5 STEPS TO A 5

500 AP Chemistry Q

to Know

Third Edition

Mina Lebitz MS

New York Chicago San Francisco Athens London Madrid Mexico City
Milan New Delhi Singapore Sydney Toronto

1 2 3 4 5 6 7 8 9 QFR 24 23 22 21 20 19

ISBN 978-1-260-44197-0
MHID 1-260-44197-0

e-ISBN 978-1-260-44198-7
e-MHID 1-260-44198-9

CONTENTS

ABOUT THE AUTHOR

Mina Lebitz worked in biochemistry research before transitioning to teaching and tutoring at both the high school and college level. As a teacher, she designed curricula for several science courses, mentored students with research projects and was awarded The New York Times' Teachers Who Make a Difference Award. She spent several years as the senior science tutor at one of the most prestigious private tutoring and test prep agencies in the United States. Mina continues to develop science curricula, working with students and their parents to help them achieve their academic goals, while continuing to learn everything she can about what she loves most, science.

INTRODUCTION

Congratulations! You've taken a big step toward AP success by purchasing *5 Steps to a 5: 500 AP Chemistry Questions to Know by Test Day*. We are here to help you take the next step and earn a high score on your AP Exam so you can earn college credits and get into the college or university of your choice.

This book gives you 500 AP-style multiple-choice and free response questions that cover all the most essential course material. Each question has a detailed answer explanation. These questions will give you valuable independent practice to supplement both your regular textbook and the groundwork you are already covering in your AP classroom. This and the other books in this series were written by expert AP teachers who know your exam inside and out and can identify crucial exam information and questions that are most likely to appear on the test.

You might be the kind of student who takes several AP courses and needs to study extra questions a few weeks before the exam for a final review. Or you might be the kind of student who puts off preparing until the last weeks before the exam. No matter what your preparation style is, you will surely benefit from reviewing these 500 questions that closely parallel the content, format, and degree of difficulty of the questions on the actual AP exam. These questions and their answer explanations are the ideal last-minute study tool for those final few weeks before the test.

Remember the old saying "Practice makes perfect." If you practice with all the questions and answers in this book, we are certain you will build the skills and confidence needed to do great on the exam. Good luck!

—Editors of McGraw-Hill Education

NOTE FROM THE AUTHOR

Dear Student,

Thank you for choosing this book to help you prepare for your AP Chemistry exam. Knowledge is perishable, so consistent practice is crucial to your success on the exam. Practice problems and questions are the most efficient and effective way to reinforce your knowledge, comprehension, and application of chemistry. Your mistakes will identify areas that need improvement.

The best time to use this book is a few months before your AP Chemistry exam (but even a few days prior will help). It is meant as a **comprehensive review** and is most helpful after you've learned at least half of the AP curriculum.

The AP Chemistry exam is a timed exam, so you may want to **introduce a time restriction into your practice**. You will have 90 minutes to complete 60 multiple-choice questions (an average of 90 seconds per question) and 105 minutes to complete 3 long and 4 short free-response questions. To practice questions with time restrictions, attempt a group of multiple-choice or free-response questions in a specific time period (e.g., 10 multiple-choice questions in 15 minutes). The College Board suggests approximately 20 minutes for long free-response questions (3 questions × 20 minutes = 60 minutes) and 7 minutes for short free-response questions (4 questions × 7 minutes = 28 minutes with 17 minutes "to spare".

All of the questions in this book are "AP Chemistry-like," but there are some notable differences. **There are no long free-response questions.** Instead, several short free-response questions have been grouped together when they refer to the same data or phenomenon and should be treated like a long free-response question on the AP Chemistry exam. The only difference is that instead of being numbered 1a, 1b, 1c etc., they are simply given their own number. For these questions, some answer explanations will give away the answers to later questions in the set and some questions later in the set will require answers from previous questions (as in the long free-response questions). To avoid inadvertently obtaining answers to later questions, *answer all the grouped questions before looking at the answers, as if you were working on a long free-response question.* If a question requires an answer from a previous question and you really want to be sure you've got it right before you move on, check *the answer* but *not the explanation.* Most free-response questions are at the end of the chapter; however, you will find a few interspersed with the multiple-choice questions if they refer to a common data set with multiple-choice questions.

If you guess on a question, even if you mostly worked out the answer using chemistry, mark that question for later review. If you guessed correctly, you may not remember to review the question and answer to clarify and reinforce the information. When the topic comes up again, you may not guess as well!

In conjunction with this book visit the College Board website (**www.collegeboard.org**). Make sure you do all of the practice questions the College Board has made available to students. You can download the AP Chemistry Course Syllabus and use the section on the Big Ideas 1 to 6 as a study outline to help you narrow down what the exam will focus on. Read all of the "exclusion topics." Even if you have no intention of reading any part of the curriculum, the AP Chemistry Course Syllabus will give you access to 43 multiple-choice practice questions with answers and free-response questions with scoring guidelines.

Each year the College Board releases the **free-response questions with scoring rubrics**. You should become familiar with as many scoring rubrics as you can, even if they are from exams administered prior to 2014 (when the AP Chemistry exam underwent a major overhaul). They are all free to download from the site. The free-response rubrics (even from older exams) will help you get acquainted with the nuances of the point-earning system used with the free-response questions. They provide important information to help you distinguish between a mediocre answer and an excellent one!

Be aware that the College Board has released at least one **full-length practice test** that is *only available to teachers*. Make sure you ask your teacher to administer it to your class.

Review your labs! Chemistry is a laboratory science, so expect to see a significant number of questions related to labs on the AP exam.

The AP Chemistry exam is 3 hours 15 minutes long. It is composed of two sections of equal point value. Be familiar with the exam *before* you sit for it so you can let your brain focus solely on the chemistry. You will be supplied with reference materials that you can use on both sections.

MULTIPLE-CHOICE SECTION

✓ **60 questions in 90 minutes**

✓ **No calculator**

✓ 4 answer choices

✓ No ¼ -point penalty for incorrect answers

FREE-RESPONSE SECTION

✓ **3 long free-response questions**

✓ **4 short free-response questions**

✓ **90 minutes**

✓ Any programmable or graphing calculator allowed. A four-function calculator is allowed but not recommended.

The College Board may make changes to the exam or its administration that may not be reflected in this version of the book, so please check the College

Board website for the most up-to-date information, especially regarding test timing and calculator use. The College Board wants those who are proficient in chemistry to do well on the AP Chemistry exam. There is a lot of good information on their website, so check it out. They offer some **excellent strategies for pacing yourself on the AP Chemistry exam** on this page: https://apstudent.collegeboard.org/apcourse/ap-chemistry/exam-tips.

Good luck!
☺ Mina

I would like to thank my uncle, Dr. John Simpson, for not only igniting my love of science but for answering five million chemistry questions over the years!

CHAPTER 1

Atoms

Guiding Principles

- The chemical elements are the fundamental building blocks of all matter.
- The properties of matter can be completely understood in terms of the arrangements of atoms that make it up.
- Chemical reactions rearrange atoms relative to one another, but individual atoms retain their identity.

MULTIPLE-CHOICE QUESTIONS

1. Which of the following particulate diagrams shows the formation of HCl gas from hydrogen and chlorine gases under standard conditions (25° C and 1 atm)?

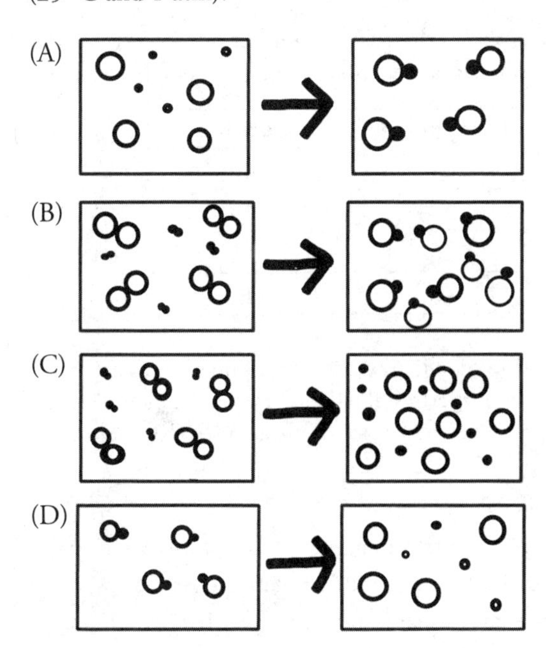

2. Which of the following particulate diagrams best represents the formation of water vapor from hydrogen gas and oxygen gas in a rigid container at 150° C and 1 atm?

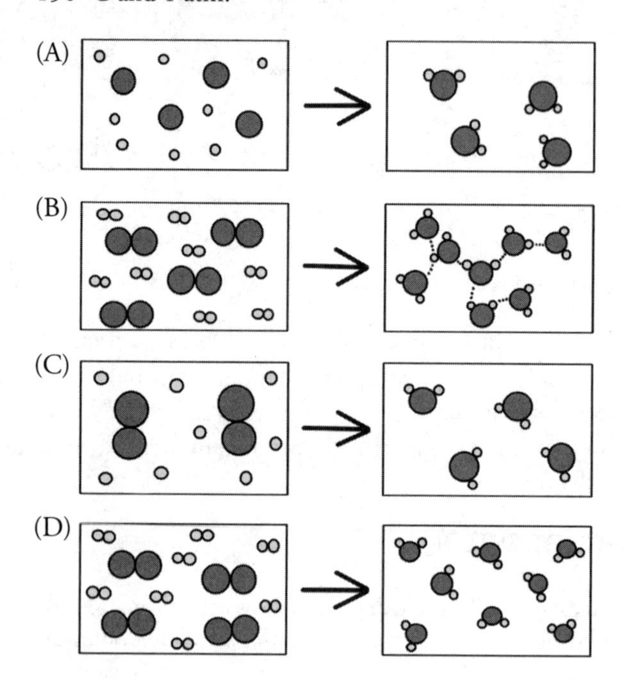

3. Which of the following shows the correct number of protons, neutrons, and electrons in a neutral cadmium −112 atom?

	Protons	Neutrons	Electrons
(A)	48	64	48
(B)	48	64	64
(C)	64	48	64
(D)	112	48	112

4. Replacing a small fraction of silicon atoms with phosphorus atoms increases the conductive properties of silicon crystals. Which of the following statements correctly identifies the reason for this increase in conductivity?

(A) P atoms are smaller than Si atoms.

(B) P atoms have a greater nuclear charge than Si atoms.

(C) P atoms introduce additional mobile negative charges.

(D) P atoms introduce additional mobile positive charges.

5. Steel is an alloy of iron and carbon with 2.1% or less carbon by mass. Which of the following statements best explains how the addition of carbon affects the properties of steel?

(A) Carbon changes the physical properties of iron atoms.
(B) Carbon changes the chemical properties of iron atoms.
(C) Carbon changes the structural arrangement of iron atoms.
(D) Carbon undergoes fusion with iron to form atoms of steel.

Question 6 refers to the following mass spectrometry data.

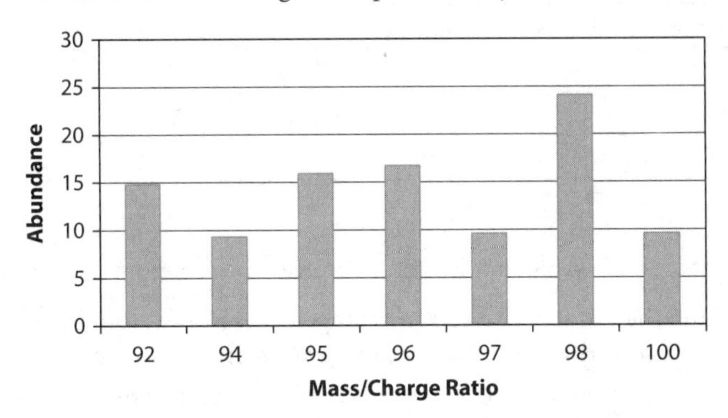

6. The atomic masses of several elements (Tc, Nb, and Ru) fall within the range of 92–100 amu. All of the following statements are true. Which of them is the best justification for concluding that this sample is pure molybdenum based on the mass spectrometry data above?

(A) The only known isotopes of ^{93}Tc and ^{93}Nb both have masses of 93 amu.
(B) The abundance of each isotope in the sample matches the known natural occurrence of Mo isotopes.
(C) The percent abundance of ^{96}Ru, ^{98}Ru, and ^{100}Ru isotopes are 5.5%, 1.9%, and 12.6%, respectively.
(D) The weighted average of the isotopes present is the atomic mass of Mo.

7. Ozone gas absorbs electromagnetic radiation in both the visible and near infrared ranges but oxygen gas only absorbs radiation in the near-infrared range. Which of the following statements best explains this observation?

(A) Ozone electrons are more easily excitable than the electrons in molecular oxygen.
(B) Infrared light matches the vibrational states of the chemical bonds in both allotropes of oxygen.
(C) The electrons in ozone require more energy to be removed than those in molecular oxygen.
(D) The electronic transitions of ozone are matched by a greater number of frequencies than molecular oxygen.

Question 8 refers to the following table of successive ionization energies ($E_{I\text{-}}$).

Element	Atomic radius (pm, 10^{-12}m)	1st E_I (kJ/mol)	2nd E_I (kJ/mol)	3rd E_I (kJ/mol)	4th E_I (kJ/mol)	5th E_I (kJ/mol)	6th E_I (kJ/mol)
Na	186	496	4,560				
Mg	160	738	1,450	7,730			
Al	143	577	1,816	2,881	11,600		
Si	118	786	1,577	3,228	4,354	16,100	
P	110	1,060	1,890	2,905	4,950	6,270	21,200
S	103	1,000	2,260	3,376	4,565	6,950	8,490
Cl	199	1,256	2,295	3,850	5,160	6,560	9,360
Ar	No data	1,520	2,665	3,945	5,770	7,230	8,780

8. The data in the table of ionization energies above can be used to support

(A) the shell model of electron arrangements in the atom because the electrons that are farther away from the nucleus are easier to remove.

(B) the shell model of electron arrangements in the atom because successive ionization energies are always increasing.

(C) the orbital model of electron arrangement because s, p, d, and f electrons always require different amounts energy to remove.

(D) the orbital model of electron arrangement because successive ionization energies are always increasing.

Question 9 refers to Rutherford's Gold Foil experiment.

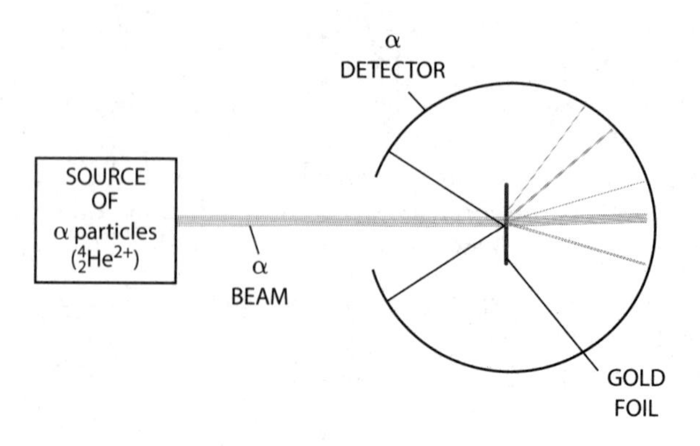

Experimental Observations:

- Most α particles passed through foil undeflected.
- A small percent were deflected from a straight-line path when passed through foil.
- Rarely but significantly, backscattering occurred.

9. In his Gold Foil experiment, Rutherford bombarded gold foil with alpha (α) particles and used a detector to visualize how their path was changed (if at all) by their interaction with the foil. Which of the following conclusions is the most accurate interpretation of Rutherford's results?

(A) Electrons are arranged in shells of increasing energy around the nucleus of an atom.

(B) The volume of an atom is mostly empty space with the positive charges concentrated in a dense nucleus.

(C) Protons and neutrons are more massive than electrons but take up less space.

(D) Discrete emissions spectrum lines are produced because only certain energy states of electrons are allowed.

Questions 10–13 are based on the following data.

Each reaction was performed at the same temperature and pressure and went to completion with no limiting or excess reagents, i.e., all hydrogen and oxygen gas reacted.

Trial	H_2 (g) (L)	H_2 (g) (g)	O_2 (g) (L)	O_2 (g) (g)	H_2O (*l*) (mL)	H_2O (*l*) (g)
1	0.122	0.01	0.061	0.08	0.09	0.09
2	0.612	0.05	0.306	0.40	0.45	0.45
3	2.440	0.20	1.220	1.60	1.80	1.80

10. Based on the data provided, which of the following observations is evidence that the volume of a gas at a given pressure and temperature is directly proportional to the total number of gas particles?

(A) In each trial, the number of grams of O_2 gas divided by the grams of H_2 gas equals a constant (8).

(B) In each trial, the volume of H_2 gas and O_2 gas that react are proportional to the volume of liquid water that forms.

(C) In each trial, the volume of each gas divided by the mass of that gas is constant (12.2 for H_2 and 0.763 for O_2).

(D) In each trial, the masses of H_2 and O_2 gases that react always sum to the mass of water that forms.

11. Which data provide the best evidence that the volume of a gas is proportional to the numbers of particles of the gas and not just the mass of the gas?

(A) In each trial, the number of liters of gas divided by the mass of the gas is always a constant (12.2 for H_2 and 0.763 for O_2).
(B) In each trial, the volume of H_2 gas is always twice the volume of O_2 gas.
(C) In each trial, the mass of H_2 gas that reacts is always 11.1% of the mass of water that forms and the mass of O_2 gas that reacts is always 89% of the mass of water that forms.
(D) In each trial, the volumes of H_2 and O_2 gases that react are proportional to the volume of liquid water that forms.

12. In addition to the data in the table above, which of the following reactions could be used to establish the diatomic nature of oxygen gas?

(A) The decomposition of H_2O_2 into H_2O *(l)* and O_2 *(g)*
(B) The decomposition of H_2O_2 into H_2 *(g)* and O_2 *(g)*
(C) The decomposition of CO_2 into C *(s)* and O_2 *(g)*
(D) The production of oxygen gas according to the reaction
$NaOCl\ (aq) + H_2O_2\ (aq) \rightarrow O_2\ (g) + NaCl\ (s) + H_2O\ (l)$

Question 13 is a **FREE-RESPONSE** question.

13. Can the data from the table and any of the observations listed in Question 12 be used to establish the diatomic nature of hydrogen gas? Justify your response.

Questions 14–16 refer to the following experiment.

The figure below is a simplified diagram of a cathode ray tube. A cathode ray strikes a detector in a straight-line path (not shown), but when a magnetic or electric field is applied (by deflecting coils), the path of the ray is deflected.

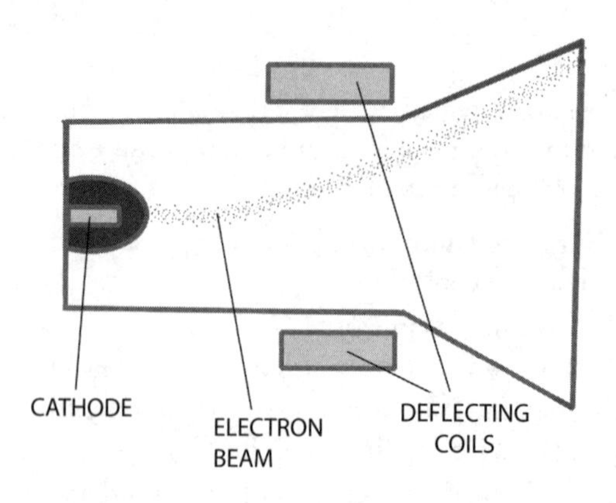

14. Which of the following statements is the most accurate interpretation of this observation?

(A) Cathode rays are composed of photons (electromagnetic radiation).
(B) Cathode rays demonstrate both wave and particle properties.
(C) Cathode ray particles carry a charge.
(D) Cathode rays have a charge-to-mass ratio that is much larger than alpha particles.

15. The experiment revealed a charge-to-mass ratio of -1.76×10^{8} C g^{-1}. The charge of an individual electron is -1.602×10^{-19} C. Which of the following expressions correctly uses this information to determine the mass of an individual electron?

(A) $\dfrac{-1.602 \times 10^{-19}}{-1.76 \times 10^{8}}$

(B) $\dfrac{-1.76 \times 10^{8}}{-1.602 \times 10^{-19}}$

(C) $\dfrac{(-1.602 \times 10^{-19})(6.02 \times 10^{23})}{-1.76 \times 10^{8}}$

(D) $(-1.602 \times 10^{-19})(-1.76 \times 10^{8})(6.02 \times 10^{23})$

16. Which of the following statements is an accurate conclusion of the observation that the deflection experienced as a result of the same force is greatest for electrons (cathode or canal rays, beta particles), intermediate for protons (hydrogen ions), and least for an alpha particle (helium nucleus)?

(A) An alpha particle has the largest charge-to-mass ratio of the three particles.
(B) The beta particle has the largest charge-to-mass ratio of the three particles.
(C) Because there are two protons in an alpha particle, the charge-to-mass ratio should be the same for the alpha particle and the proton.
(D) The neutrons in the alpha particle cause the charge-to-mass ratio to be higher in the alpha particle than the rest of the particles listed.

Questions 17–20 refer to the following experiment.

A beam of gaseous hydrogen atoms is emitted from a hot furnace and passed through a magnetic field onto a detector screen. The interaction of the electron of the hydrogen atom and the magnetic field causes the hydrogen atom to be deflected from a straight-line path. The deflection depends on the spin of the electron, which is either $+1/2$ or $-1/2$. Assume the field applied does not alter the spin of the electron. A tiny, permanent spot develops where an atom strikes the screen.

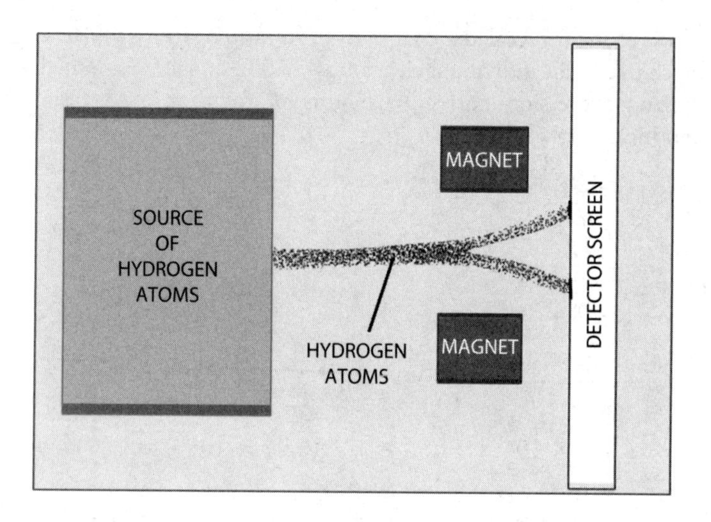

17. If the spin of the electron in each hydrogen atom were random, which of the following patterns would you expect to observe on the screen?

(A) One very small, focused spot in the center of the screen
(B) A very dark spot in the center of the screen that becomes more diffuse as the radius increases
(C) Two spots of equal intensity and distance from the center of the screen
(D) A line in the shape of a wave

18. Which of the following experimental findings was this experiment designed to further investigate?

(A) The energy of atoms is quantized.
(B) Electrons have both wave-like and particle-like properties.
(C) Electrons have a charge equal in magnitude but opposite in charge to protons, but their mass is 1/1,800 that of a proton.
(D) Emission spectra lines of hydrogen can be split when a magnetic field is applied.

19. The design of the experiment was based on a model of electrons that predicted particular behaviors of the electrons in the hydrogen atom. Which of the following illustrates the model that explains why a neutral hydrogen atom can be deflected based on the electron spin?

(A) The field was designed to deflect only one kind of charge.
(B) The mass of the proton is much greater than the mass of the electron.
(C) Only moving charges are deflected by a magnetic field.
(D) Opposite electron spins create magnetic fields of opposite direction.

20. The data produced in this experiment verify which of the following concepts of the atom?

(A) Electron energies are quantized.
(B) Electron spin is either $+1/2$ or $-1/2$.
(C) Electrons have properties of both waves and particles.
(D) The more that is known about an electron's position, the less can be known about its momentum (or velocity).

21. A beam of energy displays the properties of interference and the ability to be polarized. According to the quantum mechanical model, which of the following *must* be true of this energy beam?

(A) It behaves only as a wave.
(B) It is composed solely of particles.
(C) The beam displays both wave and particle properties.
(D) The energy of the beam is continuous.

22. Which of the following ions could be used to separate Ba^{2+} from a solution of Ba^{2+}, Fe^{3+}, and Zn^{2+} ions at room temperature?

	Ba^{2+}	Fe^{3+}	Zn^{2+}
Cl^-	soluble	soluble	soluble
OH^-	soluble	insoluble	insoluble
SO_4^{2-}	insoluble	soluble	soluble

(A) Adding dilute HCl and NaCl
(B) Adding dilute Li_2SO_4 only
(C) Adding dilute NaOH only
(D) Adding dilute NaOH or Li_2SO_4

Questions 23 and 24 refer to the following experiment.

When a discharge tube filled with pure hydrogen gas is electrified, it emits blue light, which when polarized and passed through a prism, gets dispersed into four narrow, colored bands that can be observed on a screen behind the prism. The pattern produced is the emission spectrum of hydrogen.

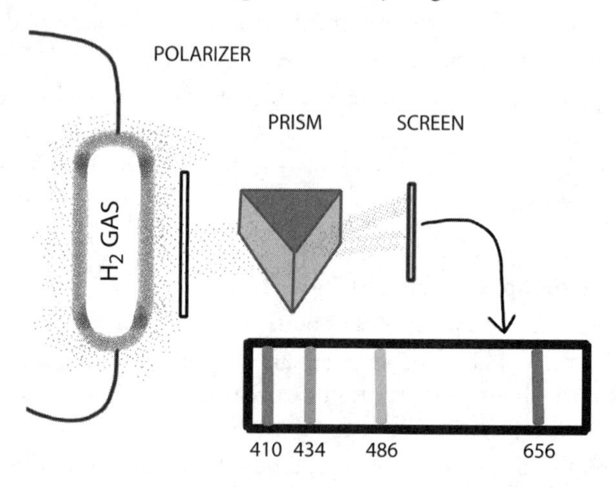

The following data correspond to the bands of the spectrum.

Band	Color	Wavelength, λ (10^{-9} m, nm)	Frequency, ν (sec^{-1})
1	Violet	410	7.3×10^{14}
2	Blue violet	434	6.9×10^{14}
3	Blue green	486	6.2×10^{14}
4	Red	656	4.6×10^{14}

You'll be given a reference document with the equation $E = h\nu$ and Planck's constant, $h = 6.63 \times 10^{-34}$ J s.

23. Which of the following statements is consistent with Bohr's interpretation of the hydrogen spectrum and accurately explains why hydrogen gas emits light when electrified?

(A) The frequencies of the photons emitted are proportional to the energies of the electronic transitions in the atom.
(B) Photons of electricity are absorbed by atoms and provide the energy needed to eject electrons.
(C) The ionized gases produced by the electric current emit cathode rays (electrons).
(D) The kinetic energy of the collisions between the energized gas particles produces photons.

24. Which of the following expressions accurately calculates the energy of the lowest energy band in the visible hydrogen spectrum?

(A) $(4.6 \times 10^{14})(656)$
(B) $(4.6 \times 10^{14})(6.63 \times 10^{-34})$
(C) $(4.6 \times 10^{14})(3 \times 10^{8})$
(D) $(656)(6.63 \times 10^{-34})$

25. A metal plate bombarded with photons of increasing frequencies gets ionized (emits electrons) at a frequency of 1.5×10^{15} sec^{-1} or higher. Refer to the table of metals below. Which of the following is most likely the identity of the metal that composed the plate?

Element	Work function (eV)
Uranium	3.6
Iron	4.5
Nickel	5.0
Platinum	6.4

An electron volt (eV) is the amount of energy gained or lost by an electron as it is moved across an electric potential difference of one volt.

$$1\ \text{J} = 6.2 \times 10^{18}\ \text{eV}$$

(A) Uranium
(B) Iron
(C) Nickel
(D) Platinum

26. Which of the following statements is *not true* regarding atomic emission spectra?

(A) Line (discrete) spectra are produced by electrified gases of the elements and continuous spectra are produced by the glow of incandescent objects.

(B) The electron configuration of the atom determines the type of spectra that will be emitted.

(C) The number of lines in the spectra is directly proportional to the number of electrons in the atom.

(D) Photons of higher and lower frequencies than those of visible light can be emitted by atoms.

27. Electromagnetic radiation in the form of X-rays can be passed through a crystal of a pure substance to produce a diffraction pattern that can be captured on photographic paper. The diffraction pattern can be analyzed to determine the molecular structure of the crystalline substance. This technique, called X-ray crystallography, allows the atoms in a structure to be visualized in three-dimensional space. Radiation of X-ray frequencies is used in this procedure because

(A) the wavelengths of X-rays are significantly greater than the size of the atoms, so they can easily be reflected off the surface.

(B) diffraction patterns emerge from crystals only when the wavelength of the radiation is comparable in size to the distance between the atoms.

(C) the energy of X-rays is high enough to break the crystal apart and scatter the atoms into a pattern on the photographic paper.

(D) the energy of X-rays is large enough to ionize the crystal but not to break the bonds between atoms.

28. A student carries out a titration using an indicator that changes color slightly past the equivalence point of pH 8. If 50.0 mL of 0.100 M KOH was added to a 100. mL sample of a monoprotic acid at the actual equivalence point, which of the following concentrations for the acid would the student obtain using the indicator?

(A) Slightly less than 0.0500 M
(B) Slightly more than 0.0500 M
(C) Slightly less than 0.200 M
(D) Slightly more than 0.200 M

29. The atomic mass of bromine is 79.904. Given that the only two naturally occurring isotopes are ^{79}Br and ^{81}Br, what is the approximate abundance of the ^{79}Br isotope?

(A) 50%
(B) 79%
(C) 80%
(D) 81%

30. Copper (Cu) has two naturally occurring isotopes, ^{63}Cu (mass = 62.9396 amu) and ^{65}Cu (mass = 64.9278 amu). The atomic mass of copper (Cu) is 63.546. Which of the following statements is a correct inference from this data?

(A) Copper does not have a whole number of neutrons.
(B) The mass of the 29 electrons in copper is 0.546 amu.
(C) Copper has an average of 63.546 neutrons.
(D) ^{63}Cu is a more abundant isotope than ^{65}Cu.

Questions 31 and 32 refer to the graph of first ionization energies below.

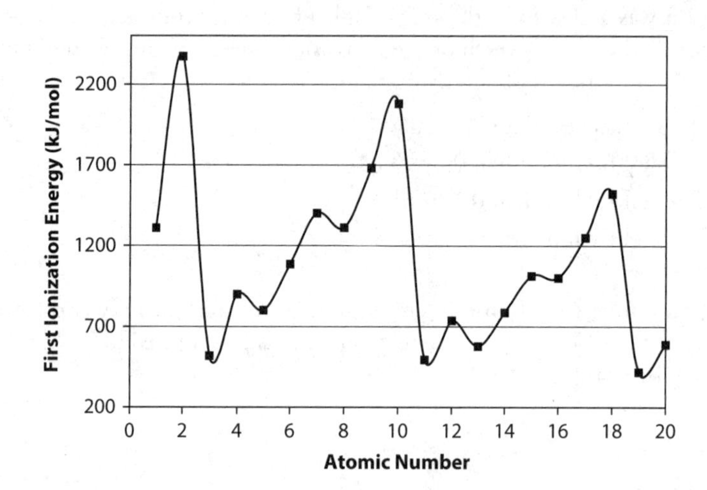

31. The large drops in ionization energies between elements of atomic number 2 and 3, 10 and 11, and 18 and 19 occur because, compared to elements 3, 11, and 19, elements 2, 10, and 18 have

(A) smaller atomic radii and a greater effective nuclear charge.
(B) smaller atomic radii and lower effective nuclear charge.
(C) larger atomic radii and a greater effective nuclear charge.
(D) larger atomic radii and a lower effective nuclear charge.

32. All of the following statements are true. Which of them is *not* a logical explanation for the decreases and increases in ionization energies between elements between atomic numbers 2 through 10, and from 11 through 18?

(A) There is repulsion between the paired electrons in the p^1 and p^4 configuration.
(B) The electrons in a filled s orbital are more effective at shielding the electrons in the p orbitals of the same energy level than each other.
(C) Electrons that are closer to the nucleus require greater amounts of energy to remove from the atom than more distant electrons.
(D) First ionization energies only measure the energy required to remove one valence electron.

33. The effective nuclear charge experienced by a valence electron of krypton (Kr) is different from the effective nuclear charge experienced by a valence electron of potassium (K). Which of the following most accurately explains this difference?

(A) K is a solid while Kr is a gas.
(B) The proton-to-electron ratio is higher for Kr than for K.
(C) Kr has a higher first ionization energy than K.
(D) The valence electrons of Kr experience less shielding by the inner electrons than the valence electrons of K.

Question 34 refers to the following photoelectron spectroscopy (PES) data.

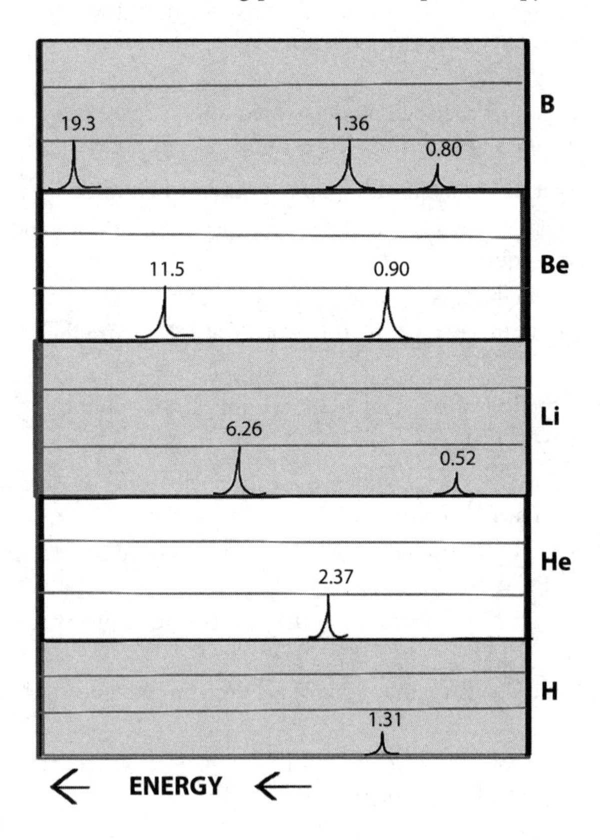

34. Which of the following statements is a correct interpretation of the photoelectron spectroscopy (PES) data shown on page 15?

(A) Helium has two electrons in the 1st shell whereas hydrogen has 1. The 2nd electron in helium's 1st shell requires slightly less energy than the 1 electron in hydrogen's 1st shell because the peak is shifted to the right.
(B) Lithium has two shells of electrons. The first shell has 2 electrons and each requires ~3.13 MJ of energy to remove whereas the 2nd shell has 1 electron, which requires 0.52 electrons to remove.
(C) Beryllium has 2 shells of electrons, each with the same number of electrons because the peak heights are equal.
(D) Boron has two 1s electrons, two 2s electrons (the first two peaks are equal size) and 1 p electron (1/2 the height of the other peaks), which requires the most energy to remove.

35. PES data support the shell model of electron energies, but which of the following data required a refinement of the shell model?

(A) Hydrogen and helium have one peak but Li and Be have two.
(B) The relative peak heights of the 2nd peaks for lithium and beryllium are not equal.
(C) Li and Be have two peaks but B, C, O, and Ne have three.
(D) The peak heights for the three peaks in carbon are the same.

36. Which of the following is *not* a direct application of photoelectron spectroscopy (PES)?

(A) PES provides direct evidence for the shell model of electron configuration in atoms.
(B) PES can quantify the amount of energy needed to eject electrons from a particular element.
(C) PES can be used to quantify the number of electrons at each energy level.
(D) PES can determine the spin of an electron in a particular orbital.

37. Which of the following factors most directly determines the number of peaks in a photoelectron spectrum?

(A) The number of electrons
(B) The number of shells
(C) The number of subshells containing electrons
(D) The number of orbitals containing electrons

38. Which of the following factors determines the height of a particular peak in a photoelectron spectra?

(A) The number of electrons in a particular orbital
(B) The quantity of energy in a particular subshell
(C) The number of electrons in a particular shell
(D) The collective quantity of energy of all the electrons represented by the peak

39. One of Bohr's interpretations of the hydrogen emission spectrum was that the energy transitions of electrons are discrete. How do atomic spectra data suggest quantized energy transitions?

(A) The dark regions between the spectral lines are frequencies at which all energy transitions occur.
(B) A finite number of lines separated by dark regions in between represent the specific frequencies in which electronic transitions occur.
(C) Atomic spectra define the limits of the continuous range of frequencies that can be emitted or absorbed by an atom. Energies above or below that range are not allowed.
(D) Each line in a spectrum represents one electron and since the number of electrons in a particular atom is discrete, the number of lines in the spectrum must also be discrete.

Question 40 refers to the information in the table below.

Element	**H**	**He**	**Li**	**Be**	**B**
Ionization energy (I_E, in eV)	13.60	24.59	5.39	9.32	8.30

40. Which of the following interpretations of the table of ionization energies above is evidence for the existence of subshells?

(A) The relative I_E of H and He. Almost twice the energy is needed to remove a valence electron from He.
(B) The relative I_E of He and Li. The energy required to remove the valence electron from lithium is lower than expected for the greater nuclear charge of Li.
(C) The relative I_E of Li and Be. The higher ionization energy of Be is lower than expected for the greater nuclear charge of Be.
(D) The relative I_E of Be and B. The ionization energy of B is lower than expected for the greater nuclear charge of B.

Question 41 refers to the diagrams below.

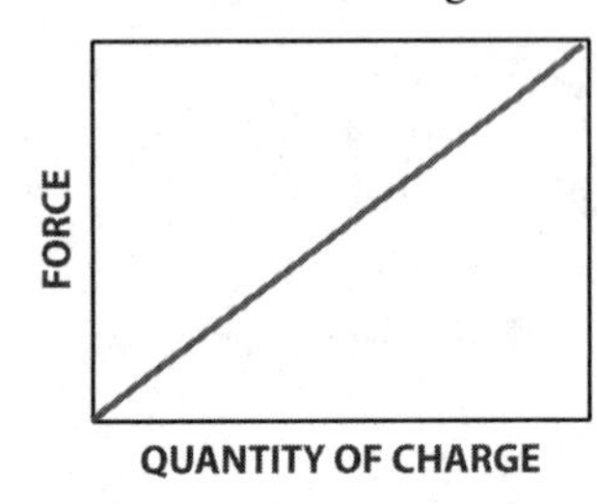

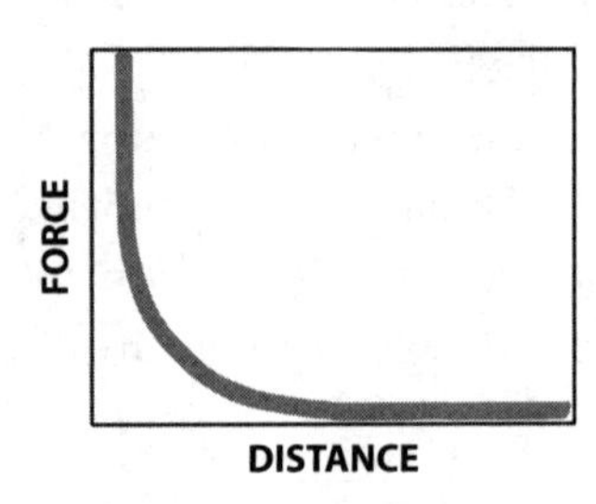

41. The graphs above show the electric force as a function of the quantity of charge on two objects (left) and the distance between charged objects (right). Which of the following statements is an accurate interpretation of these relationships?

(A) The relationship between force and charge is more significant than the relationship between force and distance.

(B) An electron that is twice the distance from the nucleus as a different electron will experience one-half the force of the nucleus.

(C) An electron that is half the distance from the nucleus as a different electron will experience four times the force.

(D) Valence electrons have the greatest potential energy because they require the greatest force to remove.

42. Equal volumes of 0.1 M Pb^{2+} and 0.1 M Cl^- solutions were mixed, forming lead (II) chloride, an insoluble compound. Which of the following particulate diagrams best represents the resulting solution?

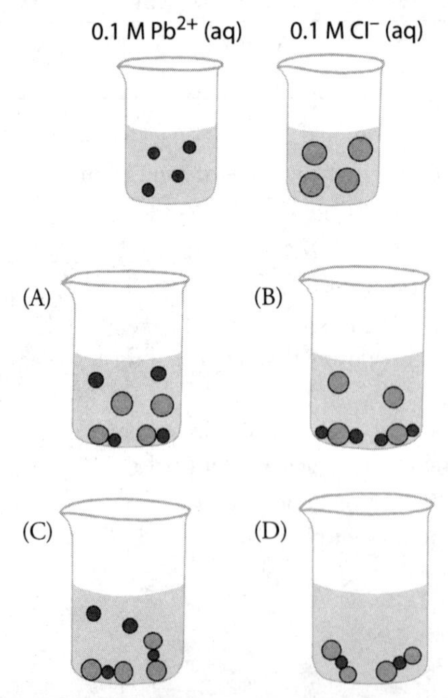

Question 43 refers to the following data tables of elements compiled in 1859.

Table 1

Simple substance	Weight	Simple substance	Weight	Common difference
Fluorine	19.0	Nitrogen	14.0	
Chlorine	35.5	Phosphorus	31.0	5
Bromine	80.0	Arsenic	75.0	
Iodine	127.0	Antimony	122.0	

Table 2

Simple substance	Weight	Simple substance	Weight	Common difference
Magnesium	12.00	Oxygen	8.00	
Calcium	20.00	Sulfur	16.00	
Strontium	43.75	Selenium	39.75	4
Barium	68.50	Tellurium	64.50	
Lead	103.50	Osmium	99.50	

43. Which pair of elements from Tables 1 and 2 do not fit the general trend of the other elements in the tables?

(A) Iodine and antimony
(B) Magnesium and oxygen
(C) Barium and tellurium
(D) Lead and osmium

Questions 44 and 45 refer to the following emission spectra.

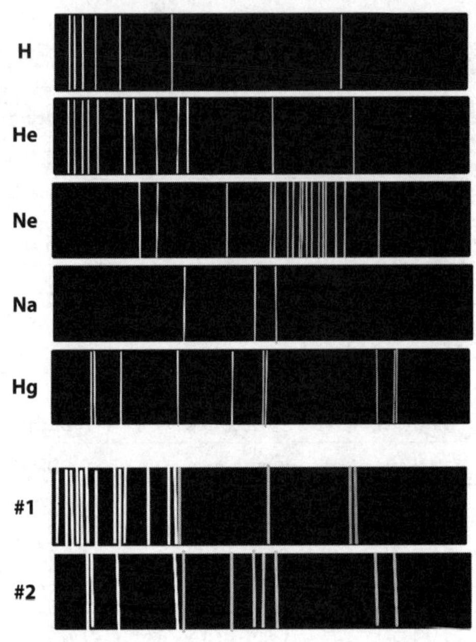

44. Which two elements are present in spectrum 1?

(A) Hydrogen and helium
(B) Helium and mercury
(C) Hydrogen and mercury
(D) Helium and sodium

45. Identify one element present in the compound that produced spectrum 2.

(A) Hydrogen
(B) Helium
(C) Neon
(D) Sodium

Question 46 refers to the Millikan's Oil Drop experiment and the data that follow.

In an experiment to determine the magnitude of the charge on a single electron, R. A. Millikan ionized oil droplets and measured their rates of motion. The size and mass of the drop along with the viscosity of the air determined the terminal velocity with which the drops would fall. When an electric field is applied, some oil drops are suspended and some rise against gravity. A single drop would be subjected to many falls and rises as the electric field was varied.

Drop	Charge
1	9.6×10^{-19}
2	1.3×10^{-18}
3	4.8×10^{-19}
4	8.0×10^{-19}
5	6.4×10^{-19}
6	1.6×10^{-18}
7	3.2×10^{-19}

46. Which of the following statements is the most accurate interpretation of Millikan's data?

(A) The average of the charges on one mole of drops is equal to the product of the elementary charge (the smallest possible charge) multiplied by Avogadro's number (6.02×10^{23}).
(B) The least common multiple of the charges on each drop reveals the smallest electric charge that occurred under these conditions.
(C) Dividing the charge on each drop by the known charge-to-mass ratio of electrons yields the charge of an electron.
(D) The frequency distribution of the charges on a large number of drops shows that the charges are quantized and the most common charge is -1.6×10^{-19} C.

47. Which of the following techniques would be the most useful in the initial determination of the molecular formula of a pure substance?

(A) mass spectroscopy
(B) photoelectron spectroscopy
(C) electron microscopy
(D) density determination

48. In 1869 two scientists (Mendeleev and Meyer) working independently produced early versions of the periodic table in which the placement of the elements was nearly identical both to each other and to the modern periodic table, even though there was no knowledge of protons or atomic numbers at the time. Which of the following is the best explanation of this occurrence?

(A) The patterns of the physical and chemical properties of the known elements were used to group the elements and the scientists made a few correct educated guesses about their specific placement.
(B) The elements were organized by oxidation number, which generally corresponds to the number of valence electrons and group number.
(C) Elements were organized by atomic mass, which generally increases with increasing atomic number.
(D) Elements were placed in categories according to their chemical properties and then placed in order on the table by their state of matter (metals on the left and gases on the right).

Questions 49–54 refer to the following diagram of the periodic table.

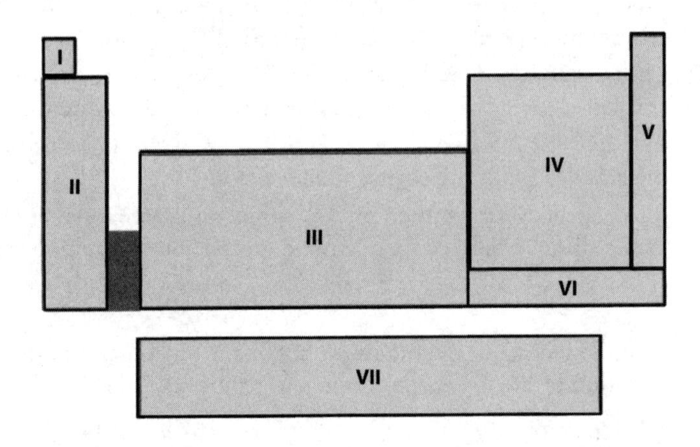

49. In which region of the periodic table would the elements that react most violently with water at 298 K be found?

(A) I
(B) II
(C) III
(D) IV

50. In which region of the periodic table would the element with the highest first ionization energy be found?

(A) I
(B) II
(C) V
(D) VI

51. In which region of the periodic table would the elements with the highest electronegativities be found?

(A) I
(B) II
(C) IV
(D) V

52. Which of the following is true regarding the numbered regions of the periodic table?

(A) None of the elements in region V are radioactive.
(B) All of the elements in region VII are radioactive.
(C) Most of the elements in region II have multiple oxidation states.
(D) Many of the elements in region III produce colored salts.

53. In which region of the periodic table would the elements with the most metallic character be found?

(A) II
(B) III
(C) IV
(D) VI

Question 54 refers to the following photoelectron spectrum (PES).

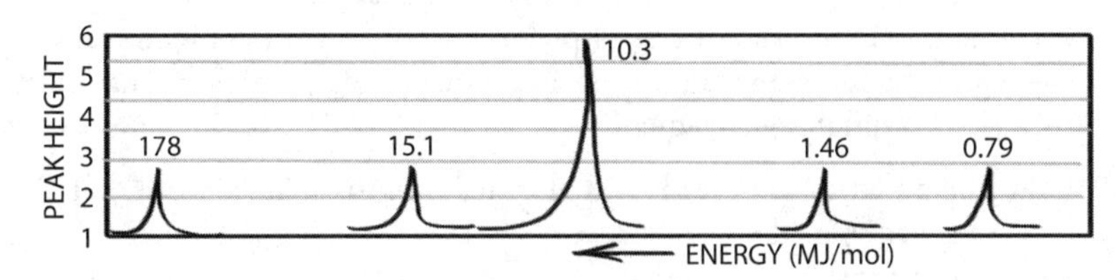

54. In which region of the periodic table would the element represented by the PES data above be found?

(A) II
(B) III
(C) IV
(D) V

55. The atomic mass of Sr is 87.62. Given that there are only three naturally occurring isotopes of strontium, ^{86}Sr, ^{87}Sr, and ^{88}Sr, which of the following must be true?

(A) ^{86}Sr is the most abundant isotope.
(B) ^{87}Sr is the most abundant isotope.
(C) ^{88}Sr is the most abundant isotope.
(D) The isotopes ^{87}Sr and ^{88}Sr occur in approximately equal amounts.

Question 56 refers to the following diagram.

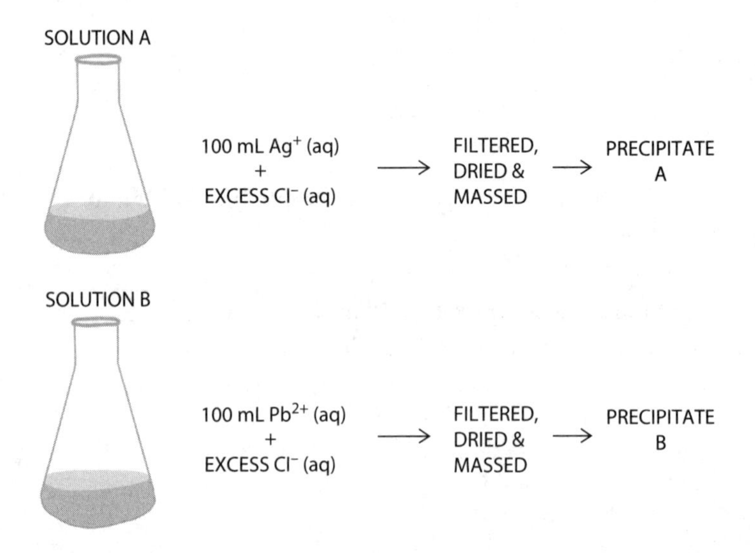

56. Two solutions were analyzed gravimetrically to determine their respective concentrations. Solution 1 was an aqueous solution of silver ions. Solution 2 was a solution of lead (II) ions. Equal volumes of a concentrated NaCl solution were added to a 100-mL sample of each solution (to provide an excess of chloride ions). The precipitates from each solution were filtered, dried, and massed. The masses of precipitates A and B were identical. Which of the following statements accurately explains why the masses of the two precipitates were equal?

(A) The molar mass of lead (II) chloride is approximately twice that of silver chloride.
(B) Solution 1 (silver ions) was twice as concentrated as solution 2 (lead ions).
(C) Solution 1 (lead ions) was twice as concentrated as solution 1 (silver ions).
(D) Twice as many chloride ions was needed to precipitate solution 1 (silverions).

57. In a process known as the Fischer esterification, a water molecule is formed from the reaction between an organic acid and an alcohol. Which of the following techniques could determine whether the hydroxyl group was from the acid (and the hydrogen of the alcohol) or from the alcohol (and the hydrogen from the acid)?

$$\underset{\text{Alcohol}}{R{-}O{-}H} + \underset{\text{Acid}}{R'{-}C({=}O)OH} \rightarrow \underset{\text{ESTER}}{R{-}O{-}C({=}O){-}R'} + \underset{\text{Water}}{H_2O}$$

(A) Mass spectroscopy
(B) Isotopic labeling
(C) Acid-base titration
(D) Hydrolysis

58. The noble gases krypton and xenon can form compounds with both oxygen and fluorine under standard conditions. Which of the following true statements most likely caused scientists to suspect this was true before these compounds were known or synthesized?

(A) Krypton and xenon have complete valence shells of electrons while fluorine and oxygen atoms do not.
(B) The ionization energies of krypton and xenon are sufficiently low while the electronegativities of fluorine and oxygen are high.
(C) Krypton and xenon have completely filled d orbitals.
(D) Krypton and xenon have electronegativity values greater than carbon.

Question 59 refers to the following information.

Ionization Energies for Element X (kJ mol^{-1})

First	Second	Third	Fourth	Fifth
786	1,577	3,228	4,354	16,100

59. Based on the ionization energies for element X listed in the table above, which of the following is most likely the identity of element X?

(A) Be
(B) Al
(C) Si
(D) As

60. Which of the following is *not* a property of the group 1 alkali metals?

(A) They have low first ionization energies.
(B) They react violently with water to form strong acids.
(C) They have strong metallic character.
(D) They are all silver solids at 1 atm and 298 K.

61. Which of the following techniques could provide a useful approximation of the ionization energy of an element?

(A) Nuclear magnetic resonance (NMR) spectroscopy
(B) Infrared (IR) spectroscopy
(C) Photoelectron spectroscopy (PES)
(D) Spectrophotometry

FREE-RESPONSE QUESTIONS

62. Photoelectron spectroscopy data for eight elements are shown below. In the table below, arrange the elements, by number, as a periodic table, by placing the number of the element into the appropriate box. Justify your arrangement.

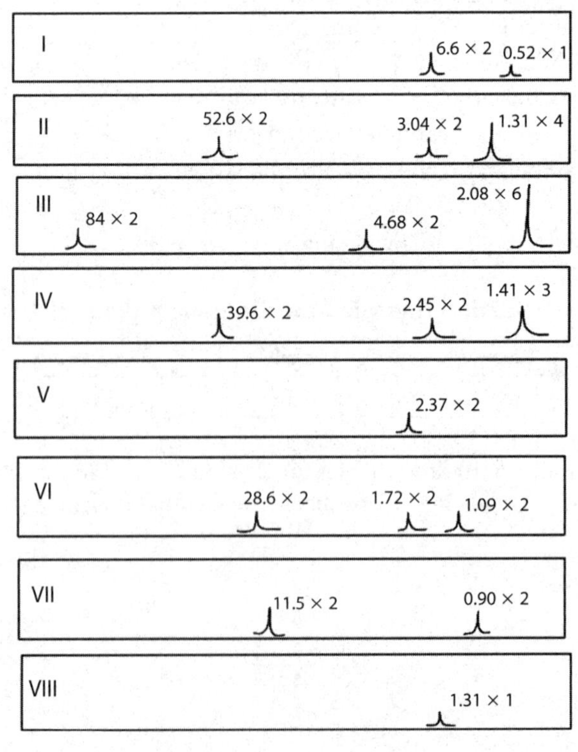

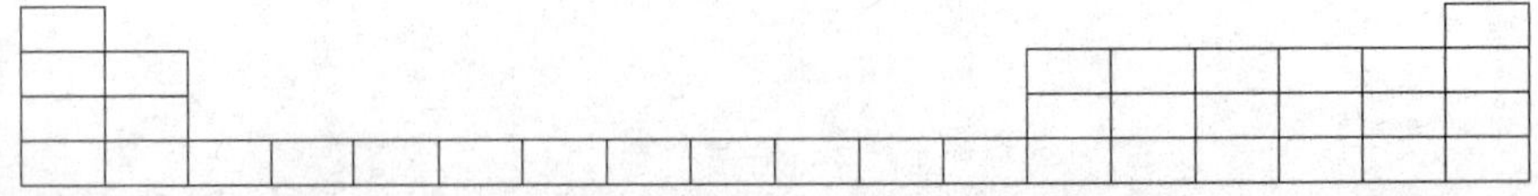

63. A photoelectron spectrum contains only three peaks of identical heights. **Identify** the element and **justify** your response.

64. **Sketch** a photoelectron spectrum for chlorine. For each peak, **identify** the energy (in eV; the energies are 1.3, 2.4, 20, 27, and 273) and **write the electron configuration** for the electrons represented by the peak (the energy level, subshell, and number of electrons in the subshell).

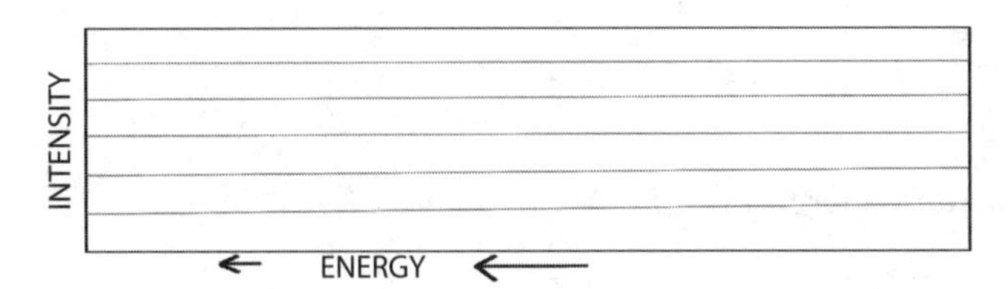

65. The photoelectron spectrum of an element is shown below. **Write an electron configuration** for the element and **identify** the element. Assume the data show all electrons present in the atom.

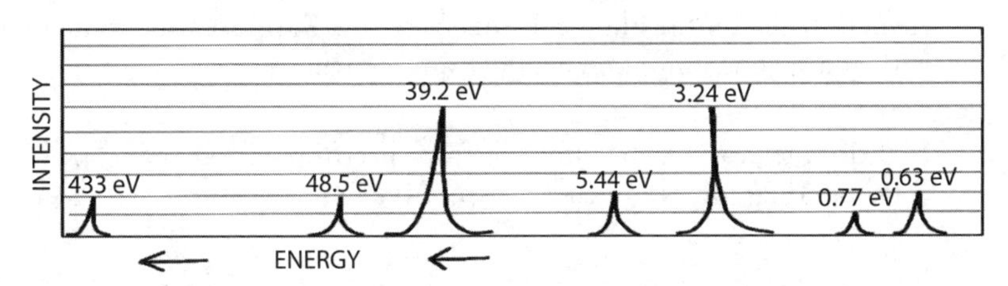

Electron configuration: ________________________________

Name of element: ____________________

Questions 66–70 refer to the photoelectron spectrum and successive ionization energies of element X shown below.

Photoelectron spectrum for element X

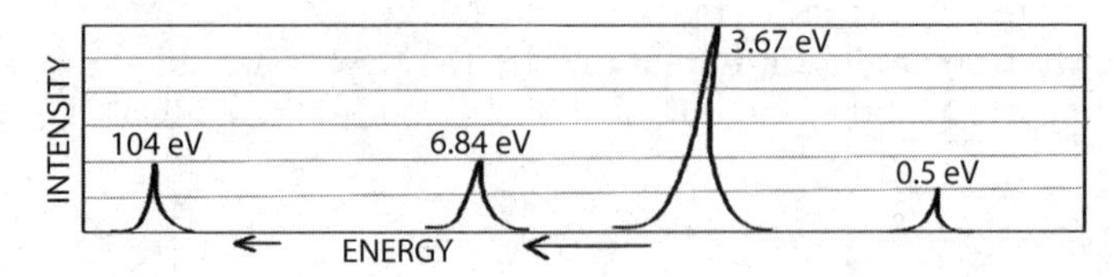

Successive ionization energies (I_E) in eV for element X

1st I_E	2nd I_E	3rd I_E	4th I_E	5th I_E	6th I_E	7th I_E	8th I_E	9th I_E	10th I_E	11th I_E
5.14	47.3	71.6	98.9	138.4	172.2	208.5	264.3	299.9	1,265	1,649

66. **Write the electron configuration** for the element and **identify** the element based on the photoelectron spectrum. **Support** your configuration with at least one ionization energy from the table.

67. **Identify** the peaks in the photoelectron spectrum with the range of corresponding ionization energies from the table.

68. Both sets of data measure electron energies in electron volts. **Explain** the discrepancy in the values obtained for each electron.

69. Use the data to **support** the existence of shells and subshells as a model to describe the electron configuration of an atom.

70. **Calculate** the range of frequencies used to obtain the photoelectron spectrum (1 J = 6.2×10^{18} eV). Use the chart below to **identify** the approximate region(s) of the spectrum of the photons.

Class of electromagnetic radiation	Range of wavelengths (m)
Gamma	$<10^{-11}$
X-ray	$10^{-11} - 10^{-8}$
Ultraviolet (UV)	$10^{-8} - 10^{-7}$
Visible	10^{-7}
Infrared	$10^{-6} - 10^{-4}$
Microwave	$10^{-4} - 10^{-1}$
Radio	$>10^{-1}$

Questions 71–72 refers to the following data.

Elemental analysis

Element	% Composition
C	40.80
H	6.16
O	43.50
N	9.52

Selected data

Molar mass	147.13 g mol^{-1}
Appearance	White, crystalline powder
Density at 20° C	1.46 g cm^{-3}
pKa values	2.1, 4.07, 9.47

71. **Calculate** the empirical and molecular formulas of the compound. Show calculations.

72. **Describe** one technique that could be used to determine the molar mass of a compound.

Questions 73–75 refer to the following experiment and data.

A student determines the concentration of two different solutions of strong base by titration using a 0.50 M HCl (*aq*) solution as the titrant. The identity and concentration of the base in each of the two analyte solutions are unknown. The two solutions may or may not contain the same strong base. The results are shown in the following table.

Analyte solution	Volume of analyte (mL)	Volume of 0.50 M acid added (mL)
A	50.	20.
B	50.	20.

73. **State** whether or not the data support the following two hypotheses and **justify** your position.

    Hypothesis I: The identity of the bases in solutions A and B are the same.
    Hypothesis II: The concentrations of the bases in solutions A and B are the same.

74. **Sketch** a theoretical titration curve of the data for solution A. **Include** important numbers on both axes. Be sure to **label** the x and y axes as well as the equivalence point.

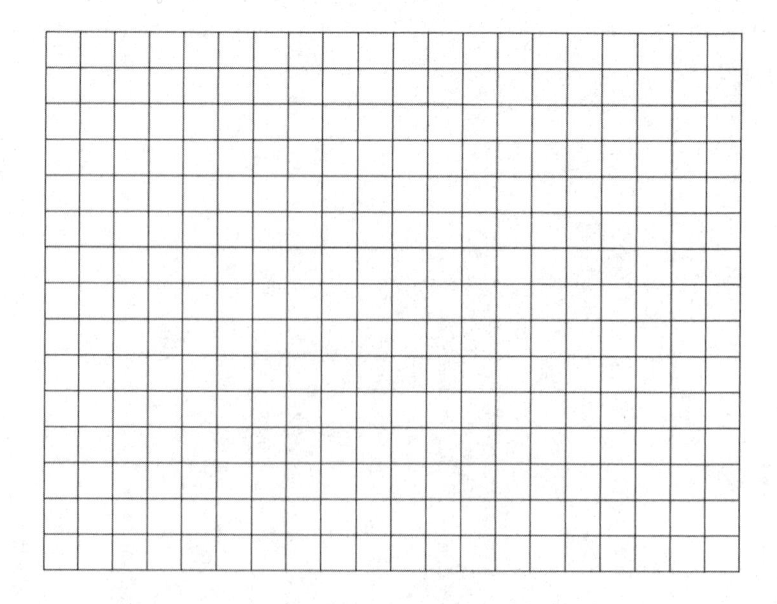

75. Draw a particulate diagram of the solution when it reaches equivalence. Use a hydroxide of a group I or II metal as the base and HCl as the acid. Include representations of each type of ion as well as the water formed by the reaction.

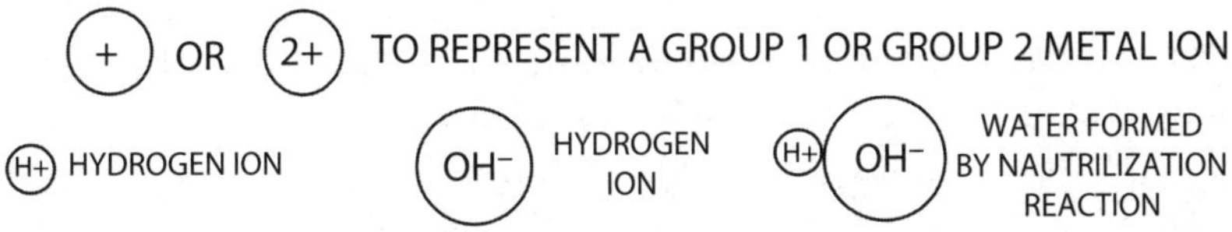

(+) OR (2+) TO REPRESENT A GROUP 1 OR GROUP 2 METAL ION

(H+) HYDROGEN ION

(OH^-) HYDROGEN ION

(H+)(OH^-) WATER FORMED BY NAUTRILIZATION REACTION

Question 76 refers to the following spectrophotometry data.

A student used an unfamiliar protocol to purify a particular substance from a mixture. The student prepared a solution (210 mL) of the substance for analysis. A 10 mL sample of the solution was taken specifically for spectrophotometric analysis. The standard curve for the absorbance versus concentration of the substance was available from a previous experiment and is shown below.

The absorbance of the student's sample was 0.547.

Concentration (mg/mL)	% Transmittance	% Absorbance
0	100.0	0
1	50.0	0.301
2	25.0	0.602
3	12.5	0.903

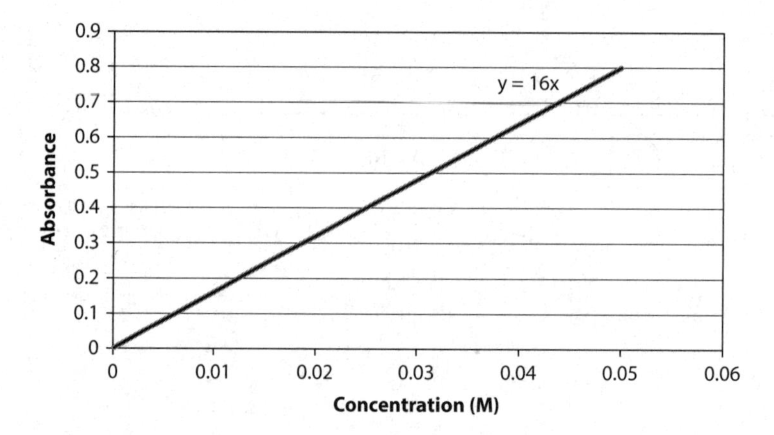

76. After filtration, the water from the 200 mL solution was completely evaporated, leaving 1.152 grams of the solid. The molar mass of the compound the student attempted to isolate is 166 g mol^{-1}. **State** whether or not the substance was pure. **Justify** your answer.

77. Combustion analysis of 3.02 grams of an organic compound yields 6.60 grams of carbon dioxide and 0.901 grams of water when reacted with excess oxygen. The compound contains carbon, hydrogen, and oxygen. **Determine** the empirical and molecular formula. The molar mass is 302.3 grams/mole.

78. The absorbance of a solution is 0.450. Use the data above to **calculate** the concentration of the solution. Show your work.

79. In August of 1875, an analysis was performed on a poorly characterized ore. A reaction was performed that precipitated out the substance of interest. The precipitate was then burned in a flame to produced spectral lines. The spectrum of the precipitate had not been observed in any of the elements previously discovered. **Explain** why this observation was evidence of the discovery of a new element.

Question 80 refers to the following experiment.

In 1895, the following data were presented in a paper that claimed that nitrogen extracted from chemical compounds is approximately 0.5% lighter than "atmospheric nitrogen." Assume the three methods of preparing N_2 chemically are 100% efficient and all three methods of oxygen removal remove 100% of the oxygen from the air. The expected mass of nitrogen in all cases is 2.3000 grams.

Source of nitrogen gas	Grams N_2
Nitric oxide	2.3001
Nitrous oxide	2.2990
Ammonium nitrate	2.2987

Method of removing oxygen from the atmosphere	Grams N_2
By hot copper	2.3103
By hot iron	2.3100
By ferrous hydrate	2.3102

80. **Propose** an explanation for the apparent difference in mass of nitrogen gas based on the data.

81. J. J. Thompson proposed the "plum pudding" model of the atom in which negative charges were embedded like plums in a "pudding" of positive charge. **Describe** what the results of Rutherford's Gold Foil experiment would have been if this had been an accurate model of the atom and **explain** why you would expect these results. (Question 9 has a diagram of Rutherford's experimental setup and his basic results.)

82. Briefly **describe** a method to determine the composition of the sun from earth.

Questions 83–85 refer to the types of spectroscopy listed in the table below.

Region of the Spectrum

Type of spectroscopy	Radiation source	Frequency (Hz)	Wavelength (m)	Energy (kcal/mol)	Type of transition
NMR	radio waves	$60–600 \times 10^{6}$	5–0.5	$6–60 \times 10^{-6}$	nuclear spin
IR	infrared light	$0.2–1.2 \times 10^{14}$	$15–2.5 \times 10^{-6}$	2–12	molecular vibrations
UV-visible	visible or UV light	$0.375–1.5 \times 10^{15}$	$8–2 \times 10^{-7}$	37–150	electronic states

83. **State** the type of spectroscopy that would be used to differentiate between an acid, an alcohol, and an aldehyde. **Justify** your choice.

84. **State** the type of spectroscopy would be used in conjunction with mass spectroscopy to determine the structure and conformation of an organic substance. **Justify** your choice.

85. **State** the type of spectroscopy would be used to detect conjugated molecules (molecules containing alternating single and double bonds). **Justify** your choice.

Structure: Property Relations

Guiding Principles:

The CHEMICAL and PHYSICAL properties of matter can be explained by

- the structure and arrangement of particles (atoms, ions, and/or molecules)
- the interparticle forces between them (within compounds and mixtures)

MULTIPLE-CHOICE QUESTIONS

Questions 86 and 87 are based on the following data.

Container	**A**	**B**	**C**
Gas	N_2	O_2	F_2
Molar mass (g/mol)	28	32	38
Temperature (°C)	22	22	22
Pressure (atm)	4.0	4.0	2.0

86. If the volume of all three containers is 1 L, in which container is the density (in g/L) of the gas greatest?

(A) A
(B) B
(C) C
(D) Both A and B

87. If the pressure in all the containers was increased at a constant temperature until condensation occurred, which gas would condense at the lowest pressure?

(A) N_2
(B) O_2
(C) F_2
(D) All the gases will condense at the same pressure.

Questions 88 and 89 are based on the following diagram and electronegativity data.

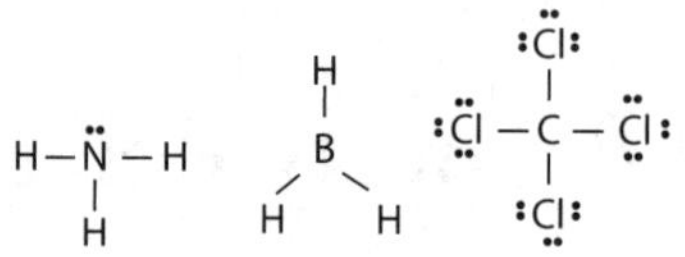

Atom	Electronegativity
H	2.20
B	2.04
N	3.04
C	2.55
Cl	3.16

88. Which of the following statements best explains why NH_3 is water soluble but BH_3 is not?

(A) The electronegativity difference is greater between N and H than it is between B and H.
(B) The molecular geometry of NH_3 is trigonal pyramidal while the molecular geometry of BH_3 is trigonal planar.
(C) The lone pair of electrons on nitrogen make it negatively charged.
(D) BH_3 has a lone pair of electrons and NH_3 does not.

89. Which of the following statements best explains why both CCl_4 and BH_3 are insoluble in water?

(A) B-H bonds and C-Cl bonds are nonpolar, making the respective molecules nonpolar.
(B) BH_3 and CCl_4 have no net dipoles.
(C) BH_3 and CCl_4 have no lone pairs of electrons.
(D) The central atom (B and C) of each molecule has an electronegativity value less than the surrounding atoms (H and Cl, respectively).

Questions 90–92 are based on the following data taken at 1 atm (760 mmHg).

	F_2	Cl_2	Br_2	I_2
Molar mass (g/mol)	38.00	70.90	159.80	253.80
Normal boiling point (°C)	−188.11	−34.04	58.80	184.30
Normal melting point (°C)	−219.67	−101.50	−7.20	113.70
Vapor pressure at 25°C (mmHg)	—	—	228.00	0.30

90. Based on the data, which substance has the weakest intermolecular forces?

(A) F_2
(B) Cl_2
(C) Br_2
(D) I_2

91. Which of the following data provides the best evidence that I_2 is a solid under standard conditions?

(A) Low vapor pressure
(B) High melting point
(C) High boiling point
(D) High molar mass

92. Which of the following data suggest that I_2 can sublimate under standard conditions?

(A) I_2 is nonpolar but has a high molar mass.
(B) There is a large difference between the melting and boiling points.
(C) The vapor pressure at 25° C is 0.3 mmHg.
(D) None of the data shown suggest that I_2 sublimates under standard conditions.

93. The synthesis of sulfur trioxide gas from sulfur dioxide and oxygen proceeds as follows:

$$2\ SO_2\ (g) + O_2\ (g) \rightarrow 2\ SO_3\ (g)$$

$$K_P = 2.63 \text{ at } 700\ °C$$

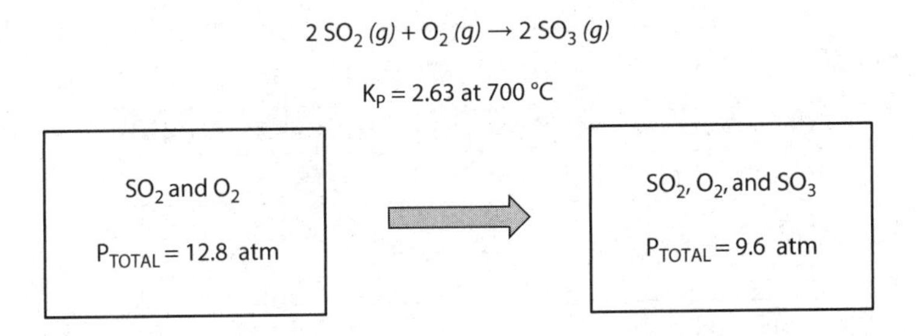

Which of the following is most likely the cause for the decrease in pressure in the container as the reaction reaches equilibrium at 700° C?

(A) The product of the forward reaction (SO_3) has more mass and therefore the particles move at a lower speed. Particles of lower speed collide with the walls of the container with less force.
(B) A decreased number of particles results in a decreased frequency of collisions with the walls of the container.
(C) An increase in the number of polar molecules causes an increased strength of intermolecular forces of attraction between the molecules in the container, which causes them to collide less with the walls of the container.
(D) As the reaction moves closer to equilibrium, the reaction rate slows, decreasing the number of collisions between the particles.

Question 94 refers to the following diagram.

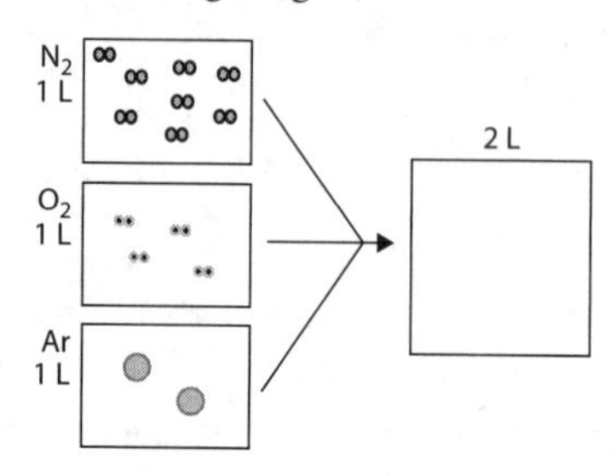

94. The figure above represents three sealed, rigid 1 L containers each containing a gas at 298 K. The pressure of the nitrogen gas is 4 atm. The ratio of the number of gas molecules is shown in the diagram. What is the total pressure at 298 K after the gases are combined into a previously evacuated 2 L container? Assume the gases will not react when combined.

(A) 3.5 atm
(B) 7 atm
(C) 14 atm
(D) 28 atm

Questions 95–97 refer to the choices in the following table.

	Compound	Vapor pressure at 25° C (torr)
(A)	$C_2H_5OC_2H_5$	545
(B)	CS_2	336
(C)	CCl_4	115
(D)	C_2H_5OH	54

95. The compound with the weakest intermolecular forces of attraction

96. The compound that can form hydrogen bonds

97. The nonpolar molecule with the lowest volatility (does not readily evaporate under standard conditions)

Question 98 refers to the following table.

Group 18 element	Normal boiling point (K)
He	4.2
Ne	27.3
Ar	87.5
Kr	120.9
Xe	166.1

98. Which of the following statements correctly accounts for the increase in boiling points of the elements going down group 18 (the noble gases)?

(A) The strength of London (dispersion) forces tends to increase with the number of electrons in a substance.

(B) At the same pressure, atoms with a large radius are closer together.

(C) At the same temperature, atoms of higher mass move more slowly on average than atoms of lower mass.

(D) At the same temperature, the kinetic energy of a substance decreases with increasing mass.

99. Which of the following does *not* occur as liquid water freezes at –1° C and 1 atm?

(A) The molecules become arranged in a crystalline structure.

(B) The density of the water decreases.

(C) Ionic bonds form between the water molecules.

(D) The average number of hydrogen bonds between the water molecules increases.

100. Which of the following statements accurately describes the change that occurs in the forces of attraction between CO_2 molecules as they change phase from a gas to a solid?

(A) Covalent bonds between CO_2 molecules are formed, creating a network solid.

(B) Weak, permanent dipoles are formed by dispersion forces.

(C) The nuclei of the carbon and oxygen atoms in the CO_2 molecules form a crystalline structure within a sea of electrons.

(D) Temporary, instantaneous dipoles form between the molecules that become more powerful than the vibrations between molecules that tend to increase the distance between them.

Question 101 refers to the following table.

Atom	Electronegativity
H	2.20
N	3.04
O	3.44
F	3.98
I	2.66

101. Which of the following bonds has the greatest polarity?

(A) H–N
(B) H–O
(C) F–O
(D) F–I

Questions 102–107 refer to the following data collected at 20° C and 1 atm. Olive oil and castor oil are both oils obtained from plants.

	Liquid	Viscosity (cP)	Surface tension ($mN\ m^{-21}$)
(A)	Water	1	72.8
(B)	Olive oil	84	32.0
(C)	Castor oil	986	?
(D)	Glycerol (propane-1,2,3-triol)	1490	63.4

102. The liquid with the steepest meniscus in a glass tube of small diameter

103. The liquid with the greatest resistance to flow in a tube

104. The liquid with the strongest intermolecular forces of attraction

105. Which of the following pair of liquids are the most miscible?

(A) Castor oil and water
(B) Olive oil and glycerol
(C) Castor oil and glycerol
(D) Water and glycerol

106. The expected surface tension (in $mN\ m^{-1}$) of castor oil is approximately

(A) <10.
(B) between 20 and 32.
(C) between 33 and 64.
(D) >64.

107. Which of the following true statements most directly explains the mechanism by which viscosity and surface tension decrease with increasing temperature?

(A) Density usually decreases with increasing temperature, increasing the distance between the molecules.
(B) Viscosity and surface tension are primarily determined by the intermolecular forces of attraction between the molecules in the liquid.
(C) Higher temperatures increase reaction rates.
(D) Higher temperatures increase the K_{EQ} of endothermic reactions.

108. Which of the following most directly explains the difference in melting points between BeO (2,507° C) and NaCl (801° C)?

(A) NaCl is an ionic compound and BeO is a covalent (molecular) compound.
(B) There is a smaller distance between the sodium and chloride ions in NaCl compared to the distance between beryllium and oxygen ions in BeO.
(C) The sodium ion is larger and less positively charged than the beryllium ion; the chloride ion is larger and less negatively charged than the oxygen ion.
(D) Beryllium and oxygen have lower atomic numbers than sodium and oxygen, respectively.

Question 109 refers to the following diagram of diatomic molecules drawn to scale.

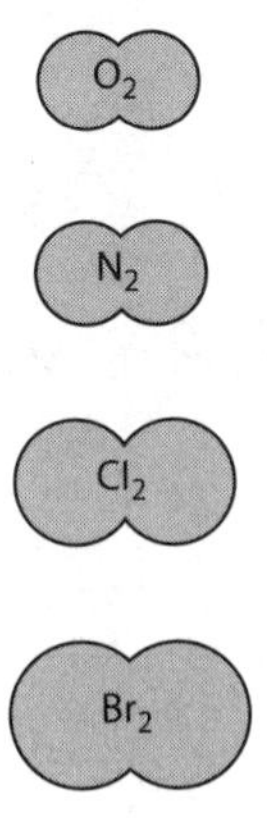

109. Which of the following is expected to have the highest boiling point?

(A) Br_2
(B) Cl_2
(C) N_2
(D) O_2

110. Which of the following *must* be true when a pure, covalent solid is heated slowly at its melting point until about half the compound has turned into liquid?

(A) The sum of the intermolecular forces holding the solid together decrease to zero as the solid continues to melt.
(B) As the temperature increases, the average kinetic energy of the molecules in the liquid phase increases.
(C) The volume increases as the substance becomes a liquid.
(D) The average kinetic energy of the particles remains the same.

111. Under which of the following sets of conditions would the most N_2 (*g*) be dissolved in H_2O (*l*)?

	Pressure of N_2 (*g*) above H_2O (*l*) (atm)	Temperature of H_2O (*l*) (°C)
(A)	2.5	75
(B)	2.5	15
(C)	1.0	75
(D)	1.0	15

112. In which of the following would sodium chloride be *least* soluble?

(A) CH_3COOH (*l*)
(B) CH_3OH (*l*)
(C) CCl_4 (*l*)
(D) HBr (*aq*)

113. Which of the following gases, when collected over water, would produce the greatest yield?

(A) CH_4
(B) HCN
(C) SO_2
(D) NH_3

114. The value of the enthalpy of solvation changes when mixing different proportions of ethanol and water. The solvation of small volumes of water in large volumes of ethanol and the solvation of small volumes of ethanol in large volumes of water are exothermic. However, the solvation of equal volumes of ethanol and water is endothermic. Which of the following statements is the most logical interpretation of this observation?

(A) The ratio of breaking hydrogen bonds between the molecules in the pure liquids and forming hydrogen bonds with the molecules in the other liquid vary with the ratios in which the liquids are combined.
(B) At low concentrations of ethanol or water, fewer hydrogen bonds are formed than when mixing them in equal amounts.
(C) No enthalpy change occurs when mixing equal volumes of liquids that form the same type of intermolecular forces.
(D) Ethanol is capable of forming more hydrogen bonds than water.

115. Ethanol, CH_3CH_2OH *(l)*, and water, H_2O *(l)*, are mixed in equal volumes at 25° C and 1 atm. Which of the following answer choices contains *only* exothermic processes regarding the preparation of this solution?

(A) Ethanol molecules move away from other ethanol molecules and water molecules move away from other water molecules as they move into solution.
(B) The hydrogen bonds between molecules in a solution are labile, continually forming and breaking as molecules move in the solution.
(C) Ethanol molecules form hydrogen bonds with water molecules as they move into solution.
(D) Ethanol molecules separate from each other and form hydrogen bonds with water molecules as they move into solution.

Question 116 refers to the following diagram.

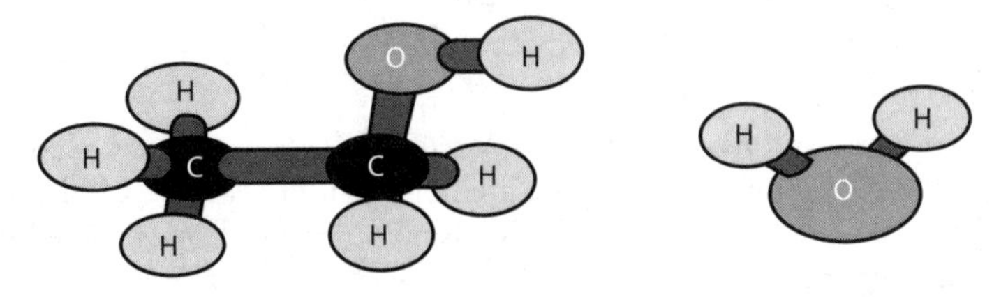

116. Intermolecular forces of attraction that occur between ethanol and water include

(A) hydrogen bonding only.
(B) ionic forces only.
(C) dispersion forces only.
(D) hydrogen bonding and dispersion forces.

117. Diamond is an extremely hard substance. This quality is best explained by the fact that a diamond crystal is

(A) an ionic compound with a large lattice energy.
(B) made completely of carbon, a very hard, dense, and stable atom.
(C) formed only under extremely high heat and pressure.
(D) one extremely large molecule in which each carbon atom forms strong bonds with each of its neighbors.

118. In which of the following processes are covalent bonds broken?

(A) $C_{10}H_8$ *(s)* → $C_{10}H_8$ *(g)*
(B) C *(diamond)* → C *(graphite)*
(C) NaCl *(s)* → NaCl *(molten)*
(D) NH_4NO_3 *(s)* → NH_4^+ *(aq)* + NO_3^- *(aq)*

119. Which of the following contributes the *least* in determining the bulk physical properties of a substance?

(A) The elemental composition of the particles in the substance
(B) The shape of the particles (atoms, ions, or molecules) that make up the substance
(C) The spacing between and the orderliness of the particles that make up the substance
(D) The forces of attraction between the particles in the substance

120. Under which of the following conditions would the predictions made by the ideal gas law be most accurate for a real gas?

(A) Low pressure and temperature
(B) High pressure and temperature
(C) High pressure and low temperature
(D) Low pressure and high temperature

121. Which assumptions of the Kinetic Molecular Theory of Gases accounts for the observation that real gases are not as compressible as ideal gases (in other words, the volume of real gases is often higher than predicted for a given temperature at high pressures)?

(A) Gases consist of many particles moving randomly and continuously through space.
(B) The volume of gas particles is negligible compared to the distances between them.
(C) Collisions between particles are completely elastic.
(D) The average kinetic energy of the molecules depends only on the temperature.

122. All gases can be condensed to form liquids. Which of the following assumptions of the Kinetic Molecular Theory of Gases is most directly challenged by this observation?

(A) Collisions between particles are completely elastic.
(B) The volume of gas particles is negligible compared to the distances between them.
(C) Attractive forces are weak and negligible.
(D) The average kinetic energy of the molecules depends only on the temperature.

123. Steel is an alloy made by adding a small percentage of carbon to iron. The addition of carbon makes the metal lattice more rigid. The alloy is less malleable and ductile than pure iron. Which of the following properties of carbon most accurately explains the behavior of the alloy?

(A) Carbon atoms are harder and more stable than iron atoms.
(B) Carbon has four valence electrons, allowing each carbon atom to form bonds with four iron atoms.
(C) Carbon nuclei are smaller than iron nuclei and can fit in the spaces between them.
(D) The higher electronegativity of carbon attracts electrons away from iron, strengthening metallic bonding.

124. The macroscopic properties of a substance, such as viscosity, surface tension, volumes of mixing for liquids, and hardness for solids, can all be explained by

(A) the organization of and strength of attraction between the particles.
(B) the temperature and pressure of the substance.
(C) the elemental composition of the substance.
(D) the origin of the substance.

Question 125 refers to the following data.

Bond	Bond energy (kJ mol^{-1})	Average bond length (pm)
O–H	464	96
F–H	568	92
S–H	366	134
Cl–H	432	127
I–H	298	161

125. Based solely on the data in the table, which of the following statements is the most logical conclusion about the relationship between bond energy and bond length?

(A) Generally, bond energy is directly proportional to bond length.
(B) Generally, bond length is inversely proportional to bond energy.
(C) The energy of a bond may be related to its length.
(D) The length of hydrogen bonds is inversely related to the bond strength.

Questions 126 and 127 are based on the following data.

Compound	Boiling point (°C)	Solubility in water at 20° C (g/100 g H_2O)
CH_3OH	65	completely miscible
CH_3CH_2OH	78	completely miscible
$CH_3CH_2CH_2OH$	97	completely miscible
$CH_3CH_2CH_2CH_2OH$	118	7.9
$CH_3CH_2CH_2CH_2CH_2OH$	138	2.7
$CH_3CH_2CH_2CH_2CH_2CH_2OH$	156	0.6

126. Which of the following most accurately and directly explains the difference in boiling points among the compounds?

(A) A greater number of hydrogen bonds increases the boiling point.
(B) Dispersion forces are stronger in molecules of higher molar mass.
(C) As water solubility increases, boiling point decreases.
(D) The greater the polarity of the compound, the lower the boiling point.

127. Which of the following most accurately explains the difference in water solubility among the compounds?

(A) The compounds of lower mass can form more hydrogen bonds with water.
(B) Molecules of high molar mass are generally less water soluble.
(C) Dispersion forces are stronger in molecules of higher molar mass.
(D) The high density of the higher molar mass compounds causes them to separate from water when mixed.

Question 128 refers to the following graph of normal boiling points versus carbon chain length of unbranched hydrocarbons.

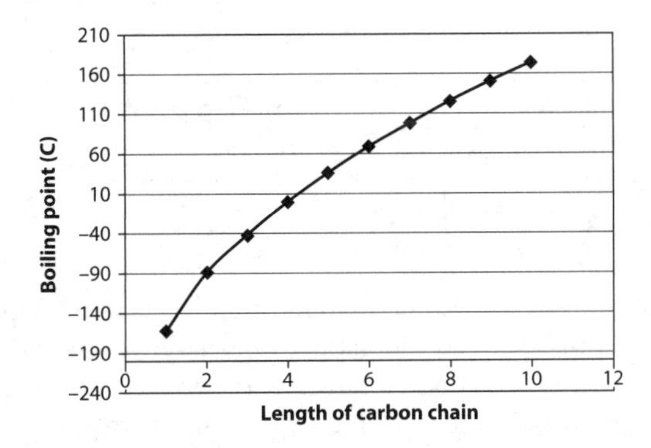

128. Which of the following statements is *not* supported by the graph?

(A) Methane (CH_4)_ is a gas at 0° C.
(B) Propane (C_3H_8) is a gas at 0° C.
(C) Hexane (C_6H_{14}) is a liquid under standard conditions.
(D) Decane ($C_{10}H_{22}$) is a solid under standard conditions.

Question 129 is a **FREE-RESPONSE** question that refers to the following diagrams and data.

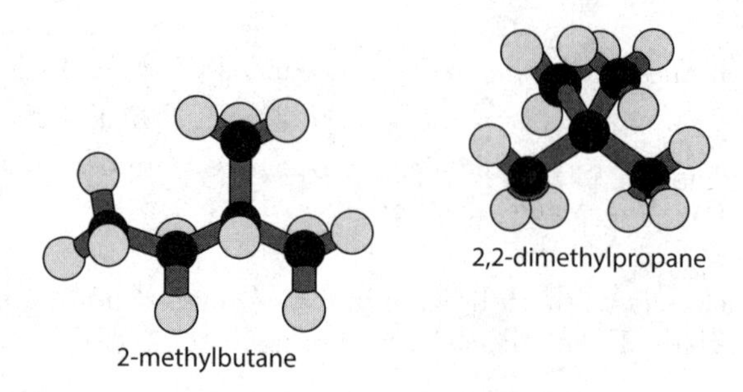

Isomer	Boiling point (°C)
2-methylbutane	28
2,2-dimethylpropane	10

129. Two isomers of pentane, 2-methylbutane and 2,2-dimethylpropane, have boiling points that vary considerably from that of an unbranched five-carbon compound (36° C; you can refer to the graph from question 128). Propose an explanation for the deviation from the expected boiling points for these molecules.

130. There is a progressive decrease in the bond angle in the series of molecules CCl_4, PCl_3, and H_2O. According to the VSEPR model, this is best explained by

(A) increasing polarity of bonds.
(B) increasing electronegativity of the central atom.
(C) increasing number of unbonded electrons.
(D) decreasing size of the central atom.

Questions 131–134 refer to the following molecules. The answer choices can be used once, more than once or not at all.

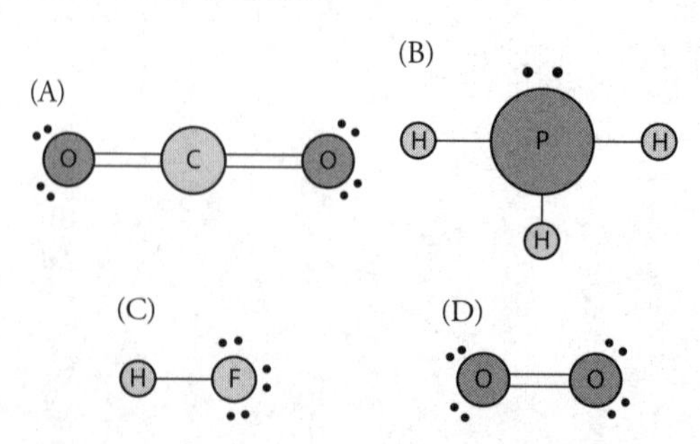

131. The molecule that exerts the greatest intermolecular coulombic (electrostatic) forces

132. The molecule stabilized by delocalized pi electrons

133. A necessary reactant for combustion reactions

134. The molecule that contains the most sigma bonds

Questions 135 and 136 refer to following tables.

Table 1

Ion	Radius (pm)
Li^+	90
Na^+	116
K^+	152
Rb^+	166

Table 2

Ion	Radius (pm)
Be^{2+}	59
Mg^{2+}	86
Ca^{2+}	114
Sr^{2+}	132

Table 3

Ion	Radius (pm)
O^{2-}	126
S^{2-}	170
Se^{2-}	184
Te^{2-}	207

Table 4

Ion	Radius (pm)
F^-	119
C^-	167
Br^-	182
I^-	206

135. Based on the data in Table 1 above, which of the following correctly predicts the relative strength of attraction of Li^+, Na^+, K^+, and Rb^+ ions to water molecules in a solution, from strongest to weakest, and provides the correct reason?

(A) $Li^+ > Na^+ > K^+ > Rb^+$ because the smaller ions have less mass so the pull of gravity on the ions in solution is smaller.

(B) $Li^+ > Na^+ > K^+ > Rb^+$ because smaller ions have a stronger electrostatic attraction to water.

(C) $Rb^+ > K^+ > Na^+ > Li^+$ because the larger ions experience greater dispersion forces.

(D) $Rb^+ > K^+ > Na^+ > Li^+$ because the larger ions experience greater gravitational pull so they must exert a stronger attraction to water in order to stay in solution.

136. Based on the data in Tables 1–4, which of the following pairs of ions would exhibit the greatest electrostatic force of attraction?

(A) Li^+ and F^-
(B) Rb^+ and I^-
(C) Be^{2+} and O^{2-}
(D) Sr^{2+} and Te^{2-}

137. Identify the hydrogen bond in the following diagram.

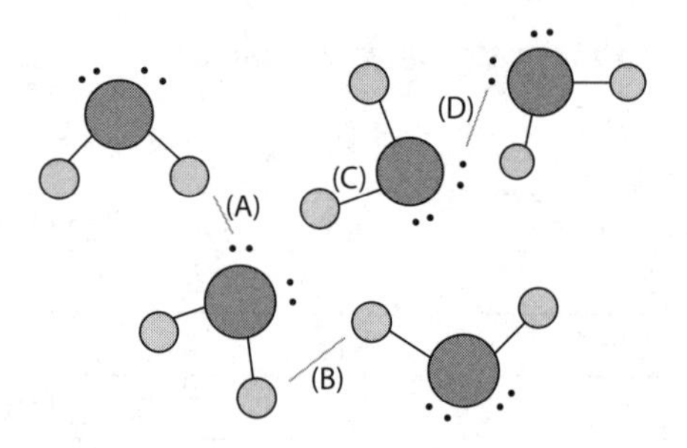

Question 138 refers to the following diagram.

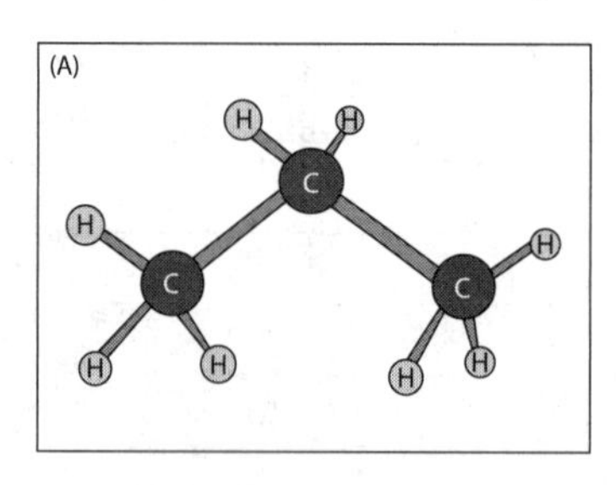

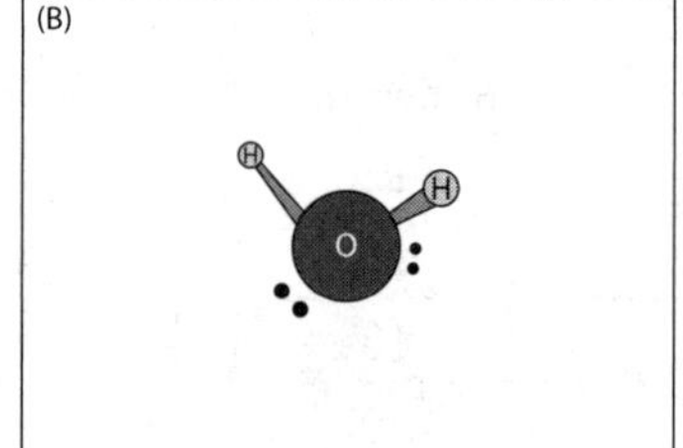

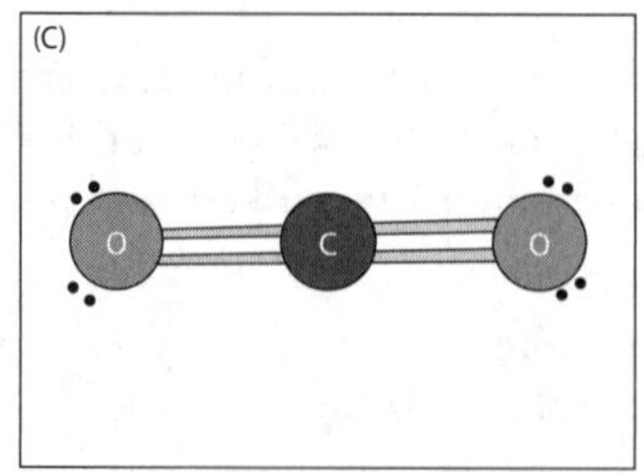

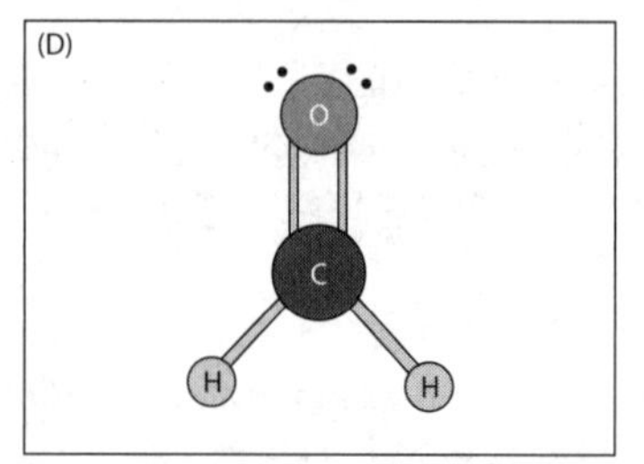

138. Which of the following choices ranks the molecules above in order of greatest to least polarity?

(A) B > D > C > A
(B) D > B > C > A
(C) B > C > D > A
(D) C > B > D > A

139. Which of the following choices correctly ranks the bond angles of the molecules from largest to smallest?

(A) $SCl_2 > PCl_3 > CCl_4$
(B) $PCl_3 > CCl_4 > SCl_2$
(C) $PCl_3 > SCl_2 > CCl_4$
(D) $CCl_4 > PCl_3 > SCl_2$

140. According to the VSEPR model, the difference in bond angles between SCl_2, PCl_3, and CCl_4 is best explained by the

(A) difference in bond polarity between the central atom and chlorine.
(B) number of unbonded electrons around the central atom.
(C) size of the central atom.
(D) difference in bond strength.

141. Which of the following is true regarding the strengths of HOCl and HOBr?

(A) HOBr is a stronger acid because Br is a larger atom than Cl.
(B) HOCl is a stronger acid because Cl is more electronegative than HOBr.
(C) The strengths of HOCl and HOBr are dependent on the concentrations of their respective solutions.
(D) HOCl is a stronger acid because HCl is a stronger acid than HBr.

Question 142 refers to the following diagrams.

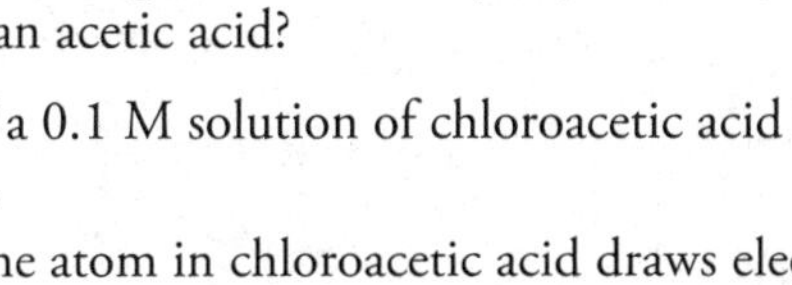

```
    H  :O:                 H  :O:
    |  ||   ..             |  ||   ..
H - C - C - O - H     Cl - C - C - O - H
            ..                     ..
    |                      |
    H                      H
  Acetic acid         Chloroacetic acid
```

142. Which of the following most accurately explains why chloroacetic acid is a stronger acid than acetic acid?

(A) The pH of a 0.1 M solution of chloroacetic acid is lower than that of acetic acid.
(B) The chlorine atom in chloroacetic acid draws electron density away from the carboxyl group, increasing the O–H bond polarity.
(C) The chloroacetic acid molecule is larger, so its electrons are more delocalized and less attracted to the nuclei.
(D) There are more hydrogen atoms on acetic acid, making it more difficult for hydrogen ions to dissociate.

143. All of the following increase the strength of an oxyacid *except*

(A) a strongly electronegative central atom.
(B) electronegative atoms bonded to the central atom.
(C) an increased number of oxygen atoms bonded to the central atom.
(D) an increased number of hydrogens.

Questions 144 and 145 refer to the following experiment.

HCl and NH_3 gases are released into opposite ends of a 1 meter (100 cm), vertical glass tube at 25° C. Their reaction quickly produces a white fog of ammonium chloride (NH_4Cl).

144. If the two gases are released at exactly the same time, which of the following most closely approximates where the ammonium chloride fog would form?

(A) 40 cm from the side where NH_3 was released
(B) 50 cm (the center of the tube)
(C) 65 cm from the side where NH_3 was released
(D) 80 cm from the side where NH_3 was released

145. Which of the following gases when released at the same time as HCl would produce a solid closest to the center of the tube?

(A) N_2H_4
(B) CH_3NH_2
(C) NH_4OH
(D) $(CH_3)_2NH$

Questions 146 and 147 refer to four samples of unknown substances shown below. The box indicates identical volumes of space sampled from containers that may or may not be the same volume; however, the conditions in the box are the same as the conditions from which the sample was taken.

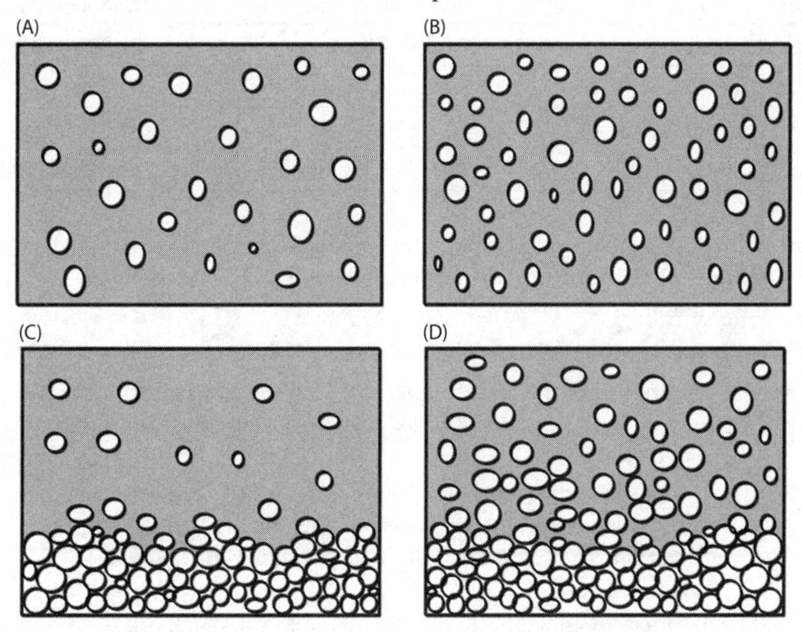

146. If all the containers are at the same temperature, which contains the substance with the strongest interparticle forces of attraction?

147. Suppose that unknown volumes of the liquids of substances A and B are placed in separate containers at the same temperature and pressure and then sealed. After equilibrium has been established, there is still a small amount of liquid at the bottom of each container. The upper portion of the containers are represented in diagrams A and B, respectively. Which of the following statements is true regarding substance A?

(A) Substance A has weaker interparticle forces of attraction.
(B) Substance A has stronger interparticle forces of attraction.
(C) The number of moles of substance A is less than the number of moles of substance B.
(D) The original container of substance A must have a larger volume than the original container of substance B.

Questions 148–153 *may* refer to information in the following table. The average kinetic energy of molecules at 25° C is 2.5 kJ mol^{-1}.

Force of attraction	Strength (kJ mol^{-1})	Distance (nm)
Van der Waals	0.4–4.0	0.300–0.600
Hydrogen bonds	12–30	0.300
Covalent bond (single)	150–560	0.074–0.267
Ionic bonds	~ 630–16,000	0.030–0.200

148. Which of the following is a logical consequence of the relative energies between the forces of attraction and the kinetic energy of molecules at 25° C?

(A) Covalent bonds cannot be broken at 25° C.
(B) Hydrogen bonds can form molecules at very low temperatures but only ionic and covalent bonds can form compounds and molecules above 25° C.
(C) The dissociating tendencies of molecules is large enough relative to the weak attractive forces between them that Van der Waals attractions and hydrogen bonds are constantly breaking and reforming at 25° C.
(D) The high melting and boiling points of molecular and ionic substances are due to the relatively high bond strengths of covalent and ionic bonds.

FREE-RESPONSE QUESTIONS

149. Use Coulomb's law to **explain** qualitatively why weak interactions are generally readily reversible.

Questions 150–153 refer to proteins. Proteins are composed of one or more linear polymers of amino acids that take on three-dimensional shapes due to complex folding of the amino acid chain. The 20 amino acids that are used to make proteins can be classified by the chemistry of their R-groups. They can be classified as polar, nonpolar, positively charged, negatively charged, neutral, basic, or acidic. Many R-groups fall into more than one category.

150. **Propose** an explanation of how weak forces such as Van der Waals and hydrogen bonding can impart stability to large molecules like proteins.

151. The following diagram represents a pentapeptide (a sequence of five amino acids joined by peptide bonds). The carbon-nitrogen backbone is highlighted. For simplicity, hydrogen atoms bonded to carbon atoms are not shown.

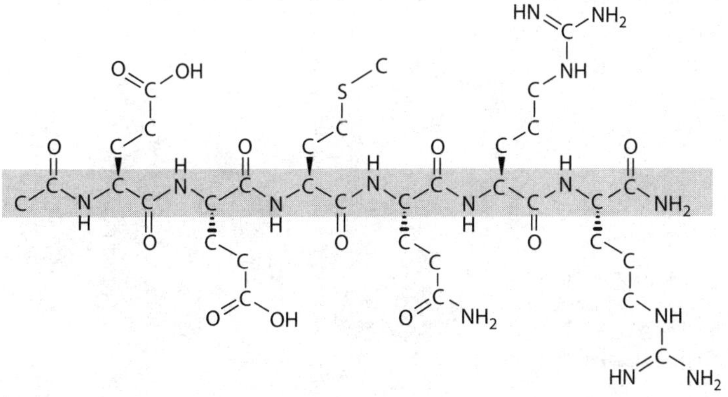

Identify three types of interactions that could occur between any of the amino acids in the chain (their place in the chain can be ignored) and /or between the amino acids in the chain and water that could contribute to the three-dimensional folding of a protein.

Circle the atom or group of atoms involved with the interaction directly on the diagram and **explain** how the interaction occurs. If the interaction involves water, draw the orientation of the water molecule relative to the amino acid(s) involved.

152. Enzymes are biological catalysts. They are typically proteins (see question 150) that have a groove or port with a specific shape to which a specific substrate binds. **Propose** a mechanism by which substrates can reversibly bind to an enzyme and **explain** how the conversion of substrate to product allows the release of the product from the active site.

Question 153 refers to the following information regarding proteins in cell membranes.

Cell membranes are composed of a bilayer of phospholipids arranged with their hydrophobic tails facing each other and their hydrophilic heads outward facing the aqueous intracellular and extracellular fluids (shown below).

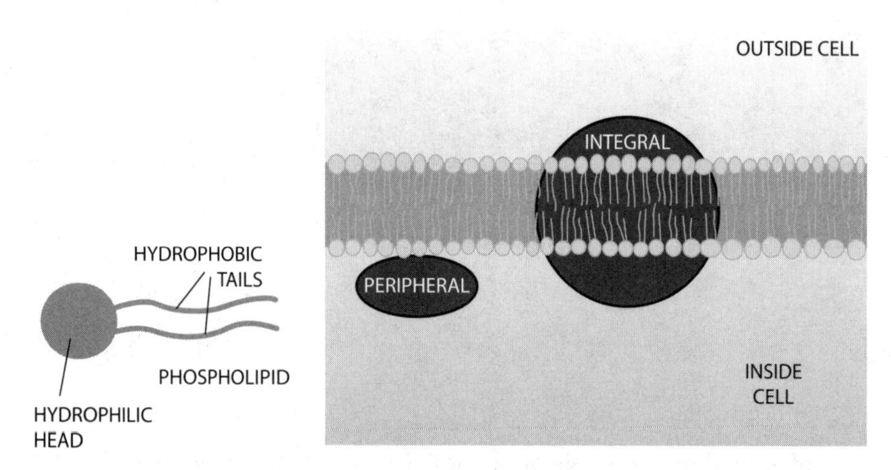

Integral membrane proteins are embedded in the bilayer with parts of the proteins facing the aqueous compartments of the tissue and parts that interface with the phospholipid tails. Peripheral proteins interact with the phospholipid heads.

153. Propose an explanation of protein structure that accounts for the ability of proteins to form stable, three-dimensional shapes in solution and in membranes, both peripherally and embedded (integral). Only the specific structural features that allow the protein to exist in their specific environments need to be addressed.

154. Use the data that follows to **explain** why water is a liquid but ammonia is a gas under standard conditions. Include in your explanation:

- A structural comparison
- A mechanism based on intermolecular attractions
- A thermodynamic comparison. Assume the average kinetic energy of molecules at 25° C is 2.5 kJ mol^{-1}.

Type of hydrogen bond **R = rest of molecule 1** **R′ = rest of molecule 2**	**Strength of hydrogen bond ($kJ\ mol^{-1}$)**	**Length of hydrogen bond (nm)**
R–O–H ••• N–R’	29	0.29
R–O–H ••• O–R’	21	0.27
R–N–H ••• N–R’	13	0.31
R–N–H ••• O–R’	8	0.29

Formula	**NH_3**	**H_2O**
Molar mass ($g\ mol^{-1}$)	17.031	18.015
Boiling point (°C)	−33.340	100.000
Melting point (°C)	−77.730	0.000
Molecular shape	Trigonal pyramidal	Bent

Atom	**Electronegativity**
H	2.20
N	3.04
O	3.44

155. Use the thermodynamic data for water that follows to (a) **explain** why the ΔH_{sub} is not simply the sum of the ΔH_{fus} and the ΔH_{vap}.
(b) **Explain** why the heat of sublimation is so high relative to the heat of fusion and vaporization.

Thermodynamic data for water:

$\Delta H_{sub\ @\ 273} = 3{,}009\ kJ\ kg^{-1}$ or $54.153\ kJ\ mol^{-1}$
$\Delta H_{fus} = 333.5\ kJ\ kg^{-1}$ or $6.003\ kJ\ mol^{-1}$
$\Delta H_{vap} = 2{,}257\ kJ\ kg^{-1}$ or $40.626\ kJ\ mol^{-1}$

156. Real gases deviate significantly from ideal behavior under certain conditions. (a) **State** two conditions that cause real gases to deviate from ideal behavior. (b) **State** two assumptions of Kinetic Molecular Theory (KMT) that are not applicable under these conditions. (c) **Describe** the specific deviation from ideal behavior that is expected for each assumption.

Question 157 refers to the following information.

The Van der Waals equation is a modification of the ideal gas law that takes into account the nonzero volume of gas particles and their interparticle forces of attraction.

One form of the Van der Waals equation is:

$$\left[(\mathrm{P} + n^2)\left(\frac{a}{V^2}\right)\right](\mathrm{V} - n\mathrm{b}) = n\mathrm{RT}$$

Values of *a* and *b* for several gases are given in the table that follows.

Substance	a ($L^2atm\ mol^{-2}$)	b ($L\ mol^{-1}$)
He	0.03412	0.02370
Ne	0.21070	0.01709
O_2	1.36000	0.03803
CO_2	3.59200	0.04267
NH_3	4.17000	0.03707

157. (a) **Describe** the deviations for which constants *a* and *b* are correct.
(b) **State** and **describe** the expected deviations in pressure and volume of He and CO_2 from ideal behavior.

Questions 158 and 159 refer to the complete Lewis electron-dot diagram for ethanoic acid shown below.

158. In the molecule, the C-O-H bond (angle x) is not 180°. **Estimate** the observed bond angle and **justify** your answer.

159. **Explain** in terms of electron domains why, in the actual molecule, the H–C–H (angle y) is not 90°.

160. Formaldehyde (methylaldehyde, methyl aldehyde) has the formula CH_2O. In the box provided, **draw** the complete Lewis electron-dot diagram for this molecule.

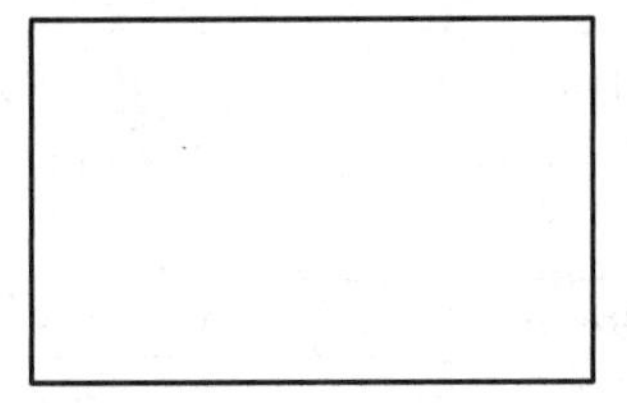

161. The structures of water and a crystal of NaCl are shown below. A student prepares an aqueous solution of NaCl and H_2O. In the space below, show the interactions between the components of NaCl *(aq)* by **drawing** the different particles present in the solution, based on the particle representations shown below. Include only one formula unit of NaCl and no more than 10 water molecules. Your drawing must (a) **identify** the ions by symbol and charge and (b) **show** the proper arrangement and orientation of the particles in the solution

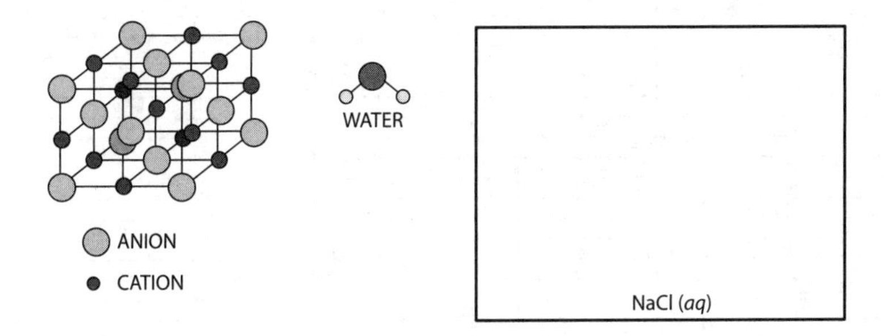

162. Suppose solid $AgNO_3$ is added to the solution above. **Draw** the resulting solution if enough $AgNO_3$ is added to produce a precipitate (AgCl has a K_{sp} of 1.8×10^{-10}).

Base your drawing on the particle representations shown below. Your drawing must (a) **identify** all the ions with symbol and charge and water molecules in the solution; (b) **show** the proper arrangement and orientation of the particles in the solution; (c) **identify** the precipitate with the correct ratio of ions; and (d) **show** the proper arrangement and orientation of the particles in the solution.

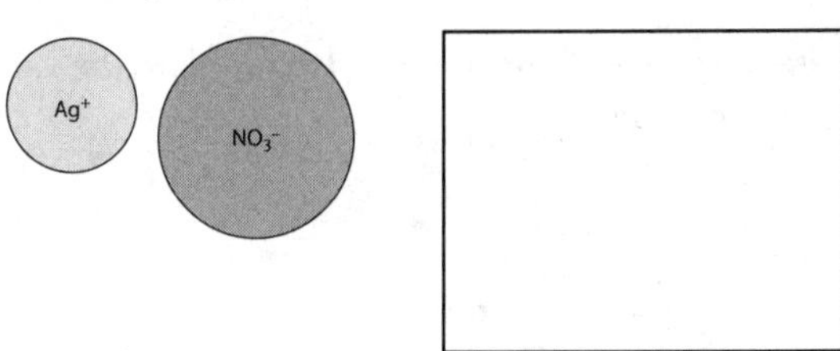

Questions 163 and 164 refer to the following experiment.

Water was heated to boiling and the vapor was collected into a previously evacuated 2 L flask. After 10 minutes of boiling and vapor collection, the temperature of the collected water was measured (99.92° C), and the flask was immediately sealed and then cooled in a large ice bath until all the water had condensed. The water was poured into a cylinder and its volume was recorded.

Temperature (°C)	Volume
99.92	2 L *(g)*
10.00	482.1 mL *(l)*

163. Graph the relationship between temperature and volume.

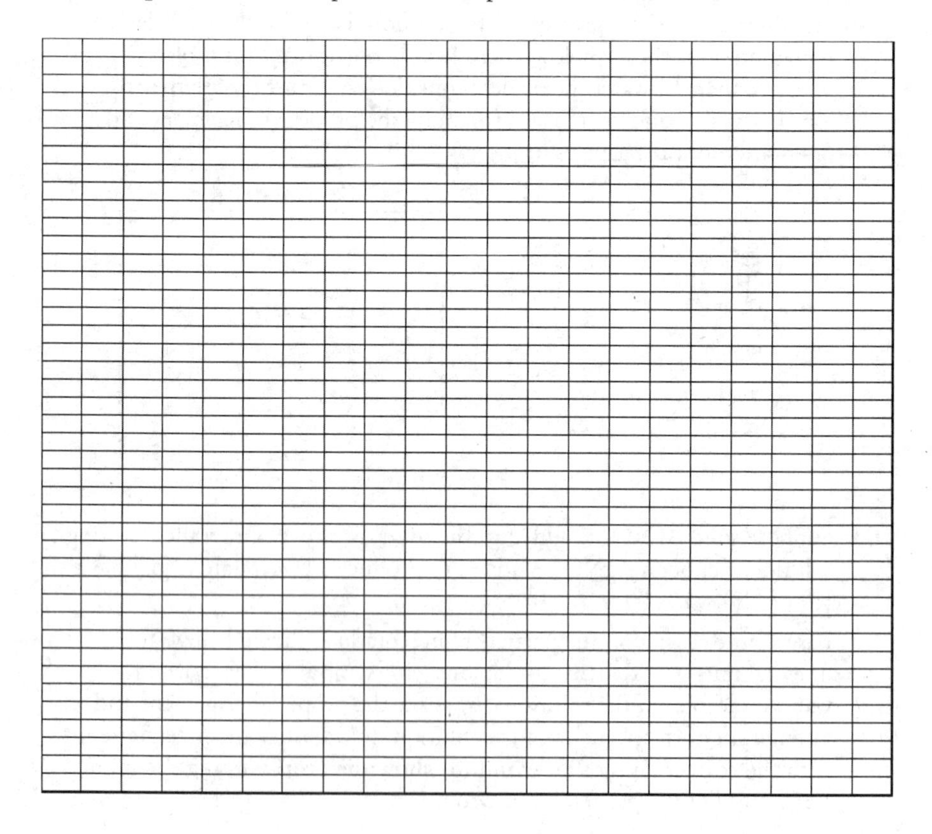

164. (a) **Determine** the slope of the line. (b) **Extrapolate** the value of absolute zero in degrees Centigrade.

165. Use Kinetic Molecular Theory to **explain** the effect of low temperatures on condensation.

Questions 166–167 refer to the chromatography procedure represented below. Samples of three mixtures, A, B, and C, were dropped onto chromatography paper. Acetone (shown below) was used as the solvent.

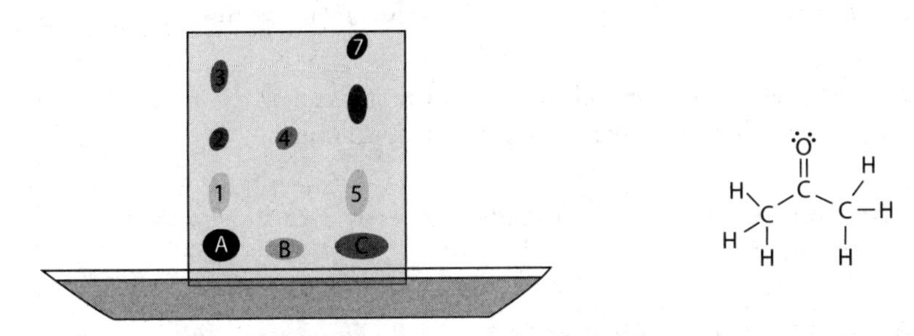

166. **Identify** which component of all three mixtures (1–7) is the *most soluble* in acetone? **Justify** your answer.

167. **Identify** which component of all three mixtures has the *lowest* R_f value? **Explain** what the R_f value describes on a molecular level.

168. The formula C_2H_6O has two isomers. **Draw** the Lewis electron-dot diagrams for both and, based on their structure, **suggest** a technique for distinguishing the two molecules or separating them. **Justify** your choice.

Questions 169 and 170 refer to the cycling of oxygen and carbon dioxide by the life-support systems aboard spacecraft and submarines. For every 1 L of oxygen gas a person breathes, 0.82 liters of carbon dioxide are released into the environment. Many life-support systems utilize the reaction below to remove carbon dioxide from the air and produce oxygen gas.

$$4\ KO_2\ (s) + 2\ CO_2\ (g) \rightarrow 2\ K_2CO_3\ (s) + 3\ O_2\ (g)$$

169. Suppose a typical person requires 550 liters of oxygen per day (at 1 atm of pressure and 298 K). (a) **Calculate** the minimum mass and volume of oxygen gas needed to keep a 10-person crew breathing for 14 days. (b) **Calculate** the mass and volume of CO_2 produced by the crew in this time.

170. (a) **Calculate** the mass of KO_2 required to remove the CO_2 from the environment. (b) **Calculate** the mass of K_2CO_3 produced. (c) **Explain** the benefit of removing carbon dioxide in this manner.

Questions 171–174 refer to the lattice structures. Lattice structures are common to both metals and ionic compounds.

171. State the different chemical and physical properties of metals and ionic compounds.

172. Compare the microscopic structures of metal lattices and ionic lattices.

173. Briefly **describe** a method to distinguish between an ionic substance and a metal that does not involve its physical appearance.

174. Explain why their properties differ so greatly even though they are both lattice structures.

Questions 175 and 176 are based on the following graph showing the boiling point versus the volume of distillate for two samples.

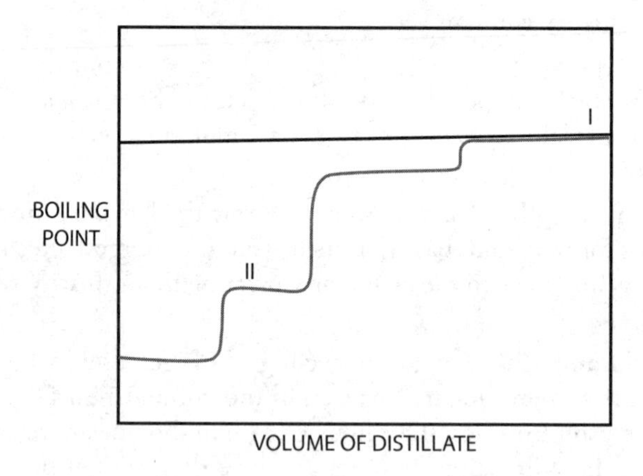

175. Identify which sample is pure and which is a mixture? **Justify** your answer.

176. State the number of pure substances that compose the mixture. **Justify** your answer. **State** whether or not there is a method to quantify the relative volumes of each substance in the mixture. **Justify** your answer.

Questions 177–180 are based on the following information.

Substance	Appearance	MP (°C)	Solubility in H_2O @ 25° C	Elements (M = metal N = nonmetal)	Type of bond
NaCl	Crystalline white solid	801	Yes	M – NM	Ionic
α - Glucose	White solid	146	Yes	NM – NM	Polar covalent
Sulfur (S_8)	A soft, bright-yellow solid with a faint odor of matches	115	No	NM – NM	Nonpolar covalent
Iodine (I_2)	Dark, gray solid	114	Slightly	NM – NM	Nonpolar covalent

Alcohol	No. of carbons	Boiling point (°C)	Solubility in H_2O (g/100 g) @ 20° C
methanol	1	65.00	completely miscible
ethanol	2	78.50	completely miscible
1-propanol	3	97.00	completely miscible
1-butanol	4	117.70	7.90
1-pentanol	5	137.90	2.70
1-hexanol	6	155.80	0.59

177. Suppose you are given a clear liquid with no detectable odor and one boiling point at approximately 102° C. **Describe** a method for determining the composition of the sample. Assume it is a pure substance or mixture containing one or more of (only) the substances listed above and assume that it may contain water.

178. **Describe** a technique to separate a mixture of powdered glucose and sulfur.

179. **State** whether or not it is possible to separate a mixture of NaCl and glucose in water. If so, **describe** the procedure. If not, **justify** your response.

180. For each of the structures A through D below, **choose** the substance from the table above that is most likely to be miscible with it. **Justify** your choices.

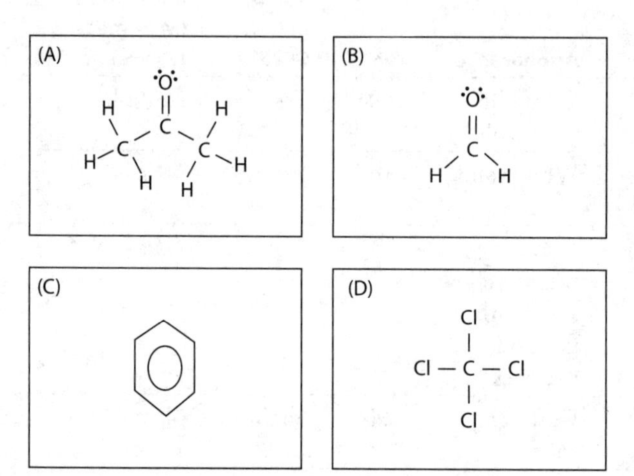

Transformation

Guiding Principles

Changes in matter involve the reorganization of atoms and/or a transfer of electrons.

MULTIPLE-CHOICE QUESTIONS

Question 181 refers to the following reactions that occur in an aqueous solution when carbon dioxide and water combine to form carbonic acid.

I. $CO_2\ (g) + H_2O\ (l) \rightarrow H_2CO_3\ (aq)$
II. $H_2CO_3\ (aq) + H_2O\ (l) \rightarrow H_3O^+\ (aq) + HCO_3^-\ (aq)$
III. $HCO_3^-\ (aq) + H_2O\ (l) \rightarrow H_3O^+\ (aq) + CO_3^{2-}\ (aq)$

181. Which of the following represent Brønsted-Lowry conjugate acid-base pairs?

(A) H_2O and H_2CO_3
(B) H_2CO_3 and H_3O^+
(C) CO_2 and HCO_3^-
(D) H_2O and H_3O^+

$$HCl + H_2O \rightarrow H_3O^+ + Cl^-$$

182. Which of the following best explains the relationship between conjugate acid-base pairs in the reaction above?

(A) The acid and the base differ by one proton (H^+ ion).
(B) The conjugate base differs from the base by one proton (H^+ ion).
(C) The conjugate base differs from the acid by one proton (H^+ ion).
(D) The conjugate acid and conjugate base differ by one proton (H^+ ion).

Questions 183 and 184 refer to the following reactions.

(A)	$HCl\ (aq) + NH_3\ (aq) \rightarrow NH_4Cl\ (aq) + H_2O\ (l)$
(B)	$Ag^+\ (aq) + Cl^-\ (aq) \rightarrow AgCl\ (s)$
(C)	$Mg\ (s) + O_2\ (g) \rightarrow MgO_2\ (s)$
(D)	$3\ Cl_2\ (aq) + 6\ OH^-\ (aq) \rightarrow 5\ Cl^-\ (aq) + ClO_3^-\ (aq) + 3\ H_2O\ (l)$

183. A reaction in which the same reactant undergoes both an oxidation and a reduction

184. The synthesis of an ionic compound that is also a redox reaction

Questions 185–187 refer to the following reaction.

$$CS_2\ (l) + 3\ O_2\ (g) \rightarrow CO_2\ (g) + 2\ SO_2\ (g)$$

185. According to the reaction above, what is the total number of moles of products when 0.400 moles of $CS_2\ (l)$ has completely reacted with 1.20 moles $O_2\ (g)$?

(A) 0.40
(B) 0.80
(C) 1.20
(D) 1.60

186. How many moles of $O_2\ (g)$ must react to form 6.30 moles of gaseous products?

(A) 2.10
(B) 4.20
(C) 6.30
(D) 9.45

187. How many moles of $CS_2\ (l)$ must react to produce 33.6 L of product at STP?

(A) 0.50
(B) 1.50
(C) 3.00
(D) 3.36

Questions 188–190 refer to the following reaction and the diagrams below.

$$2\ SO_2\ (g) + O_2\ (g) \rightarrow 2\ SO_3\ (g)$$

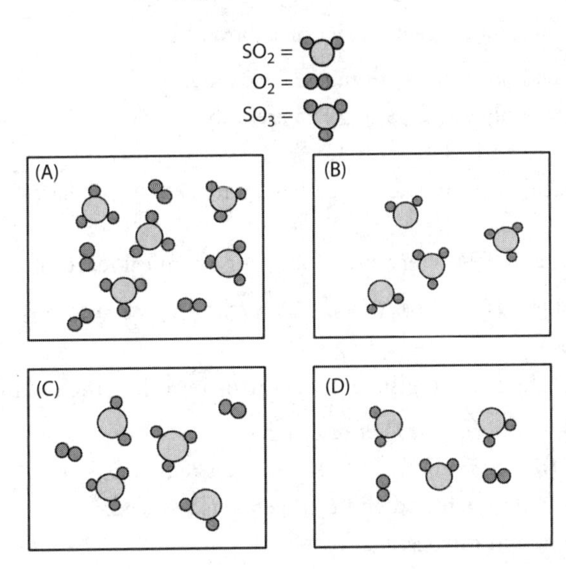

188. The ratio of reactants that would produce 3 moles of SO_3 and leave 0.5 mole of O_2 in excess

189. The ratio of products if 4 moles of SO_2 combine with 1 mole of O_2

190. The ratio of products if 5 moles of SO_2 reacts with 6.5 moles of O_2

$$F^-\ (aq) + H_2O\ (l) \rightleftharpoons HF\ (aq) + OH^-\ (aq)$$

191. Which of the following statements is true of the reaction represented above?

(A) OH^- is the conjugate acid of H_2O.
(B) HF is the conjugate base of F^-.
(C) HF and H_2O are conjugate acid-base pairs.
(D) HF and H_2O are both Brønsted-Lowry acids.

$$CH_3COOH + H_2O \rightleftharpoons CH_3COO^- + H_3O^+$$
$$NH_3 + H_2O \rightleftharpoons NH_4^+ + OH^-$$

192. In the forward *and* reverse reactions above, H_2O acts as

(A) an acid only and a conjugate base only.
(B) a base only and a conjugate base only.
(C) an acid and a base only.
(D) an acid, a conjugate acid, a base, and a conjugate base.

Questions 193 and 194 refer to the following chemical reaction.

$$3\ Cu\ (s) + 8\ H^+\ (aq) + 2\ NO_3^-\ (aq) \rightarrow 3\ Cu^{2+}\ (aq) + 2\ NO\ (g) + 4\ H_2O\ (l)$$

193. Which of the following statements is true regarding the reaction above?

(A) Cu *(s)* and H^+ *(aq)* get reduced.
(B) Cu *(s)* gets oxidized and H^+ *(aq)* gets reduced.
(C) Cu *(s)* gets oxidized and nitrogen gets reduced.
(D) Cu *(s)* and nitrogen get reduced.

194. Which of the following reactions most accurately represents the above reaction?

(A) $Cu\ (s) \rightarrow Cu^{2+}\ (aq) + 2\ e-$
(B) $8\ H^+ + 8e- + 4\ O \rightarrow 4\ H_2O$
(C) $3\ Cu + 2\ N^{+5} \rightarrow 3\ Cu^{2+} + 2\ N^{2+}$
(D) $CuNO_3\ (s) \rightarrow Cu^{2+}\ (aq) + NO\ (aq)$

Questions 195 and 196 refer to the following table of standard reduction potentials.

Half-reaction			E° (V)
$F_2\ (g) + 2\ e^-$	→	$2\ F^-$	**2.87**
$Ag^+\ (aq) + e^-$	→	Ag *(s)*	**0.80**
$Cu^{2+}\ (aq) + 2\ e^-$	→	Cu *(s)*	**0.34**
$2\ H^+\ (aq) + 2\ e^-$	→	**$H_2\ (g)$**	**0.00**
$Fe^{2+}\ (aq) + 2\ e^-$	→	Fe *(s)*	**−0.44**
$Zn^{2+}\ (aq) + 2\ e^-$	→	Zn *(s)*	**−0.76**
$Li^+\ (aq) + e^-$	→	Li *(s)*	**−3.05**

195. Which of the following representations most accurately represents the organization of the table of standard reduction potentials?

(A)

Most easily oxidized	→	
$2\ H^+ + 2\ e^-$	→	$H_2\ (g)$
	→	Strongest oxidizer

(B)

Strongest oxidizer	→	
$2\ H^+ + 2\ e^-$	→	$H_2\ (g)$
	→	Strongest reducer

(C)

	→	Strongest oxidizer
$2\ H^+ + 2\ e^-$	→	$H_2\ (g)$
	→	Most easily reduced

(D)

Most easily oxidized	→	
$2\ H^+ + 2\ e^-$	→	$H_2\ (g)$
Strongest reducer	→	

196. Which of the following statements is true based on the data in the table?

(A) The reduction of hydrogen ions and the oxidation of H_2 *(g)* require no work and can do no work.

(B) The reduction of F_2 *(g)*, Ag^+ *(aq)*, and Cu^{2+} *(aq)* as well as the oxidation of Fe *(s)*, Zn *(s)*, and Li *(s)* have the potential to do work.

(C) The reduction of Li^+ *(aq)* is more spontaneous than the reduction of F_2 *(g)*.

(D) Both of the reduction half-reactions must be thermodynamically favorable for the full reaction to proceed.

$$Li + Ag^+ \rightarrow Li^+ + Ag$$
$$Ni + 2\ Ag^+ \rightarrow Ni^{2+} + 2\ Ag$$

197. Silver ions (Ag^+) are reduced in both of the reactions above. In calculating the cell potential, the same E° value, +0.80 V, will be used for both reactions even though the second reaction requires twice the number of moles of silver. Which of the following statements most accurately explains why changing the stoichiometric coefficient in a half-reaction does not affect the value of the standard reduction potential?

(A) The values used for calculating the cell potential are based on the standard hydrogen electrode under standard conditions. Since all concentrations are 1 M, the stoichiometry doesn't matter.

(B) The standard reduction potential is an intensive property of a substance, so it does not depend on the amount of the substance (or the number of times the reaction occurs).

(C) Because the oxidation and reduction reactions need to be balanced, increasing the amount of substances to be oxidized requires an increased amount of substances to be reduced, which causes the net increase in voltage to be zero.

(D) Reaction stoichiometry is taken into account only when non-standard cell concentrations are used.

198. In a voltaic (galvanic) cell operating under standard conditions, the reaction at the cathode always has a more positive E°_{red} than the reaction at the anode. Which of the following best explains why this is true?

(A) The reaction at the anode is an oxidation described by a value given by E°_{ox}, not a reduction described by the value E°_{red}.

(B) The more positive E°_{red} value of a redox reaction, the more spontaneous the reaction.

(C) Because the anode is negative in a voltaic cell, the value of the E°_{red} at the anode must also be negative in order for an oxidation to occur.

(D) The more positive the value of E°_{red} at the cathode, the greater the driving force for reduction.

199. In a voltaic (galvanic) cell, the cathode and the anode are contained in separate cells (they are not in contact with one another). Which of the following best explains the purpose of this arrangement?

(A) Physical separation of the reduction and the oxidation forces electrons to flow through a wire where electrical energy can be harnessed.

(B) The oxidation at the anode creates positive ions that would interfere with the reduction at the cathode, causing a neutralization in the solution.

(C) The anode and cathode are maintained in separate cells so that the only connection between them is by the salt bridge, which completes the circuit and allows ions to flow between the half-cells.

(D) Reductions cause an increased driving force for oxidation, which would cause the battery to get used up too quickly.

200. A student wants to compare the efficacy of two kinds of antacids, carbonates and hydroxides (in the forms of $CaCO_3$ and $Mg(OH)_2$, respectively), on the neutralization of stomach acid. Which of the following would be the best way to do an accurate comparison?

(A) Equal masses of each compound

(B) Equal numbers of moles of each compound

(C) Equal numbers of moles of CO_3^{2-} and OH^-

(D) Equal masses of carbonate and hydroxide

201. The pressure of PCl_5 in a sealed, rigid, previously evacuated 2 L container is initially 2 atm. The PCl_5 undergoes decomposition according to the following reaction:

$$PCl_5\ (g) \rightarrow PCl_3\ (g) + Cl_2\ (g)$$

If the reaction proceeds to completion, the pressure in the container would be:

(A) 2 atm
(B) 3 atm
(C) 4 atm
(D) The partial pressures cannot be determined without knowing *either* the volume of the container *or* the pressure in the container after the reaction.

202. Which of the following substances would be present in the reaction vessel if equal masses of solid sodium and chlorine gas were reacted according the following reaction?

$$Na\ (s) + \frac{1}{2}\ Cl_2\ (g) \rightarrow NaCl\ (s)$$

(A) NaCl only
(B) NaCl and Na only
(C) NaCl and Cl_2 only
(D) NaCl, Na, and Cl_2

203. Which of the following substances would be present in the reaction vessel if equal masses of solid potassium and chlorine gas were reacted according the following equation?

$$K\ (s) + \frac{1}{2}\ Cl_2\ (g) \rightarrow KCl\ (s)$$

(A) KCl only
(B) KCl and K only
(C) KCl and Cl_2 only
(D) KCl, K, and Cl_2

Questions 204–209 refer to galvanic cells made from different combinations of the half-cells described below.

Half-cell A: Strip of Al *(s)* in 1.00 mol/L $Al(NO_3)_3$ *(aq)*

Half-cell B: Strip of Cu *(s)* in 1.00 mol/L $Cu(NO_3)_2$ *(aq)*

Half-cell C: Strip of Fe *(s)* in 1.00 mol/L $Fe(NO_3)_2$ *(aq)*

Half-cell D: Strip of Zn *(s)* in 1.00 mol/L $Zn(NO_3)_2$ *(aq)*

Galvanic cell	Half-cells	Cell reaction	E°_{cell} (V)
I	A and B	$2\,Al\,(s) + 3\,Cu^{2+}\,(aq) \rightarrow 2\,Al^{3+}\,(aq) + 3\,Cu\,(s)$	2.00
II	A and C	$2\,Al\,(s) + 3\,Fe^{2+}\,(aq) \rightarrow 2\,Al^{3+}\,(aq) + 3\,Fe\,(s)$	?
III	A and D	$2\,Al\,(s) + 3\,Zn^{2+}\,(aq) \rightarrow 2\,Al^{3+}\,(aq) + 3\,Zn\,(s)$	0.90
IV	B and C	$Fe\,(s) + Cu^{2+}\,(aq) \rightarrow Fe^{2+}\,(aq) + Cu\,(s)$	0.78
V	B and D	$Zn\,(s) + Cu^{2+}\,(aq) \rightarrow Zn^{2+}\,(aq) + Cu\,(s)$	1.10
VI	C and D	$Zn\,(s) + Fe^{2+}\,(aq) \rightarrow Zn^{2+}\,(aq) + Fe\,(s)$	?

204. What is the standard cell potential of galvanic cell II?

(A) 0.90 V
(B) 1.22 V
(C) 1.88 V
(D) 2.78 V

205. What is the standard cell potential of galvanic cell VI?

(A) −1.88 V
(B) −0.32 V
(C) 0.32 V
(D) 1.88 V

206. In galvanic cells II and IV, which of the following takes place in half-cell 3?

(A) In cell II, the oxidation of Fe and in cell IV, the reduction of Fe^{2+}
(B) In cell II, the reduction of Fe^{2+} and in cell IV, the oxidation of Fe
(C) In cell II, the oxidation of Fe^{2+} and in cell IV, the reduction of Fe
(D) In cell II, the reduction of Fe and in cell IV, the oxidation of Fe^{2+}

207. In galvanic cells I and V, which of the following takes place in half-cell 2?

(A) In cell I, Cu^{2+} is reduced, and in cell V, Cu is oxidized.
(B) In cell I, Cu is oxidized, and in cell V, Cu^{2+} is reduced.
(C) In both cells, Cu is oxidized.
(D) In both cells, Cu^{2+} is reduced.

208. If half-cell 2 containing 1.00 M $Cu(NO_3)_2$ *(aq)* in galvanic cells I and V is replaced with a half-cell containing 6.00 M $Cu(NO_3)_2$ *(aq)*, what will be the effect on the cell voltage of the two galvanic cells?

(A) The voltage will increase in both cells.
(B) The voltage will decrease in both cells.
(C) The voltage will increase in cell I and decrease in cell IV.
(D) The voltage will increase in cell IV and decrease in cell I.

209. If the half-cell containing 1.00 M $Zn(NO_3)_2$ *(aq)* in galvanic cells III and VI is replaced with a half-cell containing 6.00 M $Zn(NO_3)_2$ *(aq)*, what will be the effect on the cell voltage of the two galvanic cells?

(A) The voltage will increase in both cells.
(B) The voltage will decrease in both cells.
(C) The voltage will increase in cell W and decrease in cell VI.
(D) The voltage will increase in cell VI and decrease in cell III.

210. Which of the following is an accurate prediction made by the reaction below, representing the complete oxidation of glucose? The molar masses of the reactants and products are $C_6H_{12}O_6$, 180 g mol^{-1}; O_2, 32 g mol^{-1}; CO_2, 44 g mol^{-1}; H_2O, 18 g mol^{-1}.

$$C_6H_{12}O_6\ (s) + 6\ O_2\ (g) \rightarrow 6\ CO_2\ (g) + 6\ H_2O\ (l)$$

(A) 108 grams of water are formed when 180 grams $C_6H_{12}O_6$ and 32 grams oxygen gas react.
(B) One gram of CO_2 is formed for every gram of O_2 that reacts.
(C) The reaction of 372 grams of $C_6H_{12}O_6$ and O_2 results in 372 grams of carbon dioxide and water.
(D) For every gram of glucose that reacts, 1.5 grams of carbon dioxide are formed.

Questions 211–214 refer to the following reaction.

$$3\ H_2\ (g) + N_2\ (g) \rightarrow 2\ NH_3\ (g) \quad \Delta H_{fus}{}^\circ = -46.11\ \text{kJ}$$

211. Which of the following is *not* an accurate interpretation of the reaction, representing the complete synthesis of ammonia? Assume the reaction occurs in a closed container under the conditions necessary for it to go to completion.

(A) If equal masses of N_2 and H_2 are combined, the amount of ammonia formed will be approximately 80% of the theoretical yield as compared with if they were combined in stoichiometrically appropriate ratios (if there was no limiting reactant).
(B) Six atoms of hydrogen and 2 atoms of nitrogen are necessary to form 2 molecules of ammonia.
(C) Upon completion of the reaction at a constant temperature, the pressure of the container will decrease by half.
(D) One gram of H_2 requires less than 5 grams of N_2 to produce 2.8 grams of NH_3.

212. Which of the following statements is an accurate description of the chemical process the reaction represents?

(A) It is a chemical synthesis that involves both oxidation and reduction.
(B) It is a neutralization reaction.
(C) The energy needed to break the bonds in H_2 and N_2 exceed the amount of energy released by the formation of NH_3.
(D) The reaction is exothermic because hydrogen bonds require less energy to break than covalent bonds.

213. Which of the following combinations of reactants would provide the most efficient reaction conditions? Assume all other reaction conditions are the same.

	H_2 *(g)* (grams)	N_2 *(g)* (grams)
(A)	1.0	6.00
(B)	2.0	8.00
(C)	3.0	14.0
(D)	4.0	20.0

214. Which of the following combinations of reactants would provide the most efficient reaction conditions? Assume all other reaction conditions are the same.

	H_2 (*g*) (L)	N_2 (*g*) (L)
(A)	10	10
(B)	25	10
(C)	50	15
(D)	100	20

215. Under which of the following sets of conditions is 1 L of a gas equal to 1 mole of a gas?

	P (atm)	T (K)
(A)	8.210	10.00
(B)	22.40	273.0
(C)	273.0	8.210
(D)	40.00	300.0

216. A student is holding a beaker in which a chemical reaction is occurring and the beaker begins to feel cold. Which of the following is a true *and* for the correct reason?

(A) The reaction is endothermic; it absorbed heat from the environment.
(B) The reaction is endothermic; it lost heat to the environment.
(C) The reaction is exothermic; it lost heat to the environment.
(D) The reaction is exothermic; it absorbed heat from the environment.

217. Which of the following pairs of substances will *not* give visible or tactile (through the production or absorption of a significant amount of heat) evidence of a chemical reaction upon mixing?

(A) HCl *(aq)* and KOH *(aq)*
(B) $CaCO_3$ *(aq)* and HF *(aq)*
(C) Mg *(s)* and HI *(aq)*
(D) NH_4NO_3 *(aq)* and HCl *(aq)*

Question 218 refers to the following heating curve.

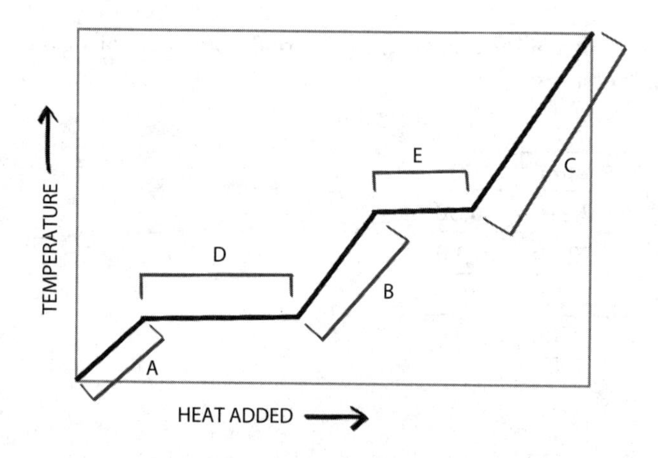

218. Which of the following statements most accurately explains why sections A, B, and C have a positive slope but D and E have a slope of zero?

(A) The processes indicated by D and E are not endothermic.
(B) Processes D and E do not change the kinetic energy of the particles.
(C) Processes D and E are not temperature dependent.
(D) Processes D and E indicate times where nothing is changing in the system.

219. Which two diagrams below accurately illustrate the processes that occur at the anode of the electrochemical cell represented by the following reaction?

$$Zn + Cu^{2+} \rightarrow Zn^{2+} + Cu$$

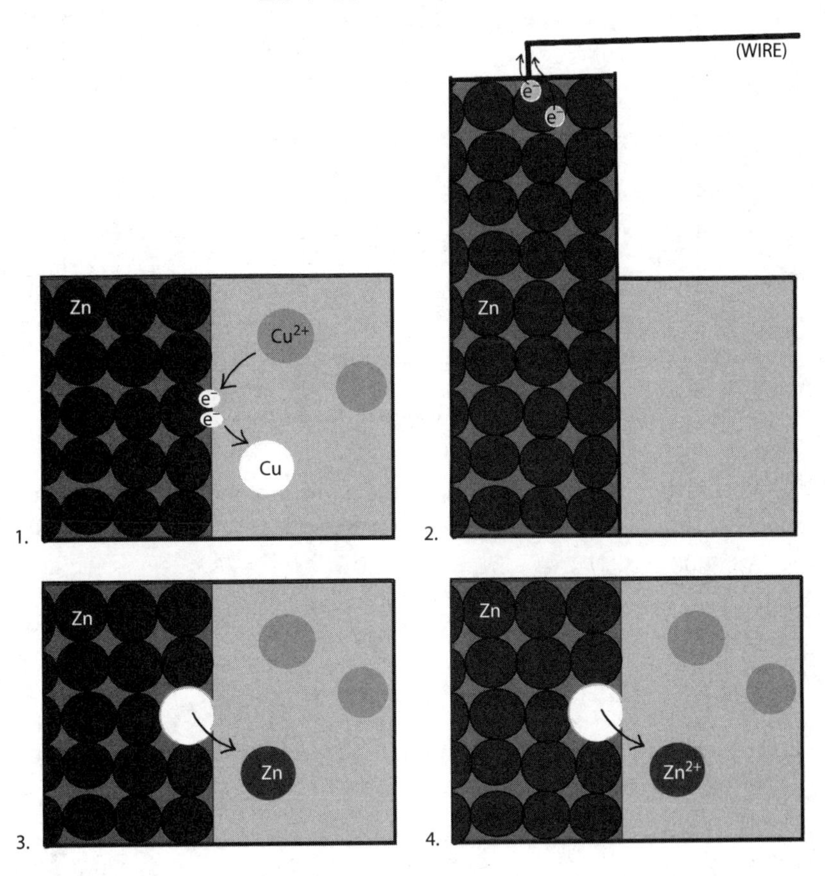

(A) 1 and 3
(B) 1 and 4
(C) 2 and 3
(D) 2 and 4

Questions 220–222 refer to the titration of a solution of a weak monoprotic acid with a 0.1 M strong base, NaOH. The titration curve is shown below.

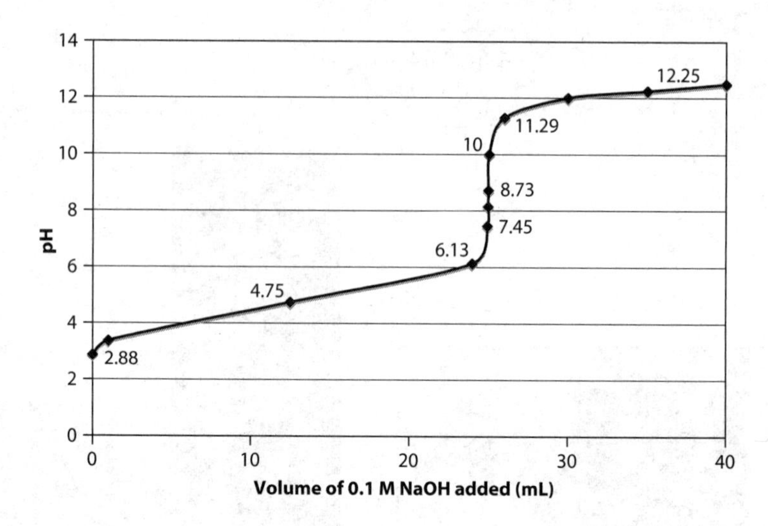

220. What is the value of the pH where the number of moles of strong base added is equal to the number of moles of weak acid in the initial solution?

(A) 7.00
(B) 7.45
(C) 8.73
(D) 10.00

221. At which pH are the concentrations of the weak acid and its conjugate base approximately equal?

(A) 2.88
(B) 4.75
(C) 6.13
(D) 7.00

222. In which of the following pH ranges does the solution act most like a buffer?

(A) 2.88 and 6.13
(B) 4.75 and 7.45
(C) 6.13 and 11.29
(D) 11.29 and 12.25

$$Cl_2\ (g) + 2\ Br^-\ (aq) \rightarrow 2\ Cl^-\ (aq) + Br_2\ (aq)$$

223. Which of the following statements most accurately explains why the reaction represented above generates electricity in a galvanic cell?

(A) Cl_2 is a stronger oxidizer than Br_2.
(B) Cl_2 loses electrons more easily than Br_2.
(C) Br_2 is more stable than Cl_2.
(D) Cl^- is more stable than Br^-.

Questions 224 and 225 refer to a galvanic cell constructed of two half-cells whose two half-reactions are represented below.

$$Zn^{2+}\,(aq) + 2\,e^- \rightarrow Zn\,(s) \qquad E° = -0.76\text{ V}$$
$$Ag^{+}\,(aq) + e^- \rightarrow Ag\,(s) \qquad E° = +0.80\text{ V}$$

224. As the cell operates, which of the following species is contained in the half-cell containing the cathode?

(A) Ag^+ only
(B) Ag and Ag^+
(C) Zn^{2+}
(D) Zn^{2+} and Zn

225. What is the standard cell potential for this galvanic cell?

(A) 0.04 V
(B) 0.84 V
(C) 1.56 V
(D) 2.36 V

$$X\,(s) + 2\,Ag^{+}\,(aq) \rightarrow 2\,Ag\,(s) + X^{2+}\,(aq) \qquad E° = +2.27\text{ V}$$
$$2\,Ag^{+}\,(aq) + 2\,e^- \rightarrow 2\,Ag\,(s) \qquad E° = +0.80\text{ V}$$

226. According to the information given above, what is the standard reduction potential for the half-reaction $X^{2+}\,(aq) + 2e^- \rightarrow X\,(s)$?

(A) +0.67 V
(B) +1.47 V
(C) −1.47 V
(D) +3.07 V

227. An electric current of 1.00 ampere is passed through an aqueous solution of $FeCl_3$. Assuming 100% efficiency, how long will it take to electroplate exactly 1.00 mole of iron metal?

(A) 32,200 sec
(B) 96,500 sec
(C) 193,000 sec
(D) 289,500 sec

Questions 228–233 refer to a galvanic cell. In one half-cell, a Zn electrode is bathed in a 1.0 M $ZnSO_4$ solution. The other cell contains a 1 M HCl solution and a hydrogen electrode.

$$Zn^{2+} + 2\,e^- \rightarrow Zn\ (s) \qquad E° = -0.76\ V$$
$$2\,H^+ + 2\,e^- \rightarrow H_2\ (g) \qquad E° = 0.00\ V$$

228. Which of the following is the correct electron configuration for Zn^{2+}?

(A) $1s^2\ 2s^2\ 2p^6\ 3s^2\ 3p^6\ 4s^1\ 3d^9$
(B) $1s^2\ 2s^2\ 2p^6\ 3s^2\ 3p^6\ 3d^{10}$
(C) $1s^2\ 2s^2\ 2p^6\ 3s^2\ 3p^6\ 3d^{10}\ 4s^2$
(D) $1s^2\ 2s^2\ 2p^6\ 3s^2\ 3p^6\ 3d^8\ 4s^2$

229. Which of the following correctly identifies and justifies which species, Zn or Zn^{2+}, has the highest ionization energy?

(A) Zn; electrons have a greater affinity for pure metals.
(B) Zn; it has a greater electron to proton ratio.
(C) Zn^{2+}; it has a greater nuclear charge.
(D) Zn^{2+}; it has a smaller radius.

230. Which of the following answer choices accurately indicates the process(es) that occur(s) at the anode?

(A) Zn *(s)* is oxidized to Zn^{2+} *(aq)*.
(B) H_2 *(g)* is oxidized to H^+ *(aq)*.
(C) Zn *(s)* is oxidized to Zn^{2+} *(aq)* and H_2 *(g)* is oxidized to H^+ *(aq)*.
(D) Zn *(s)* is oxidized to Zn^{2+} *(aq)* and 2 H^+ *(aq)* is reduced to H_2 *(g)*.

231. The salt bridge is filled with a saturated solution of KNO_3. Which of the following correctly identifies processes that occur at the salt bridge?

(A) K^+ moves into the anode half-cell only.
(B) K^+ moves into the cathode half-cell only.
(C) K^+ moves into the cathode half-cell and NO_3^- moves into the anode half-cell.
(D) K^+ moves into the anode half-cell and NO_3^- moves into the cathode half-cell.

232. Which of the following would be the ideal replacement for 1 M HCl solution in the hydrogen half-cell?

(A) NaCl
(B) HF
(C) $ZnSO_4$
(D) HNO_3

233. Which of the following changes to the cell potential, if any, would result from increasing the concentration of $ZnSO_4$?

(A) No change; the voltage of the cell is determined only by the standard reduction potentials of the electrodes.
(B) Increased voltage; the reaction becomes more favorable as the concentration of Zn^+ increases.
(C) Decreased voltage; the value of Q (the reaction quotient for the reaction) will fall below 1 and the log term in the Nernst equation will become more negative.
(D) Decreased voltage; the increased concentration of Zn^{2+} will inhibit the oxidation of Zn *(s)*.

$Zn^{2+}\ (aq) \rightarrow Zn\ (s)$	$E° = -0.76$ V
$Fe^{2+}\ (aq) \rightarrow Fe\ (s)$	$E° = -0.44$ V

234. Iron nails are often electroplated (galvanized) with zinc to prevent rusting. Which of the following statements accurately states the purpose of electroplating the iron? The standard reaction potentials for the two reactions are given above.

(A) Zn is preferentially oxidized.
(B) Zn serves as an airtight seal preventing the electrons from iron from reacting with oxygen.
(C) The galvanization process produces an alloy of Zn and Fe that is stronger and more resistant to oxidation than Fe alone.
(D) The oxidation of Zn allows the reduction of Fe^{2+} to proceed spontaneously.

235. Which of the following correctly relates the quantities needed for the conversion of chemical energy into electrical energy in a galvanic cell?

(A) The sum of the E°_{cell} values of the two half-cells
(B) The product of the cell voltage and the total charge passed through the cell
(C) The product of cell voltage and the natural log (ln) of K, the equilibrium constant for the redox reaction
(D) The difference between the voltages of the two half-cells

236. A solid, white crystalline substance is added to water to produce a basic solution. When HCl is added to the solution, a gas is liberated. Based on this information, the solid could be

(A) NaOH
(B) $NaNO_3$
(C) Na_2CO_3
(D) Na_2SO_4

Questions 237 and 238 refer to the following titration curve and list of indicators.

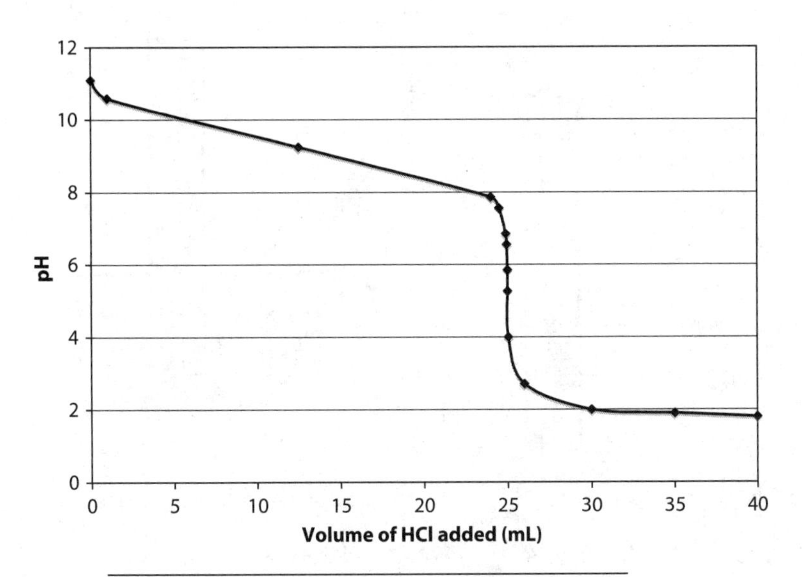

	Indicator	pH range of color change
(A)	Phenolphthalein	8.2–10.0
(B)	Bromthymol blue	6.0–7.6
(C)	Methyl red	4.4–6.2
(D)	Methyl orange	3.1–4.4

237. Which of the above indicators is the most appropriate choice for this titration?

238. Which of the above indicators would transition at the point where the concentration of conjugate acid initially exceeds the concentration of the base?

Questions 239–242 refer to the following diagram and experimental procedure.

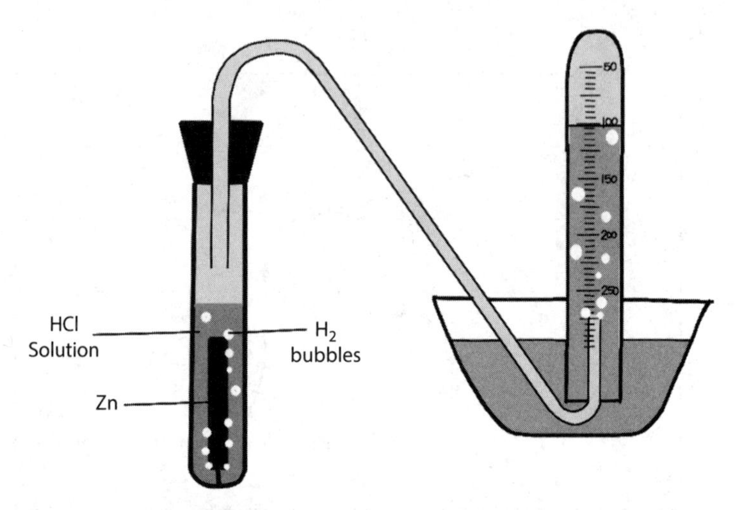

In a laboratory experiment, H_2 gas is produced by the following reaction:

$$Zn\ (s) + 2\ HCl\ (aq) \rightarrow ZnCl_2\ (aq) + H_2\ (g)$$

The H_2 gas is collected over water in a gas-collection tube (a simple eudiometer) as shown. The atmospheric pressure in the laboratory is 770 torr, the temperature of the lab and the water used in the experiment is 22°C, and the volume of the eudiometer is shown in milliliters. The vapor pressure of water at 22°C is 19.8 torr. The value of the gas constant is $0.0821\ L \cdot atm \cdot mol^{-1} \cdot K^{-1}$ or $62.4\ L \cdot torr \cdot mol^{-1} \cdot K^{-1}$.

239. Which of the following steps is necessary to correctly determine the total volume of gas inside the tube?

(A) Make sure measurements are taken under standard conditions by increasing the temperature of the water to 25° C and waiting for the pressure in the eudiometer to reach 760 torr.
(B) Measure the pH of the water to determine how much of the gas remains dissolved in water.
(C) Wait for the water level in the tube to equilibrate with the water level in the dish.
(D) Wait for the Zn electrode to be completely covered with gas bubbles.

240. Which of the following gases would have the greatest percent yield when collected by this technique?

(A) HCl
(B) NH_3
(C) SO_2
(D) CO_2

241. The partial pressure of hydrogen gas in the tube is closest to

(A) 730 torr.
(B) 750 torr.
(C) 760 torr.
(D) 770 torr.

242. The mass of the H_2 gas collected is closest to

(A) 4.0×10^{-2} grams.
(B) 4.0×10^{-3} grams.
(C) 8.0×10^{-3} grams.
(D) 1.6×10^{-2} grams.

FREE-RESPONSE QUESTIONS

Questions 243–248 refer to the following reaction (represented by an unbalanced chemical equation).

$$C_5H_{12}\ (l) + O_2\ (g) \rightarrow H_2O\ (l) + CO_2\ (g)$$

243. **Balance** the chemical reaction using a particulate representation of the reaction. Use the legend that follows as a guide to representing the molecules in the reaction.

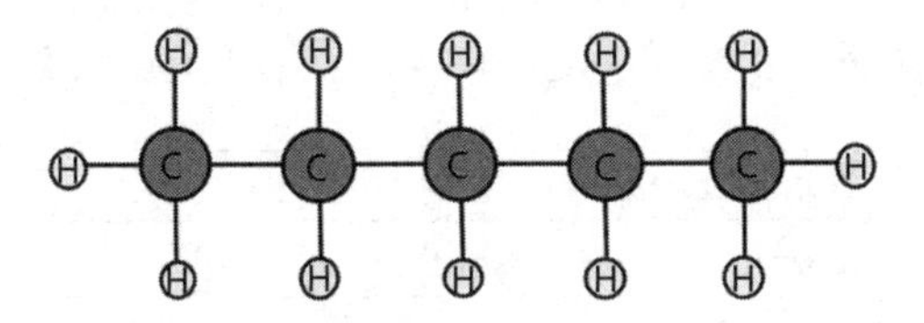

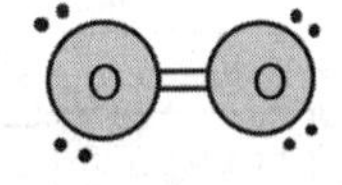

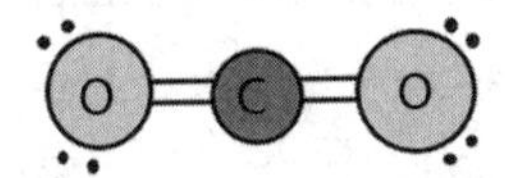

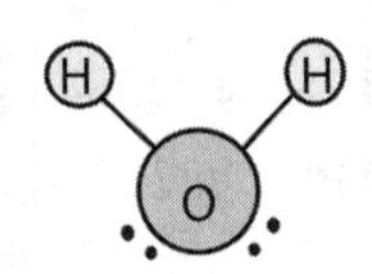

244. **Calculate** the masses of the reactants and products needed for the reaction (round to the nearest whole number). Assume the least whole number of moles of the reactants and products.

245. Calculate the mass of carbon dioxide produced if 100.00 grams of pentane reacts with excess oxygen gas.

246. Calculate the amount of heat produced if 100.0 grams of pentane reacts with excess oxygen gas. The enthalpy of combustion of pentane is −3,509 kJ/mol.

247. Calculate the theoretical yield of carbon dioxide (in mass) and the enthalpy of combustion if 100 grams reacts with 100.0 grams of oxygen gas. **Identify** the limiting and excess reagents.

248. Calculate the difference in volume (at STP) between oxygen consumed and carbon dioxide produced by the combustion of 100 grams of pentane and 100 grams of oxygen.

249. State whether this reaction is an oxidation-reduction (redox) reaction. **Justify** your answer.

Questions 250–260 refer to the galvanic cell operating under standard conditions and the data that follow. The cell contains a Cu and Ag electrode.

Standard Reduction Potentials

$$Ag^+ + 1\,e^- \rightarrow Ag\,(s) \qquad E^\circ = 0.80\text{ V}$$
$$Cu^{2+} + 2\,e^- \rightarrow Cu\,(s) \qquad E^\circ = 0.34\text{ V}$$

Solubility table	**NO_3^-**	**Cl^-**
Cu^{2+}	soluble	soluble
Ag^+	soluble	insoluble
K^+	soluble	soluble

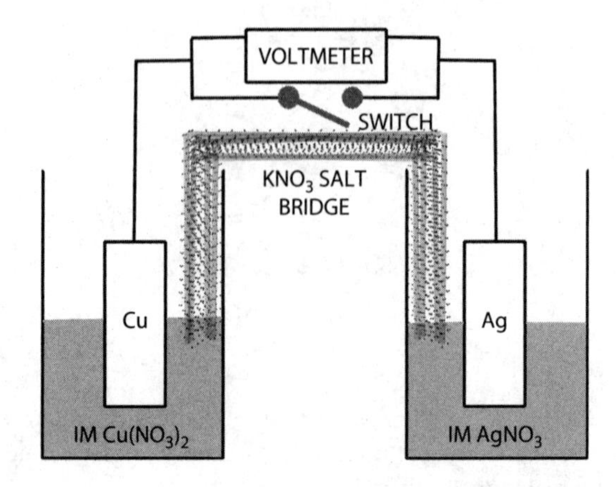

250. Identify the anode and the cathode. **Justify** your answer.

251. Calculate the voltage on the cell when the switch is closed.

252. The mass of each electrode changes as the cell functions. **State** the change in mass that occurs at each electrode. If the electrode gained mass, **explain** where the mass came from. If the electrode lost mass, **explain** where the mass went.

253. In the expanded view of the salt bridge below, **draw** the relevant ions (you can use labeled circles; you do not have to draw the Lewis structures) and **draw** arrows indicating their direction of motion. Do not include water molecules.

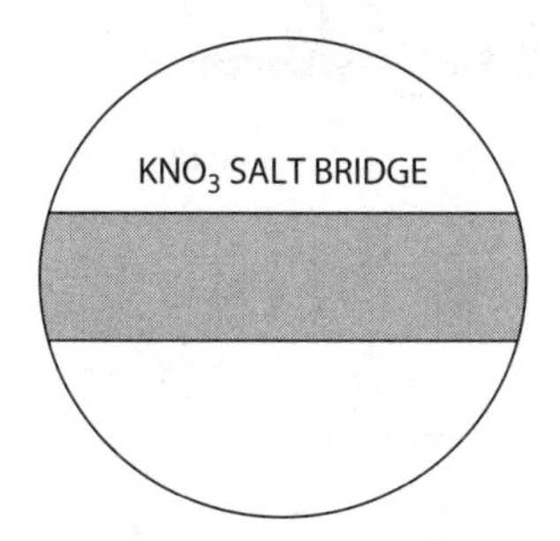

254. Explain the function of the salt bridge.

255. State whether this cell would function in an equivalent manner if KCl were used in the salt bridge instead of KNO_3? **Justify** your answer.

256. Suppose the 1 M standard solutions were replaced with equal volumes of 0.65 M solutions. **State** whether the cell potential would be greater than, equal to, or less than what you calculated in question 251. **Explain.**

257. Compare the length of time that a device could be powered by the standard cell (1 M concentration) versus a cell in which each half-cell had a concentration of 0.5 M. **Justify** your answer.

258. Write the net-ionic equation for the reaction between the Cu and Ag electrodes (in the standard cell) that would be thermodynamically favorable. **Justify** the thermodynamic favorability of the reaction.

259. Calculate the $\Delta G°$ for the reaction of the standard cell. (Make sure to include units!)

260. Imagine a cell where both the anode and the cathode were bathed in a solution of KNO_3, as shown below. **State** whether you would expect this cell to work. **Justify** your answer.

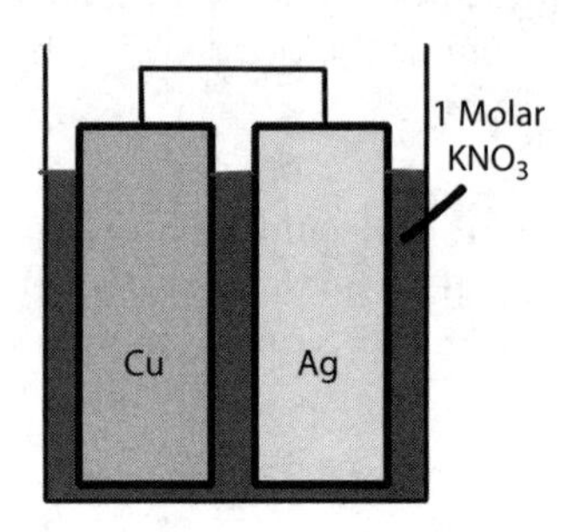

Kinetics

Guiding Principles

Reaction rates are determined by the details of the molecular collisions and influenced by several factors such as concentration and temperature.

MULTIPLE-CHOICE QUESTIONS

261. Some over-the-counter antacids contain mineral carbonates ($CaCO_3$, $MgCO_3$) as their primary ingredient. A chemistry student wants to optimize the rate of neutralization of stomach acid (HCl). Which of the following would be expected to increase the rate at which the mineral carbonates neutralize HCl *in vitro* (not in a biological system)?

(A) Increase the volume of the reaction solution.
(B) Increase the pressure on the solution.
(C) Decrease the particle size of the tablets by grinding them into a powder.
(D) Decrease the temperature at which the reaction occurs.

262. Collisions between molecules of methyl isonitrile can cause a rearrangement of atoms as follows:

$$H_3C{-}N{\equiv}C \rightarrow H_3C{-}C{\equiv}N$$

Which of the following statements is true regarding this reaction?

(A) The energy for the rearrangement is provided for by the collision between the reactants.
(B) The reaction is bimolecular because it involves two types of molecules.
(C) Because the reactant and product contain a triple bond, the reaction requires a multistep mechanism.
(D) Because the reactant and product contain a triple bond, the reaction is improbable and rarely occurs.

263. Which of the following is *not* true regarding the determination of the rate law for a specific reaction?

(A) The rate law can be determined using experimental data of initial reaction rates with varying concentrations of reactants at the same temperature.
(B) The rate law can be determined using experimental data of the initial rates of the elementary steps of a reaction.
(C) The rate law for an elementary step in a reaction is based on its molecularity.
(D) The rate law for equilibrium processes can be expressed using the stoichiometric coefficients of the reactants as reactant orders.

264. A student proposes that the following reaction proceeds via a single elementary reaction.

$$H_2\ (g) + Br_2\ (g) \rightarrow 2\ HBr$$

The experimentally determined rate law for the reaction is rate $= k\,[H_2][Br_2]^{1/2}$. Which of the following statements concerning the experimentally determined rate law is true?

(A) The experimentally determined rate law is consistent with a reaction mechanism involving one elementary reaction.
(B) The experimentally determined rate law is consistent with a reaction mechanism involving two or more elementary steps.
(C) The experimentally determined rate law is consistent with one elementary reaction that is bimolecular.
(D) The experimentally determined rate law is inconclusive because reaction orders must be whole numbers.

Questions 265 and 266 refer to the following reaction.

$$2\ NO\ (g) + O_2\ (g) \rightarrow 2\ NO_2\ (g)$$

The proposed mechanism for the above reaction is as follows:

Step 1: $2\ NO \rightleftharpoons N_2O_2$ *(fast, reversible)*
Step 2: $N_2O_2 + O_2 \rightarrow 2\ NO_2$ *(slow)*

265. Which of the following is true of the fast, reversible step (step 1)?

(A) Step 1 is the rate-determining step.
(B) Step 1 establishes an equilibrium in which the $[N_2O_2]$ is directly proportional to $[NO]^2$.
(C) Step 1 establishes an equilibrium in which the $[N_2O_2]$ is directly proportional to $[NO_2]^2$.
(D) The concentration of N_2O_2 cannot be determined because it is consumed in step 2; therefore, the rate of step 1 cannot be experimentally determined.

266. Which of the following are consistent with the proposed mechanism?

(A) The reaction is third-order.
(B) The first elementary step is unimolecular and the second is bimolecular.
(C) The reaction order of step 1 is one and the reaction order of step 2 is two.
(D) The reaction rate is directly proportional to the concentration of N_2O_2.

Question 267 refers to the following data for the decomposition of ammonia on a hot platinum wire.

$$2\ NH_3 \rightarrow 3\ H_2 + N_2$$

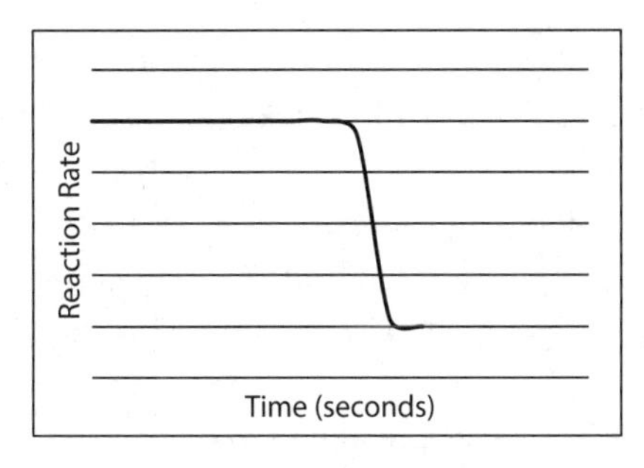

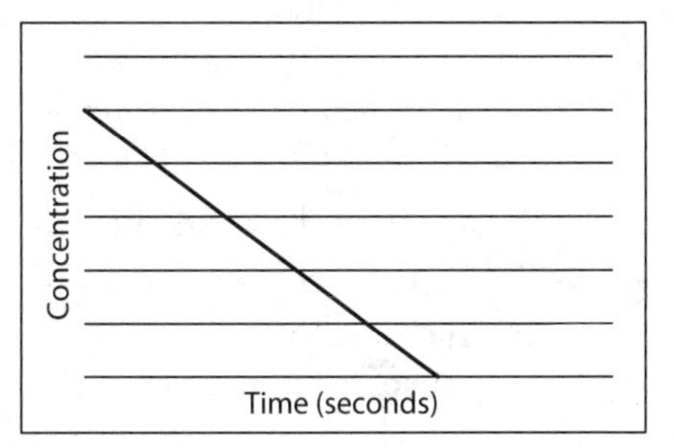

267. Which of the following statements most accurately describes the kinetics of this reaction?

(A) The reaction is first order.
(B) The rate constant is negative.
(C) The reaction rate does not depend on concentration.
(D) The rate of the appearance of products is equal to the rate of the disappearance of reactant.

268. The production of iron (II) sulfide occurs at a significantly higher rate when iron filings are used instead of blocks (volume = 0.1 mL). Which of the following best explains this observation?

(A) The iron filings are partially oxidized due to their greater exposure to oxygen.
(B) The iron in the block is Fe *(s)* and the iron in the filings is Fe^{2+}.
(C) The iron block is too concentrated to chemically react.
(D) The iron filings have a much greater area in contact with sulfur.

269. The conversion of ozone into molecular oxygen occurs in the stratosphere according to the following equation.

$$2\ O_3\ (g) \rightarrow 3\ O_2\ (g)$$
$$\text{Rate} = k\ [O_3]^2[O_2]^{-1}$$

Which of the following statements is true regarding the rate of the breakdown of ozone (O_3) into molecular oxygen (O_2) according to its rate law?

(A) The catalyst that converts ozone to oxygen is inhibited by oxygen.
(B) The rate at which ozone is converted to oxygen gas increases as the concentration of oxygen decreases.
(C) There is an inverse square relationship between ozone and oxygen concentrations.
(D) The conversion of oxygen to ozone is faster than the conversion of ozone to oxygen.

270. Which of the following correctly explains the effect of increased temperature on the rate of a chemical reaction?

(A) Increases the reaction rate of endothermic reactions only
(B) Increases the reaction rate of exothermic reactions only
(C) Increases the reaction rate of endothermic reactions and decreases the rate of exothermic reactions
(D) Increases reaction rates of both exothermic and endothermic reactions

271. Which of the following best describes the purpose of a spark in a butane lighter or a combustion engine?

(A) The spark decreases the activation energy of the combustion reaction.
(B) The spark provides the heat of vaporization for the liquid fuel.
(C) The spark supplies some of the energy to form the activated complex for the combustion reaction.
(D) The spark provides an alternative stoichiometry for the reaction, decreasing the amount of oxygen required for complete combustion.

272. Which of the following would *not* result in an increased rate of reaction in an aqueous solution?

(A) Increasing the temperature of an endothermic reaction
(B) Increasing the temperature of an exothermic reaction
(C) Increasing the surface area of a solid reactant
(D) Increasing the pressure on the solution

273. Which of the following statements is *incorrect* regarding the kinetics of radioactive decay?

(A) The length of time of a half-life is specific to a particular element.
(B) All radioactive decay displays first-order kinetics.
(C) In a sample of a pure, radioactive isotope, one half the number of radioactive atoms and one half the mass of the radioactive substance remains after one half-life.
(D) The half-life of a particular substance does not change with time or temperature.

$$A + B \rightarrow C \qquad E_A = 5.3 \text{ kJ mol}^{-1}$$

$$C \rightarrow A + B \qquad E_A = 5.3 \text{ kJ mol}^{-1}$$

274. Considering the reactions shown above, which of the following must be true of the process?

(A) A catalyst is present.
(B) The reaction order is zero.
(C) The reaction is at equilibrium.
(D) The enthalpy change of the reaction is zero.

275. All of the following are true statements about the properties of catalysts. Which one most accurately accounts for the observation that addition of a catalyst to a system at equilibrium does not change the value of the equilibrium constant?

(A) A catalyst is not consumed by the reaction it catalyzes.
(B) A catalyst speeds up both the forward and reverse reactions.
(C) Catalysts speed up chemical reactions by providing an alternate pathway for reaction in which the activated complex is of lower energy.
(D) A catalyst that works for one chemical reaction may not work for a different reaction.

Question 276 refers to the following data in which a reaction was observed for 60 minutes. Every 5 minutes, the percent of reactant remaining was measured.

Time (min)	**0**	**5**	**10**	**15**	**20**	**25**	**30**	**35**	**40**	**45**	**50**	**55**	**60**
Reactant remaining (g)	100	75	56	42	32	24	18	15	11	7	4	2	1

276. According to the data in the table above, which of the following most accurately describes the reaction order and half-life of this reaction?

	Reaction order	**Half-life (minutes)**
(A)	First	12
(B)	First	8
(C)	Second	12
(D)	Second	8

277. Which of the following most accurately explains the mechanism by which increased temperature increases reaction rates?

(A) Increased temperature increases the number of collisions between particles.
(B) Increased temperature increases the number of particles that are in the proper orientation at time of collision.
(C) Increased temperature increases the kinetic energy of the collisions between reactants.
(D) All of the above correctly explain the effect.

278. It is observed for a particular chemical reaction that the proposed rate law applies only to the first few seconds of the reaction. It does not accurately predict rates at later stages in the reaction. Which of the following is the most likely explanation for this observation?

(A) The reaction is zero-order.
(B) The products take part in the reaction.
(C) The concentrations of reactants changes as the reaction proceeds.
(D) Rate laws only apply to initial reaction rates.

279. The reaction of nitric oxide with hydrogen gas at 25° C and 1 atm is represented below. The rate law for this reaction is rate = $k\,[H_2][NO]^2$.

$$2\,NO\,(g) + 2\,H_2\,(g) \rightarrow N_2\,(g) + 2\,H_2O\,(g)$$

Which of the following choices is the most accurate prediction according to its rate law?

(A) The initial rate of disappearance of NO is the twice the initial rate of disappearance of H_2, regardless of their concentrations.
(B) The initial rate of disappearance of NO is four times as great as the disappearance of H_2, regardless of their concentrations.
(C) The initial rate of disappearance of NO is twice as great as the disappearance of H_2 if the initial concentration of NO is twice that of H_2.
(D) The initial rate of disappearance of NO is four times as fast as that of H_2 if the initial concentration of NO is initially twice that of H_2.

$$2\,NO\,(g) + Br_2\,(g) \rightarrow 2\,NOBr\,(g)$$

280. The reaction between nitrogen monoxide (commonly called nitric oxide) and bromine is represented above. The proposed reaction mechanism is as follows.

$$NO + Br_2 \rightleftharpoons NOBr_2\ \textit{(fast)}$$
$$NOBr_2 + NO \rightarrow 2\,NOBr\ \textit{(slow)}$$

Which of the following rate laws is consistent with the proposed mechanism?

(A) Rate = $k\,[NO]^2$
(B) Rate = $k\,[NO][Br_2]$
(C) Rate = $k\,[NO][Br_2]^2$
(D) Rate = $k\,[NO]^2[Br_2]$

$$NO_2\,(g) + CO\,(g) \rightarrow NO_2\,(g) + CO_2\,(g)$$

281. The reaction between nitrogen dioxide and carbon monoxide is represented above. The proposed reaction mechanism is as follows:

$$\text{Step 1: } NO_2 + NO_2 \rightarrow NO_3 + NO\ \textit{(slow)}$$
$$\text{Step 2: } NO_3 + CO \rightarrow NO_2 + CO_2\ \textit{(fast)}$$

Which of the following reaction mechanisms is consistent with the proposed mechanism?

(A) Rate = $k\,[NO_2]$
(B) Rate = $k\,[NO_2]^2$
(C) Rate = $k\,[NO][CO]$
(D) Rate = $k\,[NO]^2[CO]$

282. A chemistry student sitting around a campfire observes that the large pieces of wood burn slowly, but a mixture of small scraps of wood and sawdust added to the flame combust explosively. The correct explanation for the difference in the rates of combustion between these two forms of wood is that, compared with the wood scraps and sawdust, the large pieces of wood (assume all forms have the same composition).

(A) have a greater surface area-to-volume ratio.
(B) have a smaller surface area per kilogram.
(C) contain compounds with a lower heat of combustion.
(D) contain more carbon dioxide and water.

283. The proposed mechanism of the conversion of ozone (O_3) to molecular oxygen (O_2) by chlorine is shown below.

$$O_3\ (g) + Cl\ (g) \rightarrow O_2\ (g) + ClO\ (g)$$
$$ClO\ (g) + h\nu \rightarrow Cl\ (g) + O\ (g)$$
$$k = 7.2 \times 10^9\ M^{-1}\ s^{-1} \text{ at } 298\ K$$

Which of the following true statements is evidence in support of the proposed reaction mechanism?

(A) The rate at which ozone is destroyed increases linearly with the partial pressure of Cl *(g)*.
(B) Chlorine atoms from chlorofluorocarbon compounds photodissociate at wavelengths between 190 and 225 nm.
(C) The rate at which O_3 forms depends on both the partial pressure of O *(g)* and the frequency of molecular collisions between O_2 *(g)* and O *(g)*.
(D) The large value of *k* for the reaction suggests the reaction rate is very high.

284. Which of the following statements is *not* an accurate description of the activated complex?

(A) The energy of the activated complex determines the activation energy of the reaction.
(B) The activated complex represents the highest energy state along the transition path of a chemical reaction.
(C) The activated complex of a chemical reaction is specific to that reaction.
(D) The configuration of atoms in the activated complex of an uncatalyzed reaction is the same as that of a catalyzed reaction, except for the presence of the catalyst.

Questions 285 and 286 refer to the following graph of a chemical reaction over time at 25° C.

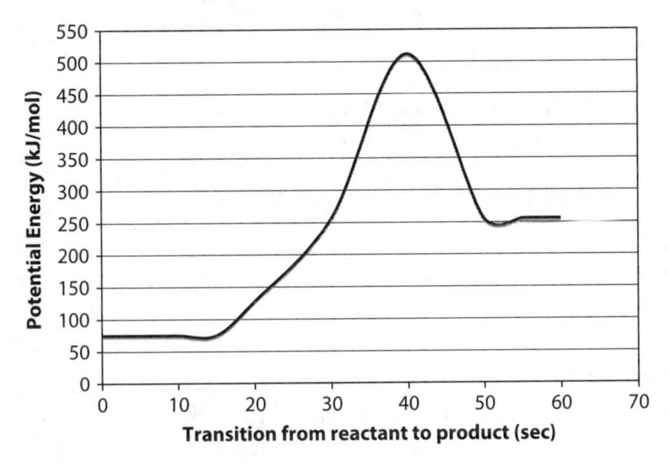

285. Which of the following statements is true concerning the above reaction at 25° C?

(A) The reaction is endothermic at 25° C but may be exothermic at higher temperatures.
(B) The activation energy (E_A) is approximately 510 kJ mol^{-1}.
(C) The activation energy (E_A) is approximately 260 kJ mol^{-1}.
(D) The enthalpy change of the reaction is approximately 175 kJ mol^{-1}.

286. Which of the following statements is the most logical prediction regarding the addition of a catalyst to this reaction?

(A) The activated complex would form in less than 40 seconds.
(B) The enthalpy change would be lower, increasing the reaction rate.
(C) More reactions would occur per second.
(D) The equilibrium would shift to favor the products.

Questions 287 and 288 refer to the data below.

Trial	[X]	[Y]	Formation of product Z
1	0.50	0.1	R
2	0.25	0.2	?

287. If the data were obtained from a reaction whose rate law is rate = $k\,[X][Y]^2$, what would be the expected rate of reaction for trial 2?

(A) R
(B) 2R
(C) $R/2$
(D) $R/4$

288. Suppose the data were obtained for a reaction whose rate law is rate = $k\,[X]^2[Y]$. What would be the expected rate of reaction for trial 2?

(A) R
(B) 2R
(C) $R/2$
(D) $R/4$

289. The proportionality (rate) constant, *k*, can vary by many orders of magnitude because

(A) the time scale over which chemical reactions proceed varies greatly.
(B) the temperature dependence of reaction rates varies greatly between reactions.
(C) the temperature dependence of a particular reaction rate can vary greatly.
(D) most reactions are second and third order, which greatly affects the magnitude of *k*.

Question 290 refers to the following reactions.

$HCl + NaOH \rightarrow NaCl + H_2O$ *(very fast, almost immediate)*
$Ag^+ + Cl^- \rightarrow AgCl$ *(within a few seconds)*
$2\,Fe^{3+} + Sn^{2+} \rightarrow 2\,Fe^{2+} + Sn^{4+}$ *(slow)*
$2\,H_2 + O_2 \rightarrow 2\,H_2O$ *(very slow)*

290. Which of the following does *not* correctly account for the rate differences observed?

(A) Reactions that require energy (endergonic) take longer than reactions that occur in the thermodynamically favored direction.

(B) A great number of collisions between hydronium and hydroxide ions occurs, each collision very likely to produce a water molecule.

(C) Reactions that do not require a transfer of electrons often occur faster than those that do.

(D) The collision between Fe^{3+} and Sn^{2+} ions is somewhat rare and a transfer of electrons does not occur with every collision.

291. A piece of solid zinc is added to a 1 M solution of HCl, and bubbles of H_2 immediately form on the metal. Within a few minutes, the reaction appears to slow down significantly. Which of the following procedures would increase the reaction rate and correctly states the reason?

(A) Brush the H_2 bubbles off the zinc because the reactant is exhausted on the surface.

(B) Brush the H_2 bubbles off the zinc to increase access to the zinc surface by the acid.

(C) Stir the solution to bring fresh acid into the vicinity of the zinc surface.

(D) Stir the solution to increase the kinetic energy in the solution.

292. Which of the following most accurately explains why radioactive decay displays first-order kinetics?

(A) The rate is proportional only to the number of particles present.

(B) Radioactive decay is not temperature dependent.

(C) The rate of decay is linear because it is inhibited by the accumulation of its product.

(D) The decay of a particular isotope is initially random, but the decay of a particular particle becomes less likely as the number of radioactive particles decreases.

Question 293 refers to the following hypothetical reaction.

$$A \rightarrow X$$
$$\text{Rate} = k\,[A]$$
$$k = 0.001\ \text{sec}^{-1}$$

293. Which of the following is true?

(A) During each second 1/1,000 of the A molecules present would be converted to X.
(B) The concentration of X increases by 0.1% each second.
(C) The number of molecules of A decreases linearly with time.
(D) The rate of the reaction does not change with time.

Questions 294–297 refer to the reaction of nitrogen monoxide (nitric oxide) and oxygen and the following data.

$$2\ NO + O_2 \rightarrow 2\ NO_2$$

Trial	[NO] (mol L^{-1})	[O_2] (mol L^{-1})	RATE (M s^{-1})
1	2.4×10^{-2}	3.5×10^{-2}	1.43×10^{-1}
2	1.5×10^{-2}	3.5×10^{-2}	5.60×10^{-2}
3	2.4×10^{-2}	4.5×10^{-2}	1.84×10^{-1}

294. The rate law for this reaction is:

(A) Rate = $k\,[NO][O_2]$
(B) Rate = $k\,[NO][O_2]^2$
(C) Rate = $k\,[NO]^2[O_2]$
(D) Rate = $k\,[NO][O_2]^{1.25}$

295. The numerical value for the rate constant (k) is closest to:

(A) 5.9×10^{-2}
(B) 170
(C) 7.0×10^{3}
(D) 1.2×10^{5}

296. Increasing the initial concentration of NO fivefold would increase the reaction rate by:

(A) 5 times
(B) 10 times
(C) 25 times
(D) 3,125 times

297. What would be the reaction rate if the initial concentration of NO was 2×10^{-2} M and the initial concentration of O_2 was 4×10^{-2} M?

(A) 3.2×10^{-5} M s^{-1}
(B) 8.0×10^{-4} M s^{-1}
(C) 2.3×10^{-2} M s^{-1}
(D) 1.1×10^{-1} M s^{-1}

298. Five trials of the same chemical reaction were performed at different concentrations (all other conditions were identical). The time it took the reactant concentration to decrease to half the initial concentration was the same in all trials. What does this indicate about the reaction order?

(A) Zero-order, because the concentrations did not affect the reaction rate.
(B) Zero-order, because the time it took to reach half the initial concentration was the same in all five trials.
(C) First-order, because the half-life was independent of the concentration.
(D) Second-order, because the half-life measures the time it takes to reach the initial concentration divided by 2.

$$2\ N_2O_5 \rightarrow 4\ NO_2 + O_2$$

299. Which of the following statements most accurately describes the reaction order for the decomposition of N_2O_5, shown above?

(A) The reaction is first-order because the reactant does not need to collide with another reactant molecule to form a product.
(B) The reaction is first-order because the rate can only be dependent on the concentration of one reactant.
(C) The reaction is second-order because two molecules of reactant must collide to form products.
(D) The reaction order is unknown. Reaction order can only be inferred from the elementary steps of a reaction mechanism.

300. Which of the following is an accurate statement about reaction order?

(A) Reaction order must be a whole number.
(B) Reaction order can be determined mathematically using only coefficients of the balanced reaction equation.
(C) Reaction order changes with temperature.
(D) Reaction order can only be determined experimentally.

301. The rate law for the reaction between O_2 and NO is rate = $k\,[O_2][NO]^2$. By what factor will the reaction rate increase if the concentrations of both O_2 and NO are increased from 2.5×10^{-4} M to 5.0×10^{-4} M?

(A) 2
(B) 4
(C) 6
(D) 8

302. The expression to calculate the half-life of a first-order reaction is half-life = $0.693/k$. It follows that

(A) the half-life of a first-order reaction must be a multiple of 0.693.
(B) the half-life of a first-order reaction is independent of the concentration.
(C) the half-life of all first-order reactions is the same.
(D) as the value of the rate constant increases, the reaction rate decreases.

$$NO + O_3 \rightarrow NO_2 + O_2$$
$$\text{Rate} = k\,[NO][O_3]$$

303. Which of the following reaction mechanisms most likely describes the elementary steps of the reaction of nitrogen monoxide (nitric oxide, nitrogen oxide) and ozone shown above?

(A)
$$NO + O_3 \rightarrow NO_2 + O$$
$$O + O \rightarrow O_2$$

(B)
$$NO + O_3 \rightarrow NO + O_2 + O$$
$$NO + O \rightarrow NO_2$$

(C)
$$NO + NO \rightarrow 2\,N + O_2$$
$$N + O_2 \rightarrow NO_2$$

(D)
$$O_3 + O_3 \rightarrow 2\,O_2 + 2\,O$$
$$2\,O \rightarrow O_2$$
$$NO + O_2 \rightarrow NO_2 + O$$
$$O + NO \rightarrow NO_2$$

FREE-RESPONSE QUESTIONS

Questions 304–306 refer to the following data for the decomposition of dinitrogen pentoxide at 65° C.

Initial concentration (M)	Initial rate (M sec^{-1})
0.01	5×10^{-5}
0.02	1×10^{-4}
0.04	2×10^{-4}
0.08	4×10^{-4}

304. Use the grid below to **plot** the initial rate versus initial concentration. **Explain** what the slope of the line represents.

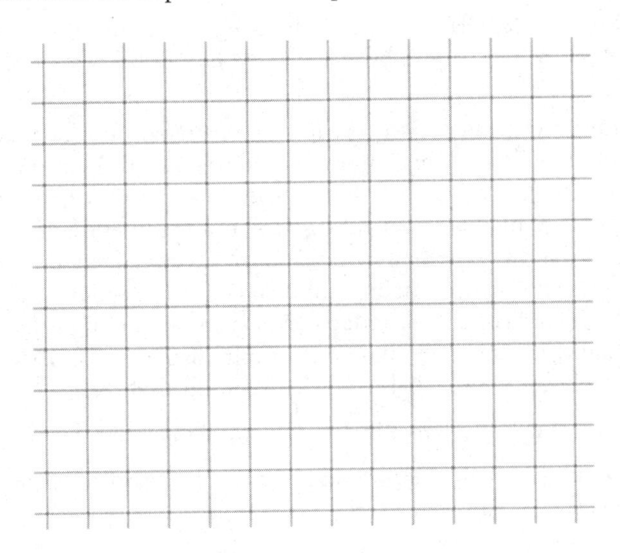

305. **Write** the rate law for the reaction. **Calculate** the value of the rate constant, k, and the **state** total reaction order.

306. (a) **Calculate** the half-life of the reaction and (b) **state** the relationship between the rate constant and the half-life.

307. Reaction rate depends on temperature and activation energy. **Rank** the activation energies, from highest to lowest, of the reactions represented by lines A, B, and C in the graph below. **Justify** your ranking.

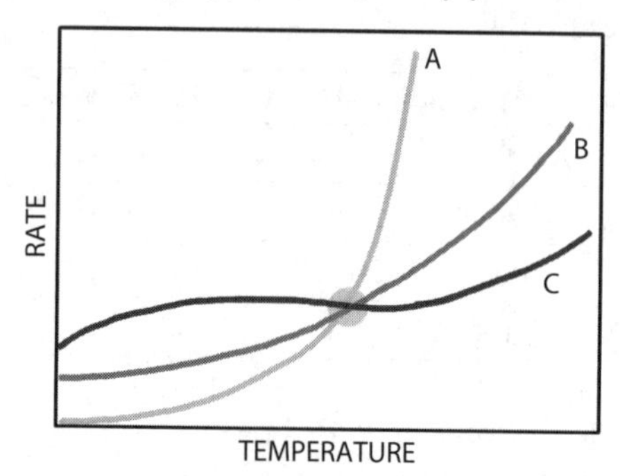

308. Use Kinetic Molecular Theory and Maxwell-Boltzmann speed distribution to **explain** the mechanism by which temperature affects reaction rates.

309. A chemist is consulted in the production of a movie in which there will be explosions. The chemist informs the filmmakers that a container that has been emptied of gasoline (but in which a small bit remains) is more explosive than one that is full of gasoline, but the filmmakers do not agree. To test this hypothesis, two identical gas containers are used. One container is filled with gasoline and then emptied by pouring the gasoline into the second container. Both containers are open to the atmosphere by the opening (a small hole about one inch in diameter). Under very safe, controlled conditions, a lighted match is dropped into each.

(a) **Describe** the difference in combustion rates between the two containers.

(b) **State** which container would produce a more violent explosion. **Justify** your answer.

310. The activation energy and rate constants of two reactions are shown in the table below (the units of the rate constant **are** L mol^{-1} sec^{-1}).
State which reaction has greater temperature dependence. **Justify** your choice.

Reaction	Activation energy, E_A (kJ mol^{-1})	Value of rate constant, *k*, at 35° C	Value of rate constant, *k*, at 37° C	Value of rate constant, *k*, at 40° C
I $C_{12}H_{22}O_{11} + H_2O \rightarrow 2\ C_6H_{12}O_6$	108	6.9×10^{-4}	1.0×10^{-4}	—
II $C_2H_5Br + OH^- \rightarrow C_2H_5OH + Br^-$	89	2.8×10^{-4}	—	5.0×10^{-4}

Energy

Guiding Principles

The laws of thermodynamics describe the essential role of energy and both explain and predict the direction of changes in matter.

MULTIPLE-CHOICE QUESTIONS

Questions 311–313 refer to the following reactions.

	Reaction	$\Delta H°$ (kJ)	$\Delta S°$ ($J \cdot K^{-1}$)
(A)	$2\ CO\ (g) + O_2\ (g) \rightarrow 2\ CO_2\ (g)$	−566.0	−173.0
(B)	$2\ H_2O\ (g) \rightarrow 2\ H_2\ (g) + O_2\ (g)$	484.0	90.0
(C)	$2\ N_2O\ (g) \rightarrow 2\ N_2\ (g) + O_2\ (g)$	−164.0	149.0
(D)	$PbCl_2\ (s) \rightarrow Pb^{2+}\ (aq) + 2\ Cl^-\ (aq)$	23.4	−12.5

311. Which reaction is spontaneous at all temperatures?

312. Which reaction is thermodynamically favored at low temperatures but becomes nonspontaneous at high temperatures?

313. Which reaction can do the most work at 298 K?

Questions 314–316 refer to the following data.

Container	Gas	Molar mass (g mol^{-1})	Temperature (K)	Pressure (atm)
(A)	O_2	32	298	1
(B)	N_2	28	298	1
(C)	H_2	2	298	3

314. Which gas has the greatest average kinetic energy?

(A) A
(B) B
(C) C
(D) The same in all three containers

315. Which container contains the gas molecules moving at the greatest speed?

(A) A
(B) B
(C) C
(D) The same in all three containers

316. The temperature of container A is lowered to 100 K. Which of the following will *not* occur to the gas in the container? Assume the container is rigid.

(A) The volume of the gas will decrease.
(B) The pressure of the gas will decrease.
(C) The speed of the particles will decrease.
(D) The kinetic energy of the particles will decrease.

Questions 317 and 318 refer to the dissolution process described below.

Dissolving an ionic solid in water is a multistep process in which three distinct processes must occur. The three steps are:

1. The separation of the ions in the solid
2. The separation of the solvent molecules (water)
3. The formation of ion-dipole interactions between the ions and the water molecules

317. What is the ΔH of each step?

(A) All of the steps are either endothermic or exothermic, depending on the ΔH of solution of the solute/solvent pair.
(B) Steps 1 and 2 are endothermic and step 3 is exothermic.
(C) Steps 1 and 2 are exothermic and step 3 is endothermic.
(D) Steps 1 and 2 are either both endothermic or both exothermic, depending on the solute and solvent, but step 3 is always the opposite.

318. The enthalpy of solvation of NaCl crystals in H_2O is + 3.87 kJ·mol^{-1} NaCl at 298 K. The conclusion is

(A) the dissolution of sodium chloride is endergonic (requires work) but occurs because the temperature is high.
(B) the separation of solid ions and water molecules requires more energy than the energy released from the attraction of water molecules to ions.
(C) every step in the dissolution process is endothermic, but only slightly.
(D) the attraction of water molecules to ions is greater than the sum of the repulsion between the sodium and chloride ions and between the water molecules for each other.

Questions 319–323 refer to the following reaction.

$$\text{Na}\ (s) + 1/2\ \text{Cl}_2\ (g) \rightarrow \text{NaCl}\ (s) \qquad \Delta H^\circ = -411\ \text{kJ}$$

The elements Na and Cl react directly to form NaCl according the equation above. Refer to the equation above and the table below to answer the questions that follow.

Process	ΔH° (kJ)
$\text{Na}\ (s) \rightarrow \text{Na}\ (g)$	v
$\text{Na}\ (g) \rightarrow \text{Na}^+\ (g) + e^-$	w
$\text{Cl}_2\ (g) \rightarrow 2\ \text{Cl}\ (g)$	x
$\text{Cl}\ (g) + e^- \rightarrow \text{Cl}^-\ (g)$	y
$\text{Na}^+\ (g) + \text{Cl}^-\ (g) \rightarrow \text{NaCl}\ (s)$	z

319. How much heat is absorbed or released when 0.10 moles of Na *(s)* are consumed to form NaCl *(s)*?

(A) 41.1 kJ is released.
(B) 41.1 kJ is absorbed.
(C) 82.2 kJ is released.
(D) 82.2 kJ is absorbed.

320. How much heat is absorbed or released when 0.05 moles of Cl_2 *(g)* are formed from NaCl *(s)*?

(A) 41.1 kJ is released.
(B) 41.1 kJ is absorbed.
(C) 82.2 kJ is released.
(D) 82.2 kJ is absorbed.

321. Which value(s) of ΔH for the process is/are less than zero (i.e., is/are exothermic)?

(A) z only
(B) y and z only
(C) w, y, and z only
(D) w, x, y, and z only

322. The reaction producing NaCl from its standard-state elements is typically observed "going to completion." Which of the following statements is true regarding the thermodynamic favorability of the reaction?

(A) The reaction is favorable and driven by both an enthalpy and an entropy change.
(B) The reaction is favorable and driven only by an enthalpy change.
(C) The reaction in unfavorable and driven only by an enthalpy change.
(D) The reaction is unfavorable due to the entropy change.

$$2\ Na\ (g) + Cl_2\ (g) \rightarrow 2\ NaCl\ (s)$$

323. Which of the following expressions is equivalent to the reaction above?

(A) $2w + x + y + z$
(B) $w + x + y + z$
(C) $2w + x + 2y + 2z$
(D) $2w + 1/2\ x + 2y + 2z$

Questions 324–326 refer to the following experiment.

A student designed a procedure to determine the heat of fusion of ice. She constructed a calorimeter using a polystyrene cup (with a removable cover with two openings) and a thermometer. She weighed the cup, filled it with 150 mL of warm water, measured the temperature of the water, and then weighed the cup and water together.

An ice cube taken from an ice bath stored in a −20°C freezer (there was still some liquid water left in the bath) was added to the cup of water and the cup, with the lid, was weighed again. The student then inserted a thermometer into one of the openings in the cover and inserted a stirring rod into the other opening to gently stir the contents of the cup until all the ice was melted. The lowest temperature reached by the water in the cup was recorded.

The student's final data is recorded below.

$T_{initial}$ of water	25°C
Mass of cup	6.40 g
Mass of cup + water	156.50 g
Mass of cup + water + ice	181.56 g
T_{final} of water	14°C

324. The purpose of weighing the cup and its contents the third (last) time was to

(A) determine the mass of water that was added.
(B) determine the mass of ice that was added.
(C) determine the mass of the calorimeter and thermometer.
(D) determine the mass of water that evaporated while the lid was off the cup.

325. Suppose a significant amount of water from the ice bath adhered to the ice cubes the student added to her calorimeter. How would this affect the value of the heat of fusion of ice she calculated?

(A) There would be no effect because the water from the ice bath would be at the same temperature as the ice cubes.
(B) The calculated value would be too large because less warm water needed to be cooled.
(C) The calculated value would be too large because more cold water needed to be heated.
(D) The calculated value would be too small because less ice melted than was weighed.

326. Which of the following pieces of information is/are necessary to calculate the heat of fusion of ice from the data that was collected?

(A) The specific heat of water
(B) The specific heat of water and the specific heat of ice
(C) The specific heat of water, the specific heat of ice, and the thermal conductivity of water
(D) The specific heat of water, the specific heat of ice, the thermal conductivity of water, and the thermal expansion coefficient of ice

Question 327 refers to the following data.

Compound	$\Delta H°_f$ (kJ mol^{-1})
CO_2 *(g)*	–393.5
CaO *(s)*	–635.5
$CaCO_3$ *(s)*	–1,207.1

$$CaCO_3\,(s) \rightarrow CaO\,(s) + CO_2\,(g)$$

327. The decomposition of $CaCO_3$ *(s)* is shown in the equation above. Using the data in the table above the reaction, which of the following values is closest to the ΔH_{rxn} of the decomposition of $CaCO_3$ *(s)*?

(A) –2,240 kJ mol^{-1}
(B) –180 kJ mol^{-1}
(C) 180 kJ mol^{-1}
(D) 2,240 kJ mol^{-1}

$$C\,(diamond) + O_2\,(g) \rightarrow CO_2\,(g) \qquad \Delta H = -395.4\text{ kJ}$$
$$C\,(graphite) + O_2\,(g) \rightarrow CO_2\,(g) \qquad \Delta H = -393.5\text{ kJ}$$
$$C\,(graphite) \rightarrow C\,(diamond) \qquad \boldsymbol{\Delta H_{rxn} = ?}$$

328. The reactions for the combustion of diamond and graphite are shown above. Which of the following values is closest to the ΔH_{rxn} for the conversion of C *(graphite)* to C *(diamond)*?

(A) –789.0 kJ
(B) –1.9 kJ
(C) 1.9 kJ
(D) 798.0 kJ

329. The standard Gibbs free-energy change, $\Delta G°_{298}$, for the conversion of C *(diamond)* into C *(graphite)* has an approximate value of –3 kJ mol^{-1}. However, graphite does *not* form from diamond under standard conditions (298 K and 1 atm). Which of the following best explains this observation?

(A) Diamond has a lower entropy (is more ordered) than graphite.
(B) The activation energy for the conversion of diamond to graphite is very large.
(C) The C–C bonds in diamond are much stronger than the C–C bonds in graphite.
(D) Diamond is significantly denser than graphite.

$$2\ CH_6N_2\ (l) + 5\ O_2\ (g) \rightarrow 2\ N_2\ (g) + 2\ CO_2\ (g) + 6\ H_2O\ (g)$$
$$\Delta H = -1{,}303\ \text{kJ mol}^{-1}\ CH_6N_2\ (l)$$

330. The combustion of methylhydrazine, a common rocket fuel, is represented above. What would be the ΔH per mole CH_6N_2 *(l)* if the reaction produced H_2O *(l)* instead of H_2O *(g)*? (The enthalpy change of the condensation of H_2O *(g)* to H_2O *(l)* is –44 kJ mol^{-1}.)

(A) –1,039 kJ
(B) –1,179 kJ
(C) –1,435 kJ
(D) –1,567 kJ

331. A student adds solid ammonium chloride to a beaker containing water at 25° C. As it dissolves, the beaker feels colder. Which of the following are true regarding the ΔH and ΔS of the dissolution (solvation) of NH_4Cl *(s)*?

	ΔH	ΔS
(A)	+	+
(B)	+	–
(C)	–	+
(D)	–	–

332. Suppose a reaction is thermodynamically favorable (spontaneous) only at temperatures below 300 K. The $\Delta H°$ for this reaction is –18.0 kJ mol^{-1}. Which of the following is closest to the $\Delta S°$ of this reaction? Assume that $\Delta S°$ and $\Delta H°$ do not change significantly with temperature.

(A) –60. J $mol^{-1}K^{-1}$
(B) –18. J $mol^{-1}K^{-1}$
(C) –0.06 J $mol^{-1}K^{-1}$
(D) 18. J $mol^{-1}K^{-1}$

Question 333 refers to the following table.

Bond	Energy (kJ mol^{-1})
H–H	432
O=O	494
H–O	459

333. Given the bond energies in the table above, which of the following statements best describes the formation of 1 mole of H_2O *(l)* from H_2 *(g)* and O_2 *(g)*?

(A) The process is endothermic with an enthalpy change of approximately 480 kJ.
(B) The process is endothermic with an enthalpy change of approximately 240 kJ.
(C) The process is exothermic with an enthalpy change of approximately 480 kJ.
(D) The process is exothermic with an enthalpy change of approximately 240 kJ.

$$CaCl_2\,(s) \rightarrow Ca^{2+}\,(aq) + 2\,Cl^-\,(aq)$$

334. The entropy change for the dissolution of calcium chloride in water shown above might be expected to be positive, but the actual value of the ΔS is negative. Which of the following is the most plausible explanation for the net loss of entropy during this process?

(A) The ions in the solution are more ordered than the ions in the solid.
(B) The ions in solution can move more freely than the ions in the solid.
(C) The decreased entropy of the water molecules in the solution is greater than the increased entropy of the ions in the solution relative to the solid.
(D) The distance between ions in solution is much greater than the distance between the ions in the solid.

335. Which of the following combinations of enthalpy and entropy changes accurately describes the combustion of wood into CO_2 *(g)* and H_2O *(g)* in a campfire?

(A) $\Delta H > 0, \Delta S > 0$
(B) $\Delta H < 0, \Delta S < 0$
(C) $\Delta H > 0, \Delta S < 0$
(D) $\Delta H < 0, \Delta S > 0$

336. Which of the following processes demonstrates a decrease in entropy ($\Delta S < 0$)?

(A) Br *(s)* → Br *(l)*
(B) Combining equal volumes of $C_2H_6O_2$ *(l)* and H_2O *(l)*
(C) The precipitation of PbI_2 from solution
(D) The thermal expansion of a helium balloon

$$Ag\ (s) \rightleftharpoons Ag\ (l)$$

337. The normal melting point of Ag *(s)* is 962° C. Which of the following is true for the process represented above at 962° C?

(A) $\Delta H = 0$
(B) $\Delta S = 0$
(C) $T\Delta S = 0$
(D) $\Delta H = T\Delta S$

338. Which of the following is true regarding the adiabatic (no heat flow between the system and its surroundings) and reversible compression of an ideal gas?

(A) The temperature of the gas remains constant.
(B) The pressure of the gas remains constant.
(C) No work can be done by the gas.
(D) The net entropy change of the gas is zero.

$$C_2H_5OH\ (l) + 3\ O_2\ (g) \rightarrow 2\ CO_2\ (g) + 3\ H_2O\ (l) \qquad \Delta H_{rxn} = -1{,}367\ \text{kJ}$$

Compound	ΔH°_f
C_2H_5OH *(l)*	–278 kJ
H_2O *(l)*	–286 kJ
CO_2 *(g)*	?

339. The equation for the combustion of ethanol, C_2H_5OH *(l)*, and selected standard heats of formation (ΔH°_f) are shown above. The standard heat of formation of CO_2 *(g)* is closest to:

(A) –1,080 kJ mol^{-1}
(B) –540 kJ mol^{-1}
(C) –510 kJ mol^{-1}
(D) –390 kJ mol^{-1}

$$2\ NO_2\ (g) \rightarrow N_2O_4\ (g)$$

340. What is the standard enthalpy change, $\Delta H°$, of the reaction represented above?

The $\Delta H°_f$ of NO_2 = 34 kJ mol^{-1} and the $\Delta H°_f$ of N_2O_4 = 9.7 kJ mol^{-1}

(A) –24.3 kJ
(B) –58.3 kJ
(C) 24.3 kJ
(D) 58.3 kJ

Questions 341–345 refer to the data in the following table.

Reaction	Equation	$\Delta H°_{298}$ (kJ)	$\Delta S°_{298}$ (J·K^{-1})	$\Delta G°_{298}$ (kJ)
X	$C\ (s) + H_2O\ (g) \rightleftharpoons CO\ (g) + H_2\ (g)$	+131	+134	+91
Y	$CO_2\ (g) + H_2\ (g) \rightleftharpoons H_2O\ (g) + CO\ (g)$	+41	+42	+29
Z	$2\ CO\ (g) \rightleftharpoons C\ (s) + CO_2\ (g)$	?	?	?

341. What is the value of $\Delta H°_{298}$ for reaction Z?

(A) 90 kJ mol^{-1}
(B) –90 kJ mol^{-1}
(C) 172 kJ mol^{-1}
(D) –172 kJ mol^{-1}

342. In which of the following reactions will the value of K_P *increase* under greater pressure?

(A) X only
(B) Z only
(C) X and Z only
(D) None of the K_P values will change with increased pressure.

343. In which of the following reactions will the value of K_P *increase* if the temperature is raised above 298 K?

(A) X and Y only
(B) Z only
(C) X, Y, and Z
(D) None of the K_P values will change with increased temperature.

344. Which of the following statements most accurately describes the entropy change of reaction Z?

(A) Positive, because there are more species of products than reactants.
(B) Positive, because there are a greater number of moles of products than reactants.
(C) Negative, because an element was formed.
(D) Negative, because a solid was formed.

345. Which of the following statements most accurately compares the rates of reactions X and Y?

(A) Y will occur more rapidly than X because the enthalpy change is less positive.
(B) X will occur more rapidly than Y because the entropy change is more positive.
(C) Y will occur more rapidly than X because the free-energy change is less positive.
(D) Reaction rates cannot be inferred from the thermodynamic data provided in the table.

346. Which of the following graphs shows the changes in a gas sample moving from higher temperature to lower temperature?

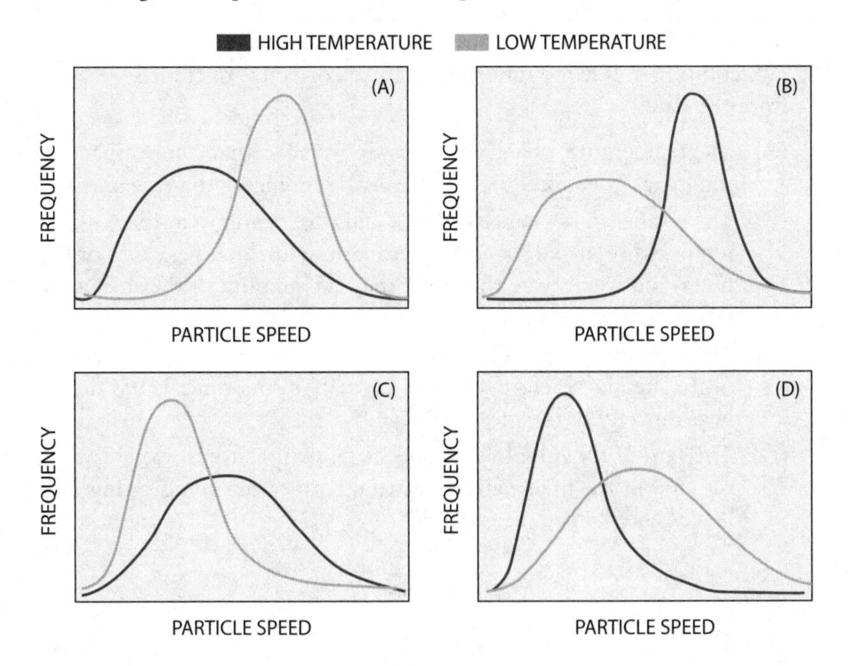

347. Which of the following is a correct interpretation of the velocity distribution of a sample of gas at a particular temperature?

(A) The height of the peak represents the particles of the highest speed.
(B) The height of the peak represents the most common velocity of the gas particles in the sample.
(C) The area under the curve can be used to approximate the number of particles in the sample.
(D) The area under the curve estimates the number of particles at a particular temperature.

Questions 348–351 refer to the following experiment and data.

Fifty grams of four different metals were heated to 100° C then immediately transferred to an insulated container containing 100 grams of water at 25° C. When thermal equilibrium was reached, the final temperature of the water was measured and recorded.

Metal	T_{metal} initial	T_{metal} final
Al	100° C	32.3
Fe	100° C	28.8
Cu	100° C	28.3
Au	100° C	26.1

348. Which of the following statements can be correctly concluded from the data?

(A) The temperature change of the water was less than the temperature of the metal; therefore, the water gained less energy than the metals lost.
(B) The temperature change of the metal was greater than the temperature of the water but because the final temperatures of the metal and water were the same, the total amount of thermal energy transferred was equal.
(C) The temperature change of the water was less than the temperature of the metal; therefore, the heat capacity of the water is less than the heat capacity of the metal.
(D) The sum of the initial water and metal temperatures was greater than the sum of the final water and metal temperatures, indicating a total loss of energy from the system.

349. Which metal has the highest specific heat (in J °C^{-1} g^{-1})?

(A) Al
(B) Fe
(C) Cu
(D) Au

350. Which metal experienced the greatest change in temperature *per atom of metal*?

(A) The metal with the highest specific heat
(B) The metal with the greatest molar mass
(C) The metal with the greatest temperature change
(D) They are approximately equal.

Questions 351–358 refer to the figure below, showing the temperature changes of 0.250 kg water when heated at a constant rate at 1 atm of pressure. Assume no mass is lost during the experiment.

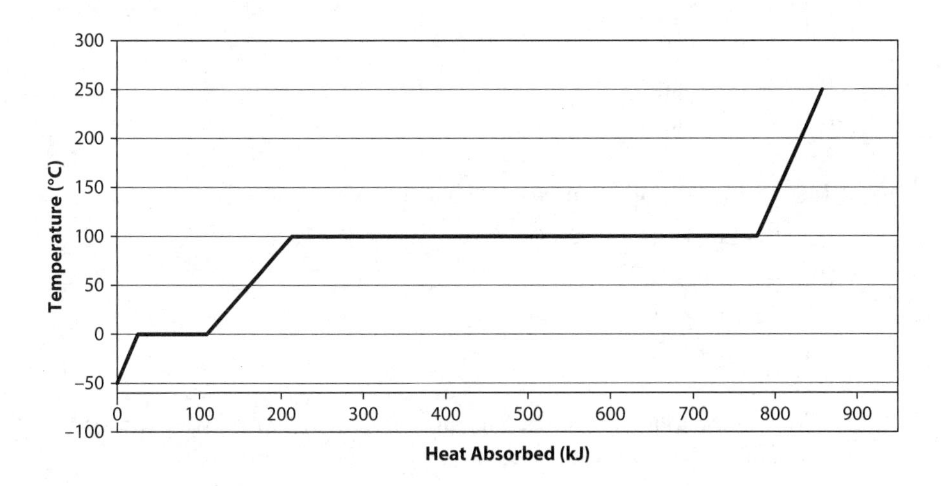

351. The sample of water requires the greatest input of energy during

(A) the melting of ice at 0° C.
(B) the heating of water from 0° C to 100° C.
(C) the vaporization of water at 100° C.
(D) The heating of steam from 100° C to 250° C.

352. Which of the following statements most accurately describes what is happening at 0° C?

(A) The average kinetic energy of the particles is increasing as heat is being absorbed.
(B) The average distance between the molecules is decreasing.
(C) The number of hydrogen bonds between the molecules is increasing.
(D) The potential energy of the substance is decreasing.

353. The heat of fusion is closest to:

(A) 75 kJ kg^{-1}
(B) 150 kJ kg^{-1}
(C) 300 kJ kg^{-1}
(D) 600 kJ kg^{-1}

354. The heat of vaporization is closest to:

(A) 1,600 kJ kg^{-1}
(B) 2,300 kJ kg^{-1}
(C) 3,200 kJ kg^{-1}
(D) 4,000 kJ kg^{-1}

355. The specific heat of steam (water vapor) is closest to:

(A) 2.0 kJ kg^{-1}°C^{-1}
(B) 4.2 kJ kg^{-1}°C^{-1}
(C) 6.0 kJ kg^{-1}°C^{-1}
(D) 8.4 kJ kg^{-1}°C^{-1}

356. Which of the following statements most accurately explains the disparity in values of the heat of fusion and the heat of vaporization?

(A) Water molecules are moving farther apart during fusion than during vaporization.
(B) Water molecules are moving closer together during fusion and farther apart during vaporization.
(C) Vaporization occurs at a higher kinetic energy than fusion.
(D) A greater number of hydrogen bonds are broken during vaporization compared to fusion.

357. The data in the heating curve graph can be used to estimate

(A) the enthalpy of formation of water.
(B) the enthalpy of hydrogen bond formation.
(C) the amount of time it takes for water to melt at 0° C.
(D) the density of water at 50° C.

358. Which of the following is *not* true regarding the energy and entropy changes in the water during the experiment?

(A) The kinetic energy of the water continuously increases as heat is added.
(B) The potential energy of water vapor is greater than that of ice.
(C) There are two points on the curve where only the potential energy and entropy of the water are increasing.
(D) The rearrangement of the water molecules during phase changes increases their potential energy.

Questions 359–361 refer to the graph below, which shows the formation of a hydrogen molecule from two hydrogen atoms.

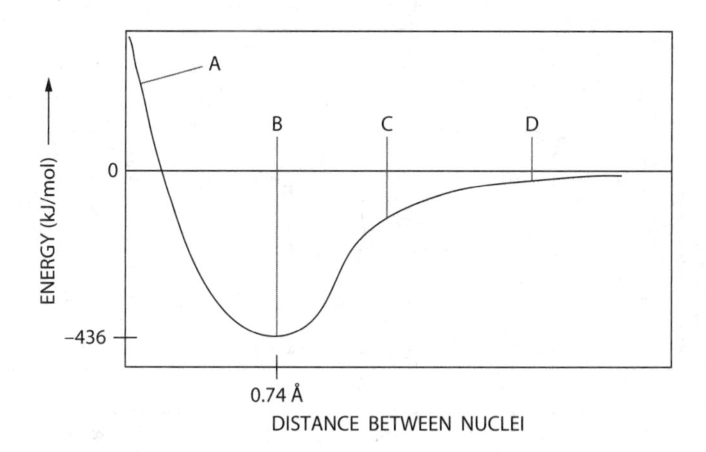

359. Which of the following quantities corresponds to the value of -436 kJ mol^{-1}?

(A) The energy change of bond formation
(B) The entropy change of bond formation
(C) The standard enthalpy of formation of hydrogen gas
(D) The ionization energy of hydrogen

360. Which of the following quantities corresponds to the value of 0.74 Å?

(A) The diameter of a hydrogen atom
(B) One-half the atomic radius of the hydrogen molecule
(C) The equilibrium bond distance
(D) The radius of the 1s orbital

361. Which of the following best explains the high energy at point A?

(A) The nuclei are very far apart.
(B) Nuclear and electronic repulsion.
(C) Nuclear and electronic attraction.
(D) The bond length is very small.

Question 362 refers to the data table and diagrams that follow. Each of the potential energy versus distance graphs correspond to one of the bonds in the table.

Bond	Bond length (Å)	Bond energy (kJ/mol)
C–C	1.54	331
N–N	1.45	159
I–I	2.67	151
Br–Br	2.28	193

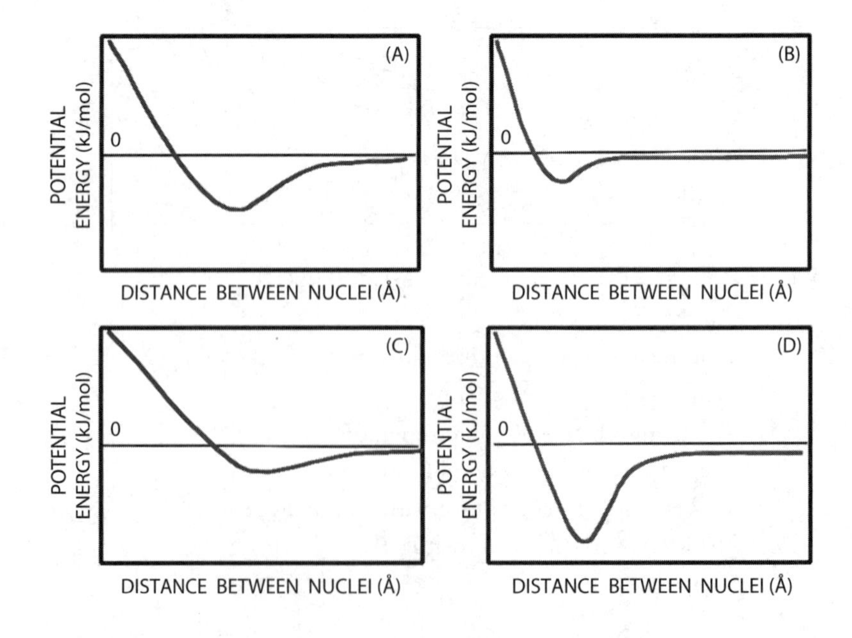

362. Which of the diagrams corresponds to the C—C bond?

363. Which of the following statements most accurately describes the limitation of the model of chemical bonding to which it refers?

(A) The VSEPR model provides a simple way to predict molecular shapes but it does not explain why bonds form between atoms.

(B) Valence-bond theory and molecular orbital theory are useful in allowing chemists to predict bond energies but are not reliable for predicting the shapes of molecules.

(C) Lewis electron diagrams are too simple to be useful in modern chemistry.

(D) The quantum mechanical model states that bond energies and lengths are inherently uncertain because they cannot be directly measured.

364. Which of the following most accurately explains how a bond can form between two atoms as the two atoms approach each other?

(A) The concentration of electron density between the nuclei causes increased electron repulsion until the distance between them is so small that they become attracted to each other.

(B) As the distance between the nuclei decreases, the electrons of each of the atoms are attracted to their own nucleus as well as the nucleus of the other atom.

(C) The electrostatic repulsion between the nuclei is negated by the electrons between them.

(D) As the electron orbitals begin to overlap, electron-electron repulsion is overcome by the attraction of the electrons of one atom for the nucleus of another until the nucleus-nucleus repulsion limits the minimum interatomic distance.

Question 365 refers to the data in the tables that follow.

Bond	Bond energy (kJ mol^{-1})
H–H	436
F–F	155
Cl–Cl	242
Br–Br	193

Reaction	ΔH
$\frac{1}{2}\ H_2\ (g) + \frac{1}{2}\ Br_2\ (g) \rightarrow HBr\ (g)$	-51 kJ mol^{-1}
$\frac{1}{2}\ H_2\ (g) + \frac{1}{2}\ Cl_2\ (g) \rightarrow HCl\ (g)$	-92 kJ mol^{-1}
$\frac{1}{2}\ H_2\ (g) + \frac{1}{2}\ F_2\ (g) \rightarrow HF\ (g)$	-269 kJ mol^{-1}

365. Which of the following statements is a logical conclusion based on the data?

(A) H–H bonds have a greater bond energy than H–F bonds.
(B) H–Br bonds have the lower bond energy than Br–Br bonds.
(C) The radii of atoms involved in bonding is inversely proportional to the energy of the bond energy.
(D) In bonds between two different elements, the greater the difference in electronegativity, the greater the strength of the bond between them.

$$NCl_3\ (g) \rightarrow N_2\ (g) + \frac{3}{2}\ Cl_2\ (g) + 235\ \text{kJ mol}^{-1}$$

366. The decomposition of nitrogen trichloride is shown above. Which of the following statements is the most likely explanation for the large, negative enthalpy change of the reaction?

(A) The nitrogen-chlorine bonds are very stable.
(B) Nitrogen gas (N_2) has a large bond energy.
(C) The electronegativity difference between nitrogen and chlorine is very large.
(D) The products are elements in their standard states.

367. Which of the following combinations of thermodynamic functions is true for the mixing of two ideal gases? Assume no reaction occurs, there are no intermolecular forces of attraction or repulsion between any of the gas particles, and the process occurs at constant temperature and pressures.

(A) $\Delta G = 0$
(B) $\Delta G = 0, \Delta S > 0$
(C) $\Delta G < 0, \Delta S = 0$
(D) $\Delta G < 0, \Delta S > 0$

FREE-RESPONSE QUESTIONS

Questions 368–372 refer to the thermal decomposition of ammonium carbonate.

Ammonium carbonate thermally decomposes into ammonia, carbon dioxide, and water according to the equation that follows.

$$(NH_4)_2CO_3\ (s) \rightarrow 2\ NH_3\ (g) + CO_2\ (g) + H_2O\ (g)$$

368. Using the standard enthalpies of formation provided in the table below (a) **calculate** the enthalpy change of this reaction, and (b) **state** whether this reaction is endothermic or exothermic. **Support** your answer with data and calculations.

Compound	**$\Delta H°_f$ (kJ mol^{-1})**
NH_3 *(g)*	−46.1
CO_2 *(g)*	−393.5
H_2O *(g)*	−241.8
$(NH_4)_2CO_3$ *(s)*	−939.0

369. Using the bond energies provided in the table below, (a) **calculate** the enthalpy change of this reaction, and (b) **state** whether this reaction is endothermic or exothermic. **Support** your answer with data and calculations.

Bond	**Bond Energy (kJ mol^{-1})**
N-H	391
C-O	360
C=O	802
H-O	463

370. **Compare** your answers from questions 368 and 369. (a) **Explain** why they do or do not agree. (b) **Explain** why you think they should or should not agree.

371. A 10.0 g sample of ammonium carbonate solid was placed in a rigid, evacuated 5.00 L container and heated to 500 K. The sample completely decomposed. **Calculate** the final pressure in the 2.00 L container after decomposition.

372. **Calculate** the partial pressures of each of the gases present.

Questions 373–376 refer to the information below.

Compound	**ΔH°_f (kJ mol^{-1})**
C_8H_{18}	−250.0
CO_2	−393.5
H_2O	−241.8

In a combustion engine, a mixture of fuel and oxygen is sprayed into a chamber where it is then ignited by a spark plug. The combustion occurs by the reaction that follows.

$$2\ C_8H_{18}\ (l) + 25\ O_2\ (g) \rightarrow 16\ CO_2\ (g) + 18\ H_2O\ (g) + \text{heat}$$

Through a series of mechanical linkages, the vertical motion of the piston is ultimately converted into rotational motion (depicted below).

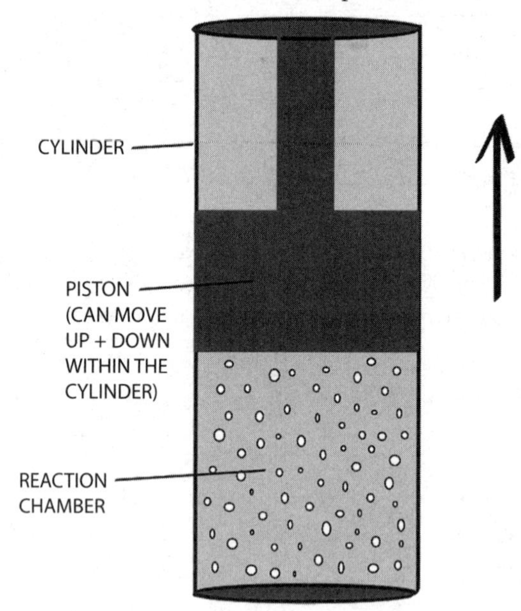

373. Use the table of bond energies that follows to **calculate** the enthalpy change of the reaction.

Bond	**Bond energy (kJ mol^{-1})**
C-C	347
C-H	413
O=O	495
C=O (in CO_2)	799
H-O	467

374. Use the table of standard heats of formation to (a) **calculate** the enthalpy change of the reaction. (b) **Compare** this value to the answer you calculated from question 373.

375. **Explain** how the combustion of the fuel results in work done on the piston.

376. **Describe** the changes in enthalpy, entropy, and free energy.

Questions 377–379 refer to the following experiment.

A coffee cup calorimeter was used to determine the heat of fusion of ice. The calorimeter was assembled using a polystyrene cup (with a removable cover with two openings) and a thermometer (shown below).

The cup was weighed, filled with 150 mL of warm water, and reweighed. The initial temperature of the water was taken. An ice cube was taken from an ice bath (assume ice cube temperature is 0° C) and added to the cup of water. The cup, lid, water, and ice cube were weighed.

The student then inserted a thermometer into one of the openings in the cover and inserted a stirring rod into the other opening to gently stir the contents of the cup until all of the ice was melted. The lowest temperature reached by the water immediately after the ice had finished melting was recorded. The data are given in the table below.

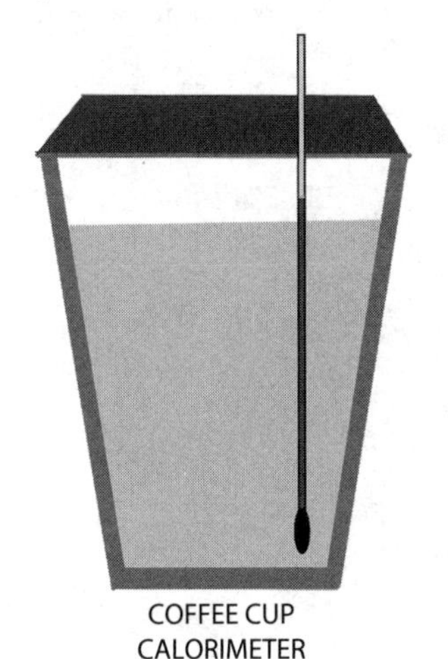

$T_{initial}$ of water	30.00° C
Mass of cup	6.40 g
Mass of cup + water	156.50 g
Mass of cup + water + ice	187.96 g
T_{final} of water	13.25° C

377. **Calculate** the heat of fusion of ice. The specific heat of water is $4.18\ J\ g^{-1°}\ C^{-1}$.

378. **State** the assumptions used in calorimetry. **State** whether or not you think each assumption is reasonable. **Justify** your answers.

Question 379 refers to the following information.

The melting of water is endothermic, but spontaneous.

$S°$ of H_2O *(l)* = 70 J mol^{-1} K^{-1}
$S°$ of H_2O *(s)* = 48 J mol^{-1} K^{-1}

379. (a) **Calculate** the molar enthalpy of fusion (in kJ mol^{-1}) for water using the accepted value for the heat of fusion of water (334 J g^{-1}).

(b) **Calculate** the ΔG (Gibbs free energy) of this process at 1°C and explain how the process can be both spontaneous and endothermic. Provide support for your reasoning.

380. Use the Kinetic Molecular Theory of Gases (KMT) to **explain** why the pressure of a gas at a constant volume increases with temperature.

381. Use the Kinetic Molecular Theory of Gases (KMT) to **explain** why the volume of a gas at a constant pressure increases with temperature.

Question 382 refers to the following situation in which two equimolar quantities of gases are separately contained in two compartments of equal volume but different temperatures, as shown below. The high-temperature compartment is 600 K and the low-temperature compartment is 300 K. The gas in box A is helium and the gas in box B is xenon. Assume there was no exchange of heat before the partition was lifted.

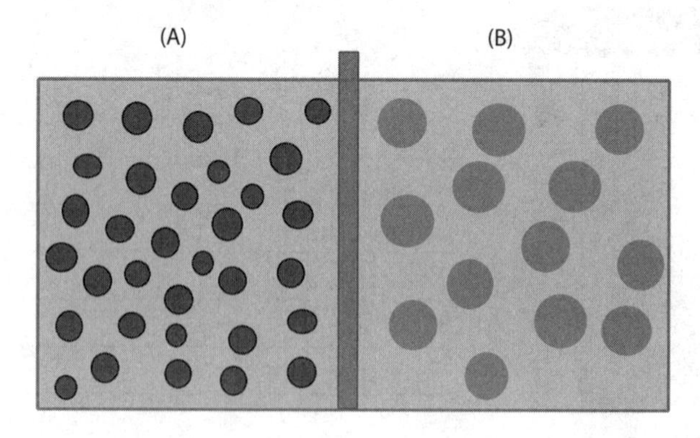

382. **Explain** what will happen after the partition is removed and the gases have mixed and reached equilibrium if (a) compartment A is at the higher temperature and (b) if compartment B is at the higher temperature. Include the expected frequency distributions of the particle speeds in your explanation.

383. When small metal cartridges of compressed gases (like N_2O or CO_2) are pierced (at room temperature at atmospheric pressure) to allow the compressed gas to quickly escape, frost forms on the cartridge exterior. **Explain** what causes the rapid decrease in cartridge temperature.

384. A particular type of glove warmer contains a supersaturated solution of sodium acetate with a metal disk containing two small slits in the middle. When the metal disk is bent, the interior of the metal disk is exposed, allowing the sodium acetate to crystallize. The crystallization process releases heat, which is used to warm the hands inside the gloves. The used glove warmers can be reused after they have been boiled for several minutes, which allows the sodium acetate crystals to dissolve. **Explain** the thermodynamic and chemical principles by which the glove warmers work.

385. Cold packs are commonly used in first aid to "ice" injuries (to reduce inflammation). They do not get as cold as ice but they can significantly lower the temperature of the surface to which they are applied. Cold packs contain ammonium nitrate in one compartment and water in a separate, adjacent compartment. When the cold pack is needed, the barrier between compartments is broken and the contents mix. Upon mixing, the temperature significantly lowers.

(a) **Explain** how the mixing of the contents of the two compartments lowers the temperature.

(b) **Calculate** the mass of NH_4NO_3 (molar mass = 80 g mol^{-1}, ΔH_{sol} = 25.7 kJ mol^{-1}) needed to lower the temperature of 250 grams of water from 25° C to 5° C. Assume the specific heat of the solution is the same as that of water, 4.18 kJ° C^{-1} kg^{-1}. You do not need to include the mass of the NH_4NO_3 in the mass of the water or solution.

Question 386 refers to the Dewar flask, also called a vacuum flask, invented by Sir James Dewar in 1892. The Dewar flask is an insulating storage device that greatly increases the time its contents remain at a different temperature from their surroundings. The flask consists of two flasks joined together at the neck with space between them (see diagram below). The space between the two flasks is evacuated of air creating a near-vacuum.

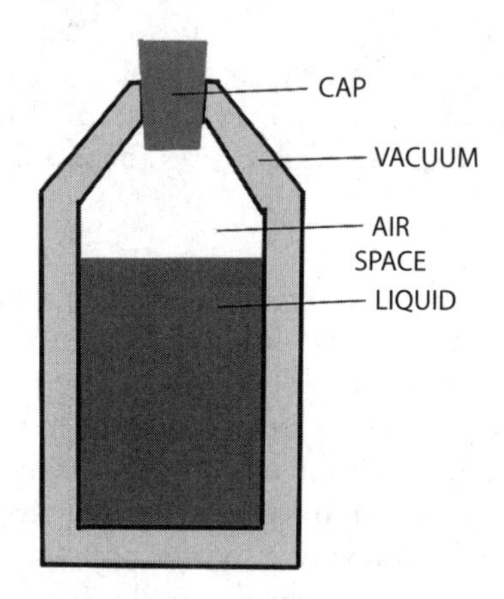

386. Explain how the Dewar flask is able to maintain its contents at a different temperature from its surroundings.

Question 387 refers to the following data.

Element	Electronegativity
H	2.20
N	3.04
F	3.98
Cl	3.16
Br	2.96
I	2.66

Compound	Standard enthalpy of formation ($\Delta H°_f$, kJ mol⁻1)
NH_3	−46.19
NF_3	−132.09
NCl_3	232.00
NBr_3	278.00
NI_3	287.00

Bond	Bond energy (kJ mol^{-1})
N–H	391
N–F	272
N–Cl	200
N–Br	243
N–I	159
N≡N	941

387. **Rank** the substances N_2, NH_3, NF_3, and NI_3 according to their predicted stability. **Justify** your ranking.

Equilibrium

Guiding Principles

- Chemical processes rearrange relationships between particles.
- These processes are reversible.
- Breakage and formation of interparticle relationships are in a dynamic competition.
- The reaction conditions determine the favorability of each opposing process.
- The relationship between the processes changes as the conditions change.

MULTIPLE-CHOICE QUESTIONS

388. Which of the following statements most accurately describes a reaction at equilibrium?

(A) The forward and reverse reaction rates are zero.
(B) At equilibrium, there is no net driving force for either the forward or reverse reaction.
(C) The concentration of products is equal to the concentration of reactants and their concentrations do not change over time.
(D) The equilibrium state is reached when both the entropy and energy of the system are at a maximum.

$$CO\ (g,\ 1\ atm) + \tfrac{1}{2}\ O_2\ (g,\ 1\ atm) \rightleftharpoons CO_2\ (g,\ 1\ atm)$$
$$\Delta G^\circ = -257\ \text{kJ at } 25^\circ\ \text{C}$$
$$\Delta H^\circ = -283\ \text{kJ at } 25^\circ\ \text{C}$$

389. Which of the following is true regarding the reaction above between carbon monoxide and oxygen?

(A) An increase in temperature will result in a decreased reaction rate.
(B) The maximum work that can be done by this reaction is 257 kJ.
(C) The reaction will proceed spontaneously under any conditions.
(D) Increased entropy is a driving force of this reaction.

Questions 390 and 391 refer to the synthesis of ammonia and the data below.

$$N_2\ (g) + 3\ H_2\ (g) \rightleftharpoons 2\ NH_3\ (g)$$

Temperature (°C)	K_{eq}
300	4.34×10^{-3}
400	1.64×10^{-4}
450	4.51×10^{-5}
500	1.45×10^{-5}
550	5.38×10^{-6}
600	2.25×10^{-6}

390. At 500° C, 3 moles each of N_2 (*g*), H_2 (*g*), and NH_3 (*g*) are combined in a 3 L container. Which of the following will occur as the system moves toward equilibrium?

(A) Net production of N_2 (*g*) + 3 H_2 (*g*)
(B) Net production of NH_3 (*g*)
(C) The total pressure will decrease.
(D) No net reaction will occur.

391. Based on the data above, the enthalpy of the reaction is most likely

(A) endothermic.
(B) exothermic.
(C) zero once equilibrium has been reached.
(D) unable to be determined from equilibrium data.

392. An unknown, monoprotic acid is titrated with a strong base. At the beginning of the titration, the pH = 2 and equivalence is reached at pH 8. Which of the following species is present in the highest concentration at pH 3 (the titration has not yet reached half-equivalence)?

(A) HA
(B) A^-
(C) H_3O^+
(D) OH^-

$$N_2O_4 \rightleftharpoons 2\ NO_2$$
$$K_P = 0.212$$

393. Pure N_2O_4 is added to a rigid, evacuated container at 100° C. Which of the following is a true statement about this system when it reaches equilibrium?

(A) The concentrations of N_2O_4 and NO_2 are equal.
(B) The partial pressure of N_2O_4 is approximately one-fifth of the total pressure.
(C) The system can do useful work.
(D) The ratio of N_2O_4 to NO_2 to is approximately 5:1.

394. Caffeine ($C_8H_{10}N_4O_2$) is a weak base ($K_b = 4 \times 10^{-4}$). The pH of a 0.1 M solution of caffeine is approximately:

(A) 8
(B) 10
(C) 11
(D) 12

Question 395 refers to the reaction below, in which thin magnesium strips were placed in a container with air at 500 K.

$$2\ Mg\ (s) + O_2\ (g) \rightleftharpoons 2\ MgO\ (s)$$
$$\Delta H°_f\ MgO\ (s) = -601.8\ kJ\ mol^{-1}$$

395. Which of the following stresses when applied to the reaction above will result in an increased amount of MgO *(s)*?

(A) Use Mg filings instead of strips.
(B) Increase the temperature.
(C) Decrease the pressure of the reaction vessel.
(D) Decrease the volume of the reaction vessel.

$$N_2\ (g) + O_2\ (g) + Cl_2\ (g) + 104\ kJ \rightleftharpoons 2\ NOCl\ (g)$$

396. Suppose the reaction represented above is at equilibrium. Which of the following changes will result in an increased amount of O_2 *(g)*?

(A) Increase the pressure.
(B) Decrease the volume.
(C) Add more N_2 *(g)*.
(D) Decrease the temperature.

Question 397 refers to the following diagram.

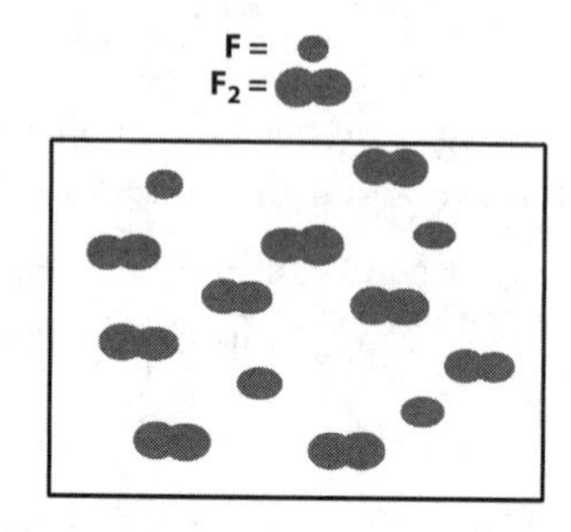

397. The diagram above represents a mixture of F_2 and F gases in a 1.0 L container at equilibrium according to the equation $F_2\ (g) \rightleftharpoons 2\ F(g)$ at a temperature above 1,200 K. Which of the following is true of the equilibrium constant for this reaction at this temperature?

(A) $K < 1$
(B) $K > 1$
(C) $K = 1$
(D) K cannot be determined because the initial concentrations are not known.

$$HX\ (aq) + Y^-\ (aq) \rightleftharpoons X^-\ (aq) + HY\ (aq)$$

398. HX and HY are monoprotic acids of different strengths. The reaction between HX and the conjugate base of HY is shown above. If the equilibrium constant for this reaction is 2.5×10^3, which of the following is a reasonable conclusion about the chemical species involved?

(A) HY is a stronger acid than HX.
(B) Y^- is a stronger base than X^-.
(C) The pH of a solution containing a 1:1 mole ratio of HX and Y^- is 7.
(D) The pK_a of HX = $-\log_{10} 2.5 \times 10^3$.

399. Suppose 50.0 mL of 0.100 M H_2SO_4 is added to 50.0 mL of a 0.100 M solution of lactic acid. Which of the following is true of the resulting solution?

The K_a of lactic acid ($HCH_3H_5O_3$) is 1.4×10^{-4}.
The K_a of hydrogen sulfate, HSO_4^-, is 1.2×10^{-2}.

(A) A smaller fraction of lactic acid and hydrogen sulfate dissociate.
(B) The pH is less than the pH of either of the two original solutions.
(C) Hydrogen sulfate behaves like a base.
(D) The strong conjugate base of the weakest acid neutralizes the pH of the solution.

Questions 400 and 401 refer to the following information.

Acid	K_a at 298 K
HOCl	2.9×10^{-8}
HOBr	2.4×10^{-9}
HOI	?

400. Which of the following is true regarding the strengths of HOCl and HOBr?

(A) HOBr is a stronger acid because Br is a larger atom than Cl and loses its proton more easily than HOCl.

(B) HOCl is a stronger acid because Cl is more electronegative than HOBr.

(C) HOBr is the stronger acid than HOCl because the negative log of its K_a, the pK_a, is larger than the pK_a of HOCl.

(D) The strength of the acids cannot be compared without knowing the concentration and pH of their respective solutions.

401. Which of the following statements is the most accurate prediction of the strength of the acid HOI?

(A) HOI is the strongest acid because I is a larger atom than Br and Cl and therefore the proton dissociates more readily in solution.

(B) HOI is the strongest acid because the O–H bond in HOI is stronger than the O–H bond in HOCl or HOBr.

(C) HOI is the weakest acid because OI^- is a stronger base than OCl^- and OBr^-.

(D) The strength of HOI depends on the pH and the concentration of the solution.

Question 402 refers to the following titration curves. In one titration (black line), 0.05 M HCl was titrated with 0.1 M NaOH. In the second titration (gray line), 0.0005 M HCl was titrated with 0.001 M NaOH.

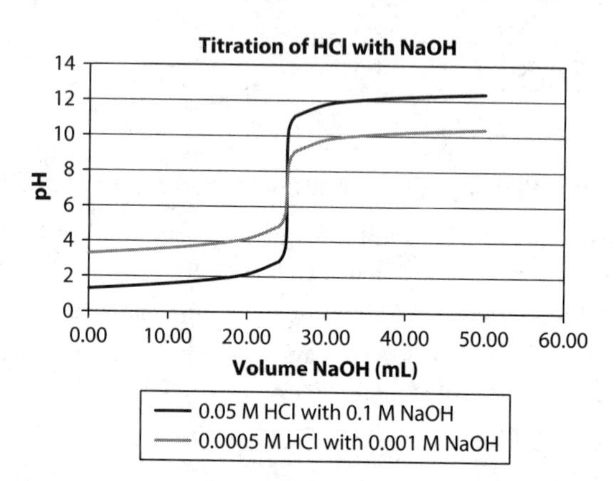

402. Which of the following statements is *not* a reasonable inference from the data?

(A) The concentration of the strong acid and strong base does not affect the pH of the equivalence point.

(B) At pH 7, the moles of titrant (NaOH) is equal to the moles of titrate (HCl) regardless of their concentrations.

(C) The pH of the buffer region of a titration in which 0.005 M HCl is titrated with 0.01 M NaOH is approximately 3.

(D) The pK_a of HCl depends on the concentration of the solution.

403. Two solutions have the same pH. One solution contains only strong acid, the other contains only a weak acid. Which of the following statements is true regarding the relative volume of base needed to neutralize them?

(A) The number of moles of OH^- needed to neutralize the acid is the same as the number of moles of H^+ ions produced by the acid.

(B) Because the pH of the two solutions is the same, the concentration of H^+ ions is the same and thus an equal volume of base is required.

(C) The weak acid has a greater concentration of acid that requires a greater volume of base to neutralize.

(D) The strongly acidic solution has a greater number of hydrogen ions per mole of acid thus requiring a larger volume of base.

Questions 404 and 405 refer to the diagram and table that follow.

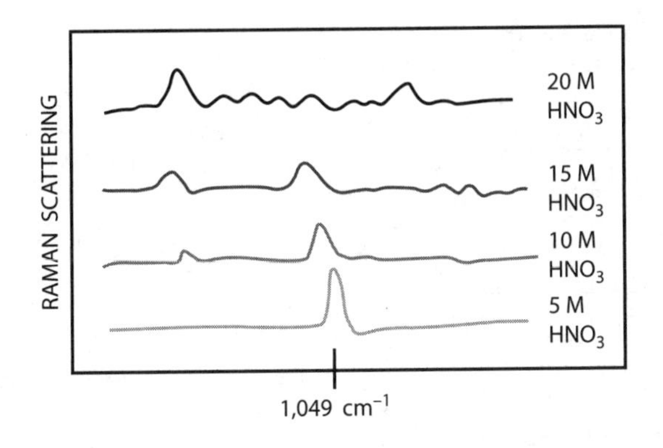

Temperature (°C)	K_a
0	46.8
25	26.8
50	14.9

404. The figure above shows the Raman spectrum of four solutions of nitric acid from 5 to 20 M. The free NO_3^- ion shows a characteristic peak at 1,049 cm^{-1}. Which of the following best explains the data?

(A) At high concentrations the dissociation of HNO_3 decreases.
(B) The pH of a 20 M HNO_3 solution is higher than the pH of a 5 M HNO_3 solution.
(C) As the concentration of HNO_3 increases, the percent ionization does not change but the higher concentration causes the relative concentration of NO_3^- ions to be lower.
(D) Raman scattering is inaccurate with solutions of high concentration and extremely low pH.

405. Under which of following sets of conditions would the percent dissociation be the greatest?

(A) High temperature and high concentration
(B) High temperature and low concentration
(C) Low temperature and high concentration
(D) Low temperature and low concentration

Questions 406–408 refer to the following data.

Temperature (°C)	pH of pure water
0	7.47
10	7.27
20	7.08
25	7.00
30	6.92
40	6.77
50	6.63
100	6.14

406. Which of the following statements is *not* true concerning the effect of temperature on pure water?

(A) The acidity of water increases with increased temperature.
(B) The K_w of water changes with temperature.
(C) The autoionization of water is endothermic.
(D) Pure water is neutral at all temperatures.

407. All of the following statements are true. Which is evidence that the autoionization of water is endothermic?

(A) Water has a high specific heat (4.18 J g^{-1} °C^{-1}) and high heat of vaporization (2,260 J g^{-1}).
(B) Water has a negative standard enthalpy of formation ($\Delta H°_f$ = −285.8 kJ mol^{-1}).
(C) The electrical conductivity of water increases with increasing temperature.
(D) The pH of water changes with increasing temperature.

408. Which of the following is *not* a logical interpretation of the observation that the pH of pure water decreases with increasing temperature?

(A) Excess hydrogen ions lower the pH and increase the acidity of water at higher temperatures.
(B) The K_w of water increases with increasing temperature.
(C) At higher temperatures there is greater ionization, increasing the concentration of both hydrogen ions and hydroxide ions.
(D) Even though the pH of water is lower at higher temperatures, pure water is neutral at all temperatures.

$$2\,X(g) \rightleftharpoons 3\,Y(g) + Z(g) \qquad \Delta H_{FORWARD\ RXN} > 0$$

409. The molar equilibrium concentrations for the reaction mixture represented above at 298 K are [X] = 4.0 M, [Y] = 5.0 M, and [Z] = 2.0 M. What is the value of the equilibrium constant, K_{eq}, for the reaction at 298 K?

(A) 0.060
(B) 2.50
(C) 16.0
(D) 62.5

Question 410 refers to two completely sealed containers, a 10 mL test tube and a 1.0 L bowl, each partially filled with methanol at 25° C as shown in the figure below.

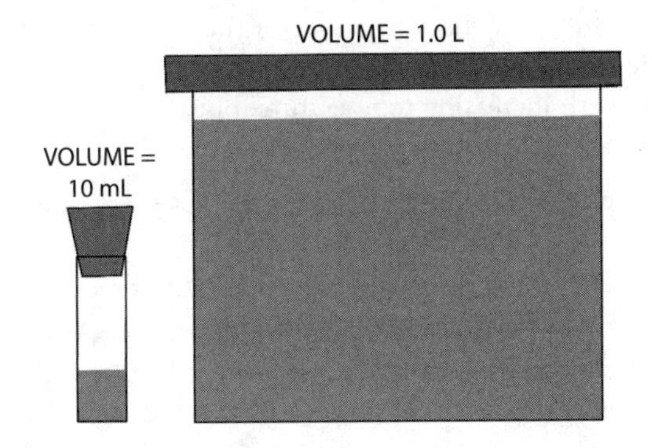

410. Which of the following statements accurately describes the relative vapor pressure in the two containers?

(A) The vapor pressure is lower in the test tube because the volume of methanol is lower.
(B) The vapor pressure is higher in the test tube because the volume of the container above the methanol is smaller.
(C) The vapor pressure is higher in the bowl because the surface area of the methanol is larger.
(D) The vapor pressure is equal in both containers because they are at the same temperature.

$$\text{liquid} + \text{heat} \rightleftharpoons \text{gas}$$

411. The vaporization of a liquid is represented in the equation above. Which of the following most accurately accounts for the fact that the vapor pressure of liquids increases with increasing temperature?

(A) A liquid at particular temperature continues to vaporize until it has completely evaporated.
(B) If a chemical system at equilibrium experiences a change in temperature, the equilibrium shifts to counteract the change.
(C) The condensation of gases takes longer than the vaporization of liquids.
(D) The condensation of gases releases more energy than is absorbed by vaporizing liquids.

412. Which of the following best accounts for the fact that, at the same altitude, the partial pressure of water vapor in the atmosphere can increase significantly with increasing temperature, but the partial pressures of N_2, O_2, Ar, and CO_2 stay relatively constant?

(A) The vapor pressures of N_2, O_2, Ar, and CO_2 are much greater than that of H_2O.
(B) The water cycle (evaporation, condensation, precipitation) causes water to constantly change between liquid and gas phases.
(C) The conditions required for atmospheric N_2, O_2, Ar, and CO_2 to be in equilibrium with their liquid phases do not exist on the earth's surface.
(D) Water in the atmosphere constantly form water droplets around condensation nuclei whereas N_2, O_2, Ar, and CO_2 only condense in or above the stratosphere.

Question 413 refers to the following chemical reaction.

$$CH_3COOH\ (aq) + HCO_3^-\ (aq) \rightleftharpoons CH_3COO^-\ (aq) + H_2CO_3\ (aq)$$

$$K_a \text{ of } CH_3COOH = 1.76 \times 10^{-5}$$
$$K_a \text{ of } H_2CO_3 = 4.3 \times 10^{-7}$$

413. If all of the species are added to water in equimolar quantities, which of the following will be true of the predominant reaction direction?

(A) Forward, because the reaction proceeds in the direction that forms a weaker acid.
(B) Forward, because the reaction proceeds in the direction that forms a stronger acid.
(C) Reverse, because the reaction proceeds in the direction that forms a weaker acid.
(D) Reverse, because the reaction proceeds in the direction that forms a stronger acid.

414. Which of the following is always true regarding the titration of a strong acid versus a weak acid?

(A) The main species present in solution at the halfway to equivalence point differs.
(B) A strong acid requires a greater volume of base to reach equivalence because a greater concentration of hydrogen ions exists for each mole of acid.
(C) A weak acid requires more time to reach equivalence because most of the ionization of acid occurs after the base is added.
(D) The buffer region of the titration curve of the weak acid is much steeper toward the equivalence point as the weak acid resists changes in pH.

Question 415 refers to the following information.

The table shows the pK_a range for the carboxylic acid and amino groups on amino acids. The pH of blood is 7.4.

Group	**pK_a range**
NH_2	8.80–10.6
COOH	1.77–2.58

```
H     H
 \    |    O
  N - C - C//
 /    |    \
H     R     OH
```

415. What is the expected state of ionization of the amino acid backbone for free amino acids in blood?

	Carboxylic acid group	**Amino group**
(A)	COOH	NH_2
(B)	COOH	NH_3^+
(C)	COO^-	NH_2
(D)	COO^-	NH_3^+

Question 416 refers to the following table of K_a values.

Acid	K_{a1}	K_{a2}	K_{a3}
Phosphoric acid **H_3PO_4**	7.5×10^{-3}	6.2×10^{-8}	4.8×10^{-13}
Citric acid **$C_3H_5O(COOH)_3$**	8.4×10^{-4}	1.8×10^{-5}	4.0×10^{-6}
Carbonic acid **H_2CO_3**	4.3×10^{-7}	5.6×10^{-11}	
Sulfuric acid **H_2SO_4**	Large	1.2×10^{-2}	
Sulfurous acid **H_2SO_3**	1.5×10^{-5}	1.0×10^{-7}	
Perchloric acid **$HClO_4$**	Large		
Chloric acid **$HClO_3$**	Large		
Chlorous acid **$HClO_2$**	1.1×10^{-2}		
Hypochlorous acid **$HClO$**	2.9×10^{-8}		
Hydrofluoric acid **HF**	6.6×10^{-4}		
Nitric acid **HNO_3**	24		
Nitrous acid **HNO_2**	7.2×10^{-4}		

416. According to the data in the table, which of the following does *not* increase the strength of an oxyacid?

(A) A strongly electronegative central atom
(B) Electronegative atoms bonded to the central atom
(C) An increased number of oxygen atoms bonded to the central atom
(D) An increased number of hydrogen atoms

417. What is the molar solubility of $BaSO_4$ *(s)* in water? The K_{sp} of $BaSO_4$ *(s)* is 1.1×10^{-10}.

(A) 5.50×10^{-11} M
(B) 1.10×10^{-10} M
(C) 1.05×10^{-5} M
(D) 5.50×10^{-5}

418. What is the molar solubility of CaF_2 *(s)* in water? The K_{sp} of CaF_2 *(s)* is 4.0×10^{-11}.

(A) $(4.0 \times 10^{-11})^{1/2}$ M
(B) $(1.0 \times 10^{-11})^{1/2}$ M
(C) $(4.0 \times 10^{-11})^{1/3}$ M
(D) $(1.0 \times 10^{-11})^{1/3}$ M

419. A student adds aqueous NH_3 to a solution of Ni^{2+} ions and observes the formation of a precipitate. When the student adds additional, excess NH_3, the precipitate dissolves and the solution develops a deep blue color. Which of the following statements most accurately explains why the precipitate that was formed by the addition of NH_3 was dissolved by excess NH_3?

(A) NH_3 is a strong base at high concentrations.
(B) Ni^{2+} forms a soluble, complex ion with NH_3.
(C) Ni^{2+} solubility decreases with increased pH.
(D) Ni^{2+} is reduced by the excess NH_3, forming soluble Ni atoms.

420. BaF_2 is sparingly soluble in water. The addition of dilute HF to a saturated BaF_2 solution at equilibrium is expected to

(A) raise the pH.
(B) react with BaF_2 to produce H_2 gas.
(C) increase the solubility of BaF_2.
(D) precipitate out more BaF_2.

$$Fe^{2+}\ (aq) + 2\ OH^-\ (aq) \rightleftharpoons Fe(OH)_2\ (s)$$

421. Iron (II) hydroxide is sparingly soluble. Its solvation in water is represented by the reaction above. Which of the following is a true statement about the solubility of iron (II) hydroxide?

(A) Increasing the pH reduces the solubility of $Fe(OH)_2$ *(s)* only.
(B) Decreasing the pH reduces the solubility of $Fe(OH)_2$ *(s)* only.
(C) A significant increase or decrease in the pH increases the solubility of $Fe(OH)_2$ *(s)*.
(D) The value of K_{sp} for the reaction is specific to the pH of the solution in which it is dissolved.

422. Citric acid ($H_3C_6H_5O_7$) is a triprotic acid with $K_{a1} = 8.4 \times 10^{-4}$, $K_{a2} = 1.8 \times 10^{-5}$, $K_{a3} = 4.0 \times 10^{-6}$. In a 0.01 M aqueous solution of citric acid, which of the following species is present in the *lowest* concentration?

(A) $H_3C_6H_5O_7$ *(aq)*
(B) $H_2C_6H_5O_7^-$ *(aq)*
(C) $H_1C_6H_5O_7^{2-}$ *(aq)*
(D) $C_6H_5O_7^{3-}$ *(aq)*

423. Which of the following compounds could produce the curve that follows?

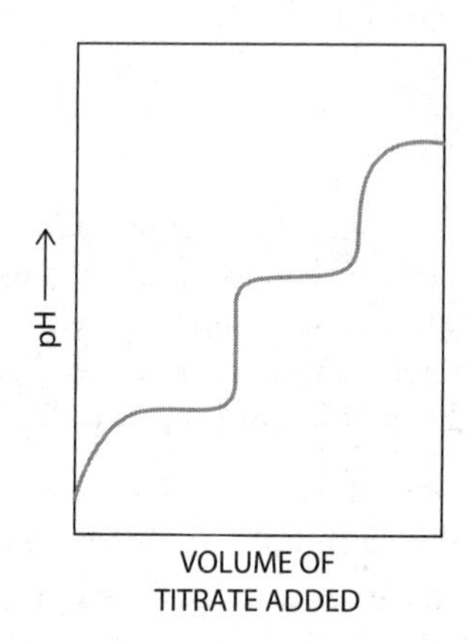

(A) HCl
(B) H_2SO_4
(C) NH_3
(D) $Sr(OH)_2$

424. In a solution of a monoprotic acid, $HA \rightarrow H^+ + A^-$, $[A^-] = [HA]$. Which of the following expressions represents the relative concentration of A^- and HA if the concentration of acid increased so that the pH decreased by 1 unit (pH point)?

(A) $\frac{2[A^-]}{[HA]}$

(B) $\frac{[A^-]^2}{[HA]}$

(C) $\frac{[A^-]}{10\,[HA]}$

(D) $\frac{10\,[A^-]}{[HA]}$

Question 425 is based on the information in the following table.

Compound	**K_{sp} at 25° C**
$Al(OH)_3$	1.0×10^{-33}
$CaCO_3$	8.7×10^{-9}
$Ca(OH)_2$	5.5×10^{-6}
$Fe(OH)_2$	1.6×10^{-14}
HgS	1.4×10^{-53}
Sb_2S_3	1.7×10^{-93}
AgCl	1.6×10^{-10}

425. Which of the following is true concerning the solubility of the compounds in the table?

(A) The solubility of AgCl is greater than the solubility of $CaCO_3$.
(B) HgS is the least soluble compound listed.
(C) $Ca(OH)_2$ is more soluble than $CaCO_3$.
(D) The dissociation of Sb_2S_3 is endothermic.

426. Neutralization of a 0.10 M NaOH solution would require the least volume of which of the following solutions?

(A) 0.1 M phosphoric acid (K_a range from $7.5 \times 10^{-3} – 4.8 \times 10^{-13}$)
(B) 0.2 M hydrochloric acid
(C) 0.3 M HCN ($K_a = 6.17 \times 10^{-10}$)
(D) 0.4 M acetic acid ($K_a = 1.76 \times 10^{-5}$)

Question 427 refers to the following table of K_a values.

Acid	K_a
HNO_2	4.00×10^{-4}
$HClO$	3.00×10^{-8}
CH_3COOH	1.76×10^{-5}
$HCH_3H_5O_3$	1.40×10^{-4}

427. Which of the following pairs of chemical species, when combined in equimolar amounts, results in a buffer with a pH closest to 4?

(A) $HClO$ and NO_2^-
(B) $HCH_3H_5O_3$ and HNO_2
(C) CH_3COOH and ClO^-
(D) $HCH_3H_5O_3$ and $CH_3H_5O_3^-$

Questions 428 and 429 refer to the following graph of concentration versus time for the reaction $A \rightleftharpoons B$.

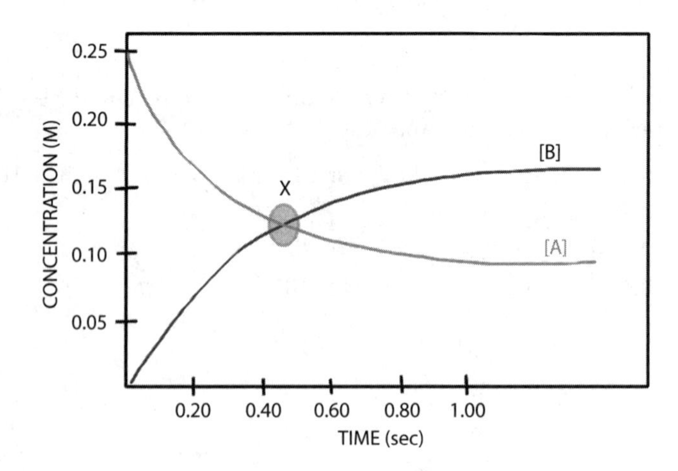

428. Which of the following is a true statement regarding the progress of the reaction?

(A) The reaction reaches equilibrium at point X.
(B) The rate of $A \rightarrow B$ is equal to the rate of $B \rightarrow A$ at point X.
(C) After 1 second, the reaction has reached equilibrium.
(D) The rate of $A \rightarrow B$ is greater than the rate of $B \rightarrow A$ until approximately 1 second.

429. Suppose at 1.5 seconds, more B was added to the system (enough to momentarily raise the concentration to 0.25 M). In which of the following ways is the system expected to respond?

(A) The forward reaction would be pushed to completion, consuming all of reactant A.
(B) The reverse reaction would consume all of reactant A.
(C) The concentration of only B would increase.
(D) The concentrations of both A and B would increase.

Question 430 refers to the following information.

$$\text{Creatine } (aq) + \text{P}_\text{i}\ (aq) \rightarrow \text{Phosphocreatine } (aq) + \text{H}_2\text{O } (l)$$

$$\Delta G^\circ \text{ (25° C and 1 M concentrations)} = 42.8 \text{ kJ mol}^{-1}$$
$$\Delta G \text{ (37° C and 1 mM concentrations, physiological conditions)} = 60.5 \text{ kJ mol}^{-1}$$

430. Which of the following statements is true regarding the reaction above?

(A) The reaction is thermodynamically favorable in the direction written.
(B) The free-energy change of the reaction is dependent upon concentrations of reactants.
(C) The reaction becomes more thermodynamically favorable at higher temperatures.
(D) Work can only be done by this reaction under physiological conditions.

Questions 431 and 432 refer to the following chemical reaction that occurs in a closed, rigid container and proceeds to equilibrium.

$2\ NH_3\ (g) \rightleftharpoons 3\ H_2\ (g) + N_2\ (g)$ $\Delta H_{FORWARD\ RXN} = +46.1$ kJ mol^{-1}

431. After the reaction has reached equilibrium, pure $N_2\ (g)$ is injected into the tank at a constant temperature and equilibrium is reestablished. Which of the following quantities will have a lower value compared to the original equilibrium?

(A) The amount of $NH_3\ (g)$ in the tank
(B) The amount of $H_2\ (g)$ in the tank
(C) K_{eq} for the reaction
(D) The total pressure in the tank

432. Which of the following changes *alone* would cause a decrease in the value of K_p for the reaction as written?

(A) Increasing the temperature.
(B) Increasing the volume.
(C) Adding a catalyst.
(D) Removing the products as they form.

433. Which of the following statements most accurately describes the synthesis of benzene from its standard-state elements under standard conditions?

$$6\ C\ (s) + 3\ H_2\ (g) \rightarrow C_6H_6\ (l) \qquad \Delta G^\circ = +124 \text{ kJ}$$

(A) Benzene is less thermodynamically stable than its elements.
(B) The decomposition of benzene into its standard-state elements is likely to occur under standard conditions.
(C) Benzene cannot be synthesized under standard conditions.
(D) The synthesis of benzene requires a catalyst to overcome the positive free energy of the reaction.

434. Which of the following best explains the relationship between ΔG° and K_{eq}?

(A) At equilibrium, ΔG° is equal to K_{eq}.
(B) At equilibrium, ΔG° and K_{eq} are equal to zero.
(C) The point at which a reaction in solution reaches equilibrium is a function of ΔG°, so it is another way of expressing K_{eq}.
(D) The free-energy change of a reaction proceeding toward equilibrium is given by ΔG°, and the point at which the system reaches equilibrium is given by K_{eq}.

FREE-RESPONSE QUESTIONS

Questions 435–443 refer to reactions involving lactic acid. Several of the questions in this group rely on answers from other questions.

Questions 435–438 refer to the reaction below in which lactic acid, a carboxylic acid, reacts with water at 25° C.

$$CH_3CH(OH)COOH\ (aq) + H_2O\ (l) \rightleftharpoons CH_3CH(OH)COO^-(aq) + H_3O^+(aq)$$

435. Identify a Brønsted-Lowry conjugate acid-base pair in the reaction above. Clearly **label** both the acid and its conjugate base.

436. The pH of a 0.250M lactic acid solution is 2.23.

(a) **Calculate** the equilibrium concentrations of the reactants and products.

(b) **Determine** the K_a of lactic acid.

437. Determine the mass of lactic acid required to make 50.0 mL of a 0.250 M solution.

438. State whether each of the following statements is *true* or *false*. **Justify** your answer.

1. Because lactic acid is a weak acid and NaOH is a strong base, the volume of 0.250 M NaOH required to neutralize 50.0 mL of 0.250 M lactic acid is less than 50.0 mL.
2. The pH of a solution prepared by mixing 50.0 mL of 0.250 M lactic acid and 50.0 mL 0.250 M NaOH is 7.0.
3. If the pH of a solution of HNO_3 is the same as the pH of a solution of lactic acid, then the molarity of the HNO_3 solution must be less than the molarity of the lactic acid solution.
4. All carboxylic acids are weak acids.

Questions 439–443 refer to the titration 50.0 mL of a solution of lactic acid with 0.200 M $Sr(OH)_2$. The volume of base needed to reach equivalence is 15.6 mL.

439. Calculate the concentration of lactic acid in the solution.

440. Calculate the pH of the $Sr(OH)_2$ solution.

441. Using the concentration you determined in question 439, **calculate** the pH of the lactic acid solution.

442. State whether the pH at equivalence is greater than, less than, or equal to 7. **Justify** your answer.

443. Choose one suitable indicator for this titration from the following table. **Justify** your choice.

Indicator	pK_a
A	2.3
B	5.1
C	7.6
D	10.4

Questions 444–449 refer to an experiment in which 2.00 moles of H_2 *(g)* and 2.00 moles of I_2 *(g)* are added to a previously evacuated 1.00 L container at 500 K. At equilibrium, 3.46 moles of HI *(g)* were present. The reaction is shown below.

$$H_2\ (g) + I_2\ (g) \rightleftharpoons 2\ HI\ (g)$$

444. Calculate the K_c of the reaction at 500 K.

445. Calculate the partial pressures of each gas at the beginning of the experiment. **Calculate** the partial pressures of each gas at equilibrium.

446. Calculate the K_p of the reaction. **Compare** it to the K_c. **Describe** how they differ and **explain** why.

447. Calculate the partial pressures of the gases if contents of the first container (at equilibrium) were transferred to a previously evacuated 2.00 L container at the same temperature and allowed to reach equilibrium.

448. The K_c of the reaction at 700 K is 54. **State** whether this reaction is endothermic or exothermic. **Justify** your prediction using Le Chatelier's principle.

449. In another experiment, 1 mole each of H_2, I_2, and HI are added to a 1.00 L container at 500 K. **Predict** whether the amount of H_2 will increase, decrease, or remain the same. **Justify** your answer with a calculation.

Questions 450–454 refer to an experiment in which a pure sample of PCl_5 *(g)* is placed in a rigid, previously evacuated 1.00 L container. The initial pressure of PCl_5 *(g)* is 1.4 atm. The temperature is held constant until the system reaches equilibrium. At equilibrium, the partial pressure of PCl_5 is 0.70 atm.

$$PCl_5\ (g) \rightleftharpoons PCl_3\ (g) + Cl_2\ (g)$$

450. (a) **Calculate** the K_P for the decomposition of PCl_5 at the temperature at which the reaction occurred.

(b) **Calculate** the total pressure of the container at equilibrium.

451. Draw a particulate diagram of the equilibrium based on the diagram below of the initial conditions shown.

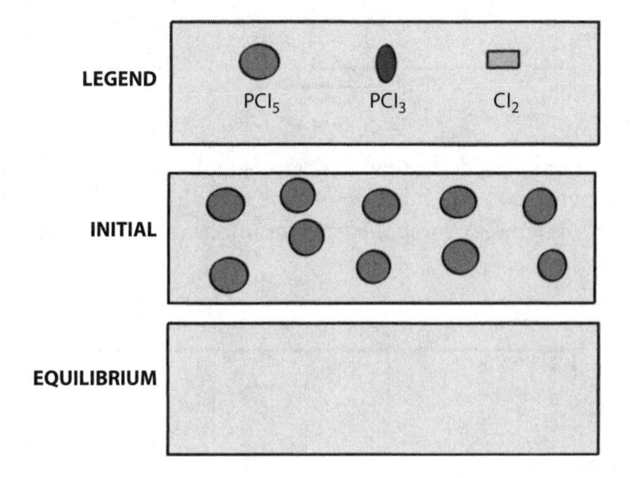

452. Graph the estimated partial pressure changes of PCl_5, PCl_3, and Cl_2 and the total pressure change of the system on the graph below. Include initial and equilibrium positions. Do not include units of time.

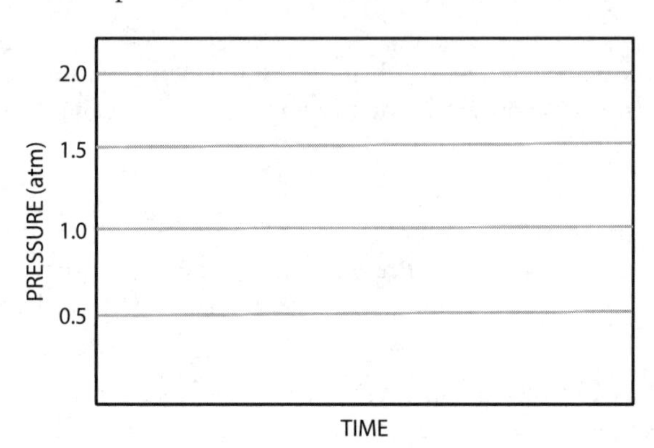

453. Use the graph provided below to **draw** a graph that represents the rate of the *forward* reaction as a function of time if additional Cl_2 were injected into the container. Assume the time for injections and mixing of additional Cl_2 is negligible.

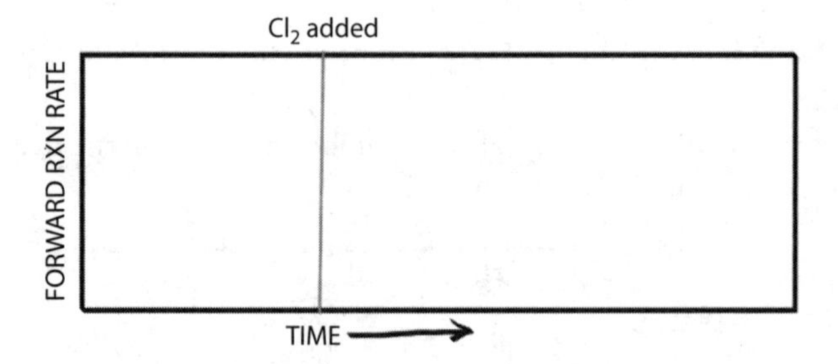

454. Use the graph provided below to **draw** a graph that represents the rate of the *reverse* reaction as a function of time if additional Cl_2 were injected into the container. Assume the time for injections and mixing of additional Cl_2 is negligible.

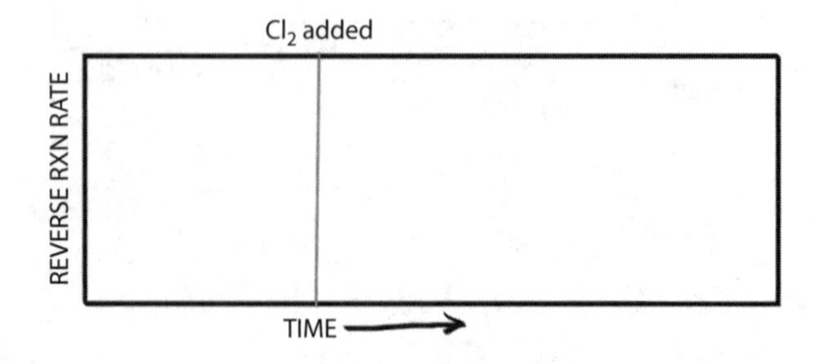

455. List three macroscopic indications that a system has reached equilibrium (or has gone to completion).

Questions 456–459 are based on an experiment where 2.0 moles of N_2O_4 *(g)* were added to a rigid, previously evacuated 5.00 L container at 298 K. At equilibrium, 0.24 moles of NO_2 were present.

$$N_2O_4\,(g) \rightleftharpoons 2\,NO_2\,(g)$$

Substance	ΔG°_f (kJ mol^{-1})
N_2O_4 *(g)*	**97.9**
NO_2 *(g)*	**51.3**

456. Calculate the K_p of the reaction as written above at 298 K.

457. (a) **Calculate** the ΔG° for the reaction as written. (b) **Predict** whether the reaction will proceed spontaneously in the direction written. **Justify** your answer.

458. Calculate the K_p of the reaction using the value of ΔG° you obtained in question 457.

459. Calculate the K_p of the reverse reaction.

Question 460 refers to the following titration curve.

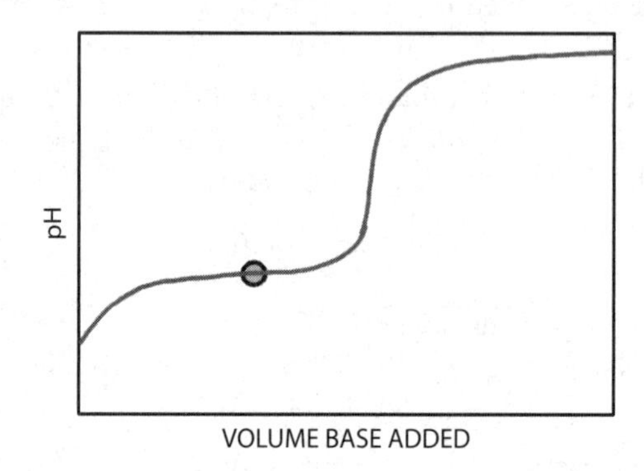

460. Suppose the titration curve represents the titration of a weak acid with a strong base. **Draw** a particulate diagram of the species present at the circled point. Represent the acid and base as shown below. Draw at least three particles of acid and three particles of base.

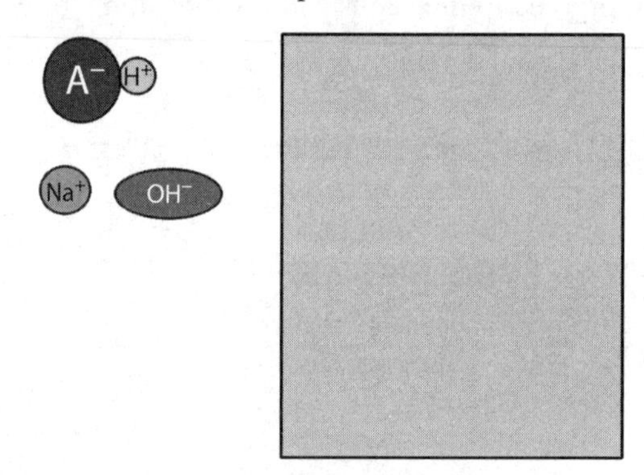

Questions 461–463 refer to the following table.

Compound	K_a	pK_a
$HClO_2$	1.1×10^{-2}	1.96
HNO_2	4.0×10^{-4}	3.39
CH_3COOH	1.8×10^{-5}	4.75
HCN	6.2×10^{-10}	9.21

461. Suppose you need a buffer with a target pH of 5. (a) **Select** the proper acid from the table and (b) **state** the name of the second compound you will need, besides water. (c) **Justify** your choices.

462. **Draw** a particle diagram of the buffer. Include at least four particles each of the two compounds you added and their ionization state. **Label** each particle.

463. **Calculate** or **qualitatively reason** the relative concentrations of the two compounds needed to obtain buffering capacity at pH 5.

Questions 464 and 465 refer to two solutions with a pH of 2.5.

Solution 1: Hydrochloric acid (HCl)
Solution 2: Acetic acid (CH_3COOH), $K_a = 1.8 \times 10^{-5}$

464. (a) **Calculate** the concentrations of each solution. (b) **Calculate** the percent ionization of each solution.

465. (a) **Demonstrate** *qualitatively* or *quantitatively* how a strong acid and a weak acid can both produce a solution of the same pH. (b) **Calculate** and **compare** the volumes of a 0.010 M solution of NaOH required to titrate 0.50 L of solutions 1 and 2.

Questions 466 and 467 refer to a solution of CH_3COOH and $NaCH_3COO$.

466. Use Le Chatelier's principle to **explain** how the buffer would respond to the addition of a small amount of HCl.

467. Use Le Chatelier's principle to **explain** how the buffer would respond to the addition of a small amount of NaOH.

468. **Select** which salt of NH_4^+, NH_4F, or NH_4Cl, would be a better conjugate acid to use to prepare a basic buffer with ammonia as the weak base. **Justify** your answer.

Questions 469 and 470 refer to the synthesis of ammonia by the Haber process, which combines hydrogen gas with the nitrogen gas found in the atmosphere to form ammonia.

$$3\,H_2\,(g) + N_2\,(g) \rightleftharpoons 2\,NH_3\,(g) \qquad \Delta H^\circ = -92.2\text{ kJ}$$
$$K_p = 6.0 \times 10^5 \text{ at } 25^\circ\text{ C}$$
$$E_a = 230 - 420\text{ kJ mol}^{-1}$$

469. Approximately 1.4×10^{12} L of hydrogen gas is produced annually. **Calculate** the theoretical yield of ammonia that can be produced by this hydrogen gas in kilograms at 298 K and 1 atm. Assume nitrogen gas is present in excess.

470. **Design** a set of conditions that will optimize the product yield. **Justify** your conditions.

Questions 471 and 472 refer to anserine, a dipeptide that is important in the maintenance of intracellular pH in some tissues.

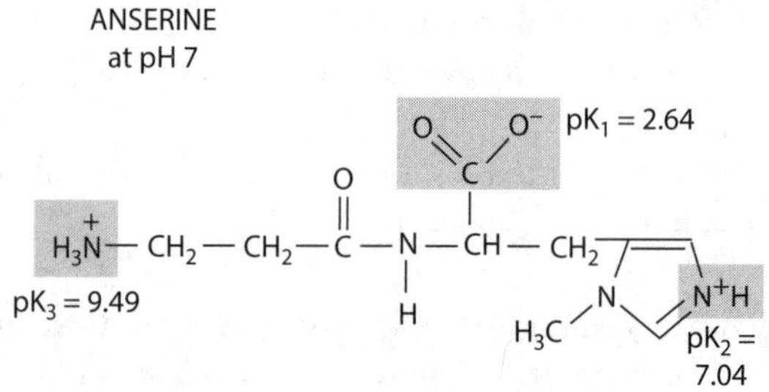

ANSERINE
at pH 7

$pK_1 = 2.64$

$pK_3 = 9.49$

$pK_2 = 7.04$

471. **Complete** the diagrams below, showing the structure at pH 2 and pH 12.

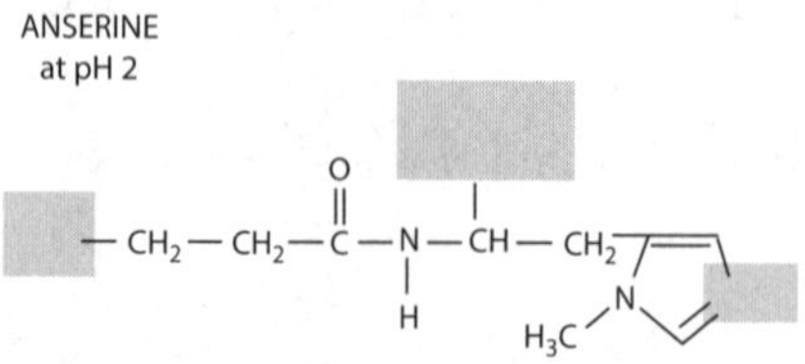

ANSERINE
at pH 2

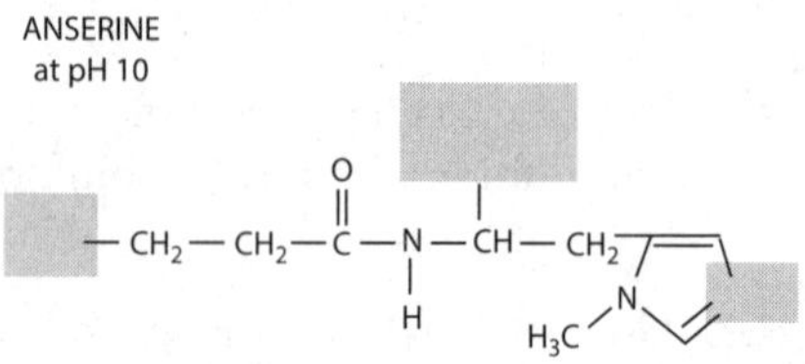

ANSERINE
at pH 10

472. The physiological pH of intracellular fluids is approximately 6.8–7.2. **Select** which of the three ionizable groups would be most effective for buffering at tissue pH. **Justify** your answer.

Questions 473 through 476 refer to the hypothetical reaction A $\rightleftharpoons$ 2 B, which proceeds to equilibrium under hypothetical conditions ($K_{eq} = 4$).

473. Using the boxes below, **sketch** a graph that compares the rates of the forward and reverse reactions and the relative concentrations of A and B over time. You can, but do not need to, quantify the relative rates and concentrations. Assume the reaction starts with reactant A only (no B is present).

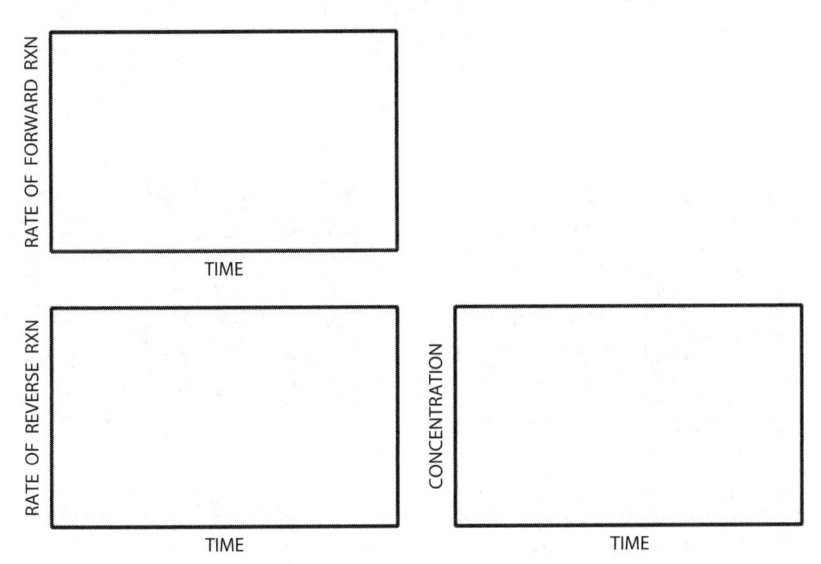

474. **Sketch** a graph that compares the rates of the forward and reverse reactions and the relative concentrations of A and B over time with additional A added at the time indicated on the graph below. Assume the reaction starts with reactant A only (no B is present).

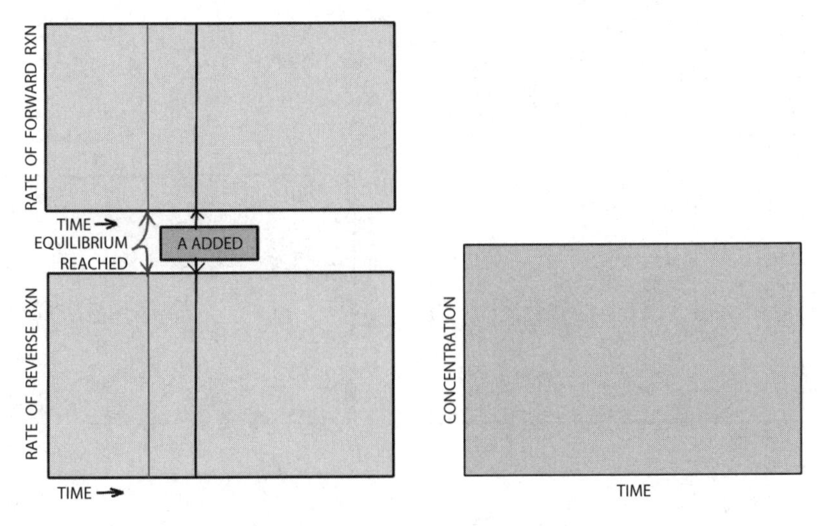

475. Sketch a graph that compares the rates of the forward and reverse reactions and the relative concentrations of A and B over time with additional B added at the time indicated on the graph below. Assume the reaction starts with reactant A only (no B is present).

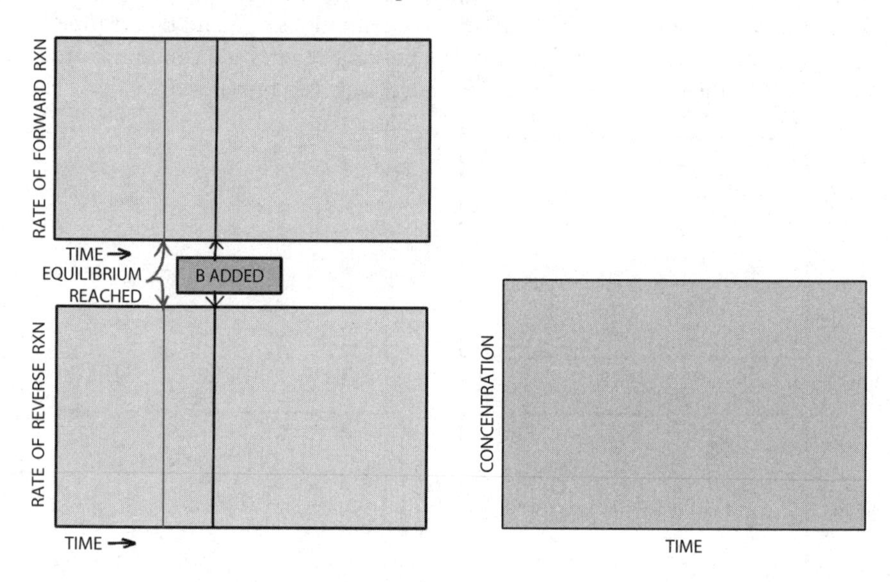

476. Sketch a graph that compares the rates of the forward and reverse reactions and the relative concentrations of A and B over time with half the amount of A removed at the time indicated on the graph below. Assume the reaction starts with reactant A only (no B is present).

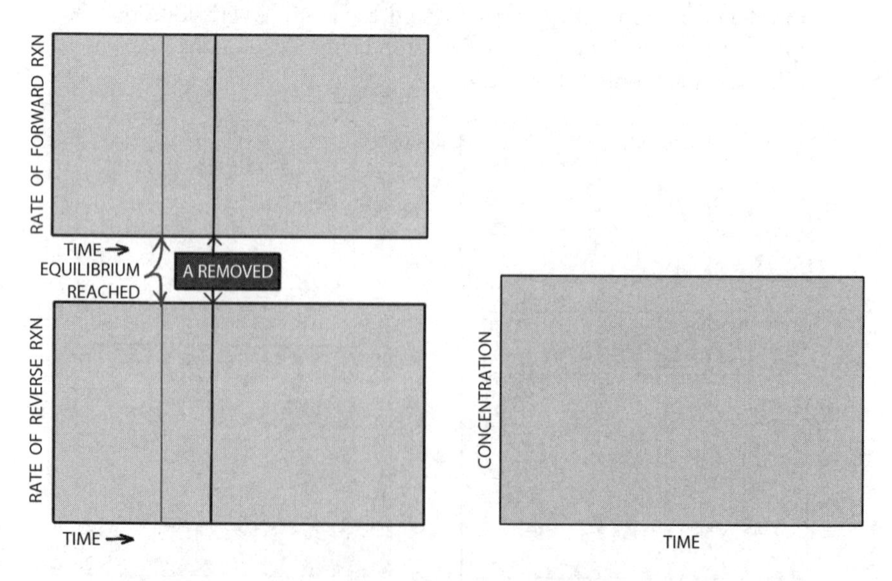

477. **Explain** what determines whether a chemical reaction will proceed to completion or reach equilibrium.

478. **State** three indications that a chemical reaction has reached equilibrium.

479. The equilibrium constant of a reaction is related to its free-energy change by the following equations.

$$K = e^{\frac{-\Delta G^\circ}{RT}}$$

$$\Delta G = -RT \ln K$$

(a) **Explain** how the magnitude of the equilibrium constant, K, affects whether the reaction is spontaneous.

(b) **Distinguish** between an *exothermic* and an *exergonic* reaction.

CHAPTER 7

Mixed Free-Response Questions

480. (a) **Calculate** the mass of $CoCl_2$ needed to make 150 mL of a 0.200 M $CoCl_2$ solution.

(b) **Calculate** the mass of H_2CO_3 needed to make 150 mL of a 0.200 M H_2CO_3 solution.

481. (a) **List** the steps of the procedure to prepare 150 mL of 0.20 M H_2CO_3 standard solution using solid, powdered reagents.

(b) Briefly **describe** the function and importance of a standard solution.

482. Briefly **explain** a procedure, other than titration, to measure the concentration of the H_2CO_3 solution.

483. (a) **Draw** a particulate diagram of the original solution of $CoCl_2$. Include 1 formula unit of $CoCl_2$ and at least 4 water molecules.

(b) **Draw** a particulate diagram of the reaction that occurs when H_2CO_3 is added to $CoCl_2$. Include 2 formula units of each and at least 4 water molecules.

Question 484 refers to the following data and information.

$CoCl_2$ is used as a desiccant because it can absorb water from the surroundings to form a deep purple hydrate. A student weighs 142.74 grams of the purple crystals then places the sample in a drying oven, re-massing the sample every 10 minutes. The data are shown below:

Time	Mass $CoCl_2$	Color
0	142.74	Deep purple
10	115.75	
20	97.78	
30	84.26	
40	77.92	
50	77.96	
60	77.94	Sky blue

484. (a) **State** the time at which the sample dried. **Explain.**
(b) **State** the expected sample color at 15 minutes. **Explain.**
(c) **Write** the formula for the hydrate.

Questions 485–492 refer to the following experiment.

A student is given the task of determining the amount of lithium in a tablet of LiCl. The tablet also contains an inert, water-soluble binding agent. One tablet is dissolved in 50.00 mL of distilled water at 25° C. After an excess of 0.2500 M $Pb(NO_3)_2$ *(aq)* is added, a yellow precipitate is formed, which is filtered, washed, and dried. The data are given below.

Mass of filter paper	1.501 g
Mass of LiCl tablet + filter paper	2.705 g
Mass of filter paper after first drying	6.314 g
Mass of filter paper after second drying	6.311 g
Mass of filter paper after third drying	6.310 g

485. Write a balanced, net-ionic equation for the precipitation reaction that occurs and **explain** why the reaction is best represented by a net-ionic equation.

486. Explain the purpose of drying and weighing the precipitate three times.

487. State whether the $[Li^+]$ is greater than, less than, or equal to the $[NO_3^{2-}]$ in the filtrate solution. **Justify** your answer.

488. (a) **Calculate** the number of moles of precipitate produced in the experiment.
(b) **Calculate** the number of moles of Li^+ in the tablet.
(c) **Calculate** the mass percent of Li^+ in the tablet.

489. (a) Suppose the tablet was dissolved in 60.0 mL of solution instead of 50.0 mL. **Predict** whether the amount of precipitate would be different. **Justify** your answer.
(b) Suppose the concentration of the $Pb(NO_3)_2$ *(aq)* was 0.20 instead of 0.25. **Predict** whether the amount of precipitate would be different. **Justify** your answer.
(c) **Calculate** the concentration of $Pb(NO_3)_2$ *(aq)* (in molarity) that would be needed for 10.00 mL of the $Pb(NO_3)_2$ solution to precipitate out the *exact* amount of Cl^- with *no excess*. Assume the solubility of $PbCl_2$ is negligible.
(d) **Calculate** the minimum volume of 0.25 M $Pb(NO_3)_2$ solution that would precipitate out the Cl^- ions.

490. (a) Use your answer from question 489(d) to **determine** the final volume of solution before filtration (assume the volume of the tablet is negligible). The K_{sp} of $PbCl_2$ is 1.2×10^{-5}.

(b) **Calculate** the mass of Pb^{2+} and Cl^- left in solution.

(c) **Quantify** the effect of the solubilized $PbCl_2$ on the theoretical yield. **State** whether or not the assumption of negligible solubility was reasonable. **Justify** your response.

(d) **Describe** what was done in the original procedure to minimize the effect of the solubility of $PbCl_2$.

491. Another experiment is performed in which the precipitation of chloride ions was done by the addition of 0.250 M $AgNO_3$ *(aq)* instead of $Pb(NO_3)_2$ *(aq)*. The K_{sp} of AgCl is 8.5×10^{-17}. **State** whether or not the substitution will work. **Justify** your answer.

492. Suppose there is only one tablet available for analysis and the lab is equipped with a balance that measures to 0.001 g. **State** whether or not the data can be expressed with four significant figures. **Explain.**

Questions 493–496 refer to the following experiment.

A new pigment has been isolated (purified). Your lab specializes in the spectrophotometric analysis of substances. You have 30 mL of a solution with a concentration of 1 g pigment mL^{-1} solution. Previous analysis showed that the molar mass is 642 g mol^{-1}.

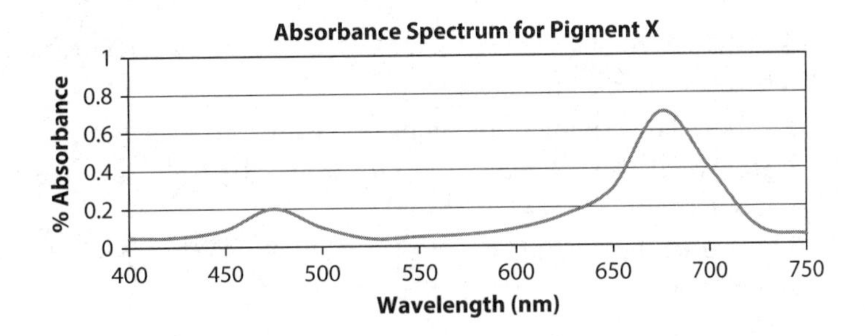

Table 1: % Absorbance vs. Concentration

Concentration of pigment X (M)	Absorbance @ 675 nm
0.0	0.00
0.1	0.14
0.2	0.36
0.3	0.51
0.4	0.62
0.5	0.81
0.6	0.98

493. Calculate the molarity of the solution.

494. Calculate the dilutions required to make the standard curve using Table 2. Dilute with water only. The final volume of each sample is 5 mL.

Table 2: Dilutions for Constructing a Standard Curve

	0 M	0.1 M	0.2 M	0.3 M	0.4 M	0.5 M	0.6 M
Volume H_2O							
Volume Solution							
Total Volume	5	5	5	5	5	5	5

495. (a) **Construct** the standard curve for 675 nm using the data in Table 1.
(b) **Write** an equation for the line.
(c) **Explain** the meaning of the slope of the line.
(d) **Calculate** the absorbance change for every 0.01 M increase in concentration.
(e) **Determine** the expected absorbance of a 0.02 M solution and a 0.36 M solution.

496. Suppose the measured absorbance of a sample is 1.
(a) **Explain** what an absorbance of 1 means.
(b) **Describe** the procedure for dealing with an absorbance of 1.

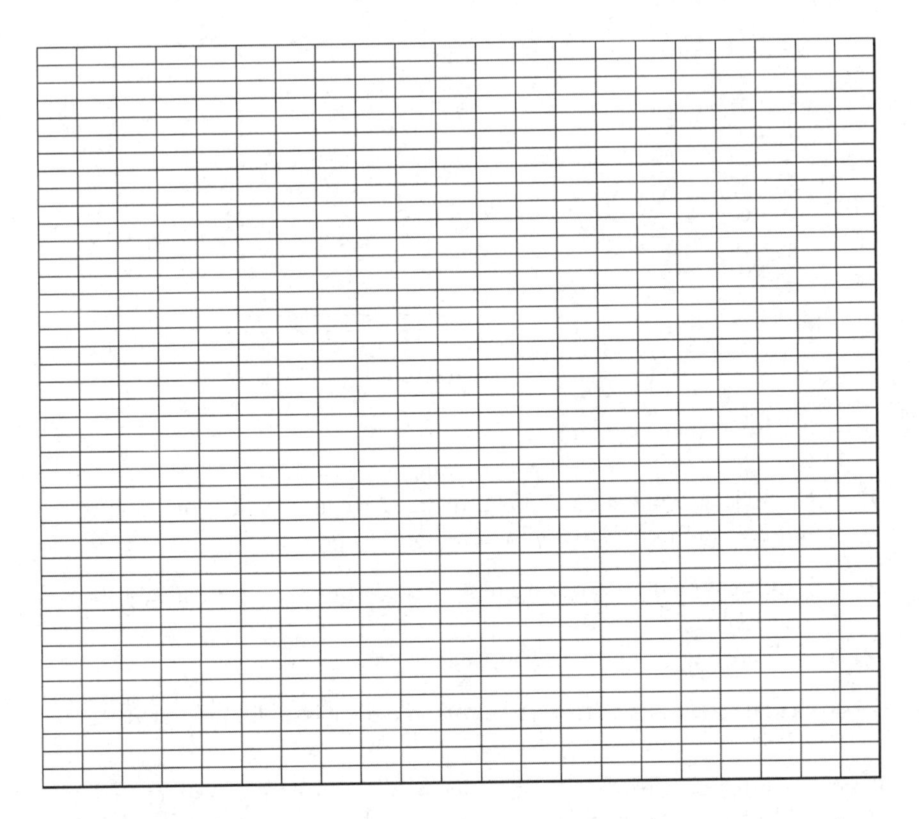

Questions 497–499 refer to the hypothetical chemical reaction and absorbance data that follow.

$$2\,A + X_2 \rightarrow 2\,AX$$

The absorbance of A can be measured at 600 nm. Figure 1 shows the calibration curve for A at 600 nm. The reaction above was performed at 600 nm. Previous tests showed that X_2 and AX had zero absorbance at this wavelength. Figure 2 shows the change in absorbance over time.

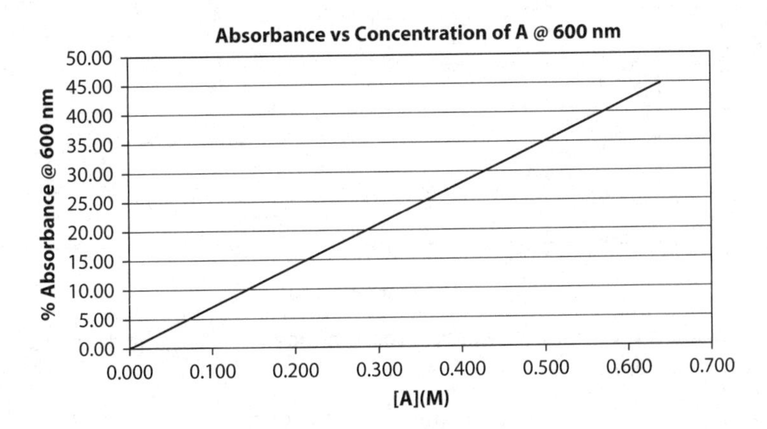

Figure 1: Absorbance vs. Concentration of A at 600 nm

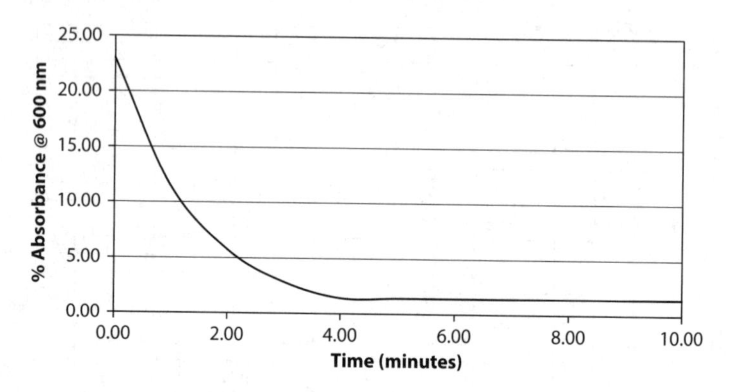

Figure 2: Absorbance vs. Time

497. Use the data from Figures 1 and 2 to **calculate** the reaction rate from 0 to 2 minutes and from 2 to 4 minutes.

498. (a) **Calculate** the absorbance of A after 1 minute if the initial absorbance was 35%.

(b) **State** any assumptions you made in your calculations.

499. **Calculate** the K_{eq} for the reaction. Assume there is no limiting or excess reactant.

Question 500 refers to the decomposition of the calcium oxalate hydrate, $CaC_2O_4 \cdot H_2O$, shown below. The graph shows the change in mass of the solid over 20 minutes. Assume the decomposition is complete by 20 minutes.

$$CaC_2O_4 \cdot H_2O\ (s) \rightarrow CaC_2O_4\ (s) + H_2O\ (g)$$
$$CaC_2O_4\ (s) \rightarrow CaCO_3\ (s) + CO\ (g)$$
$$CaCO_3\ (s) \rightarrow CaO\ (s) + CO_2\ (g)$$

500. **Calculate** the masses of water, carbon monoxide, and carbon dioxide produced during the decomposition.

Congratulations, you're done! ☺

ANSWERS

Chapter 1: Atoms

1. (B) The balanced equation for the reaction is H_2 *(g)* + Cl_2 *(g)* → 2 HCl *(g)*. The answer choices are represented by particulate diagrams, which *the College Board expects you to be proficient with.* In this case, the circles represent individual atoms that are not to scale but differ in size based on their respective atomic radii: the larger circle represents a chlorine atom and the smaller circle represents a hydrogen atom. In their standard states (at 298 K and 1 atm) both hydrogen and chlorine are diatomic gases (remember Professor $Br_2I_2N_2Cl_2H_2O_2F_2$, or, **H_2**ave **N_2**o **F_2**ear **O_2**f **I_2**ce **Col_2**d **$Beer_2$**).

Only choices B and C show the correct form of the reactants. HCl is diatomic, too, with one larger chlorine atom bonded to one smaller hydrogen atom. (**Diatomic** refers to substances that contain two atoms, which may or may not be part of the same element. **Binary**, on the other hand, refers *only to compounds* that contain two different elements, but may contain more than two atoms, like $CaCl_2$). The correct particulate diagrams for the product are only shown in choices A and B; choice D shows HCl as a reactant. **Particulate diagrams for chemical reactions need to demonstrate conservation of atoms; the number and type of atoms in the reactants are the same as the products.** Choice D violates the requirement of conservation of atoms.

Correct particulate diagrams will conserve the number and types of atoms. In other words, the products should contain the same number and type of atoms as the reactants.

2. (D) The balanced equation for the reaction is 2 H_2 *(g)* + O_2 *(g)* → 2 H_2O *(l)*. Hydrogen and oxygen are both diatomic gases (H_2 and O_2) under standard conditions (298 K and 1 atm) and so we can expect they will be gases at 150°C (423 K) and 1 atm. The correct particulate diagrams for the reactants, H_2 and O_2, are shown only in choices B and D. Answer choices B and D are both stoichiometrically correct, showing eight H_2O molecules formed from the reaction of eight H_2 and four $O_{2\ \text{molecules}}$. Particulate diagrams for chemical reactions need to demonstrate conservation of atoms; the number and type of atoms in the reactants should be the same as the products. None of the answer choices violate this requirement. However, choice B shows the water as a liquid, identifiable as such by the hydrogen bonds between them (the dotted lines connecting the hydrogen of one water molecule to the oxygen of another).

3. (A) The **mass number** of cadmium is 112, not the **atomic mass** (the weighted average of the relative abundances of naturally occurring isotopes). The mass number will always be a whole number because it is the sum of the number of protons and neutrons (collectively called the **nucleons**, referring to their location in the nucleus) in an atom. **The atomic number is the number of protons.** It determines the identity of the atom, so finding cadmium on the periodic table will tell us its atomic number (48). Choices C and D both have the incorrect atomic number). The number of electrons and protons will always be the same in a neutral atom because they are the only negatively and positively charged (respectively) particles in the atom. Neutral Cd has 48 electrons (choice B has 64 electrons, 16 too many). If we subtract the atomic number from the mass number we get the number of neutrons in that particular isotope of cadmium, ^{112}Cd, 112 protons and neutrons – 48 protons = 64 neutrons.

4. (C) Crystals are highly organized but typically rigid structures. (That is, if they're solid crystals, which is the type of crystal you're most likely to see on your AP Chemistry exam. Liquid crystals do exist and they have many amazing properties.) **The periodic trend for atomic radius is that from left to right across a period, the radius typically decreases,** so P atoms are in fact slightly smaller than Si atoms. However, the size difference is minor and it does not directly impact conductivity (so choice A is incorrect even though the statement is true, as is the case for choice B, which is the reason that P atoms are smaller than Si atoms).

Electrical conductivity is a consequence of mobile charges. In solids that conduct electricity, it is the *valence electrons* that are mobile (solid ionic crystals do not conduct electricity), not protons, so choice D is incorrect. Phosphorus atoms have five valence electrons and silicon has four, so introducing phosphorus atoms would increase the number of mobile charges. Specifically, there is one extra mobile electron for each phosphorus atom in the crystal.

5. (C) Steel is an **interstitial alloy** (as opposed to a **substitutional alloy**; see figure below). Interstitial alloys combine elements of different radii. The atoms of larger radius (in this case, iron) are present in much greater quantities and form the "main" lattice structure, whereas the atoms of smaller radius fit in between them (hence the name *interstitial*). This changes the microscopic structure of the lattice and therefore the bulk (macroscopic) properties of the alloy.

The physical properties of a particular atom cannot be changed by the presence of a different type of atom (although their combined effects can produce new, *emergent properties*), so choices A and B are incorrect. Choice D implies that nuclear fusion occurs when steel is produced, but it doesn't! Importantly, steel is not an element (it is an alloy). If you're unsure, check the periodic table (but be aware that the periodic table provided for your AP Chemistry exam will have only element symbols, not names).

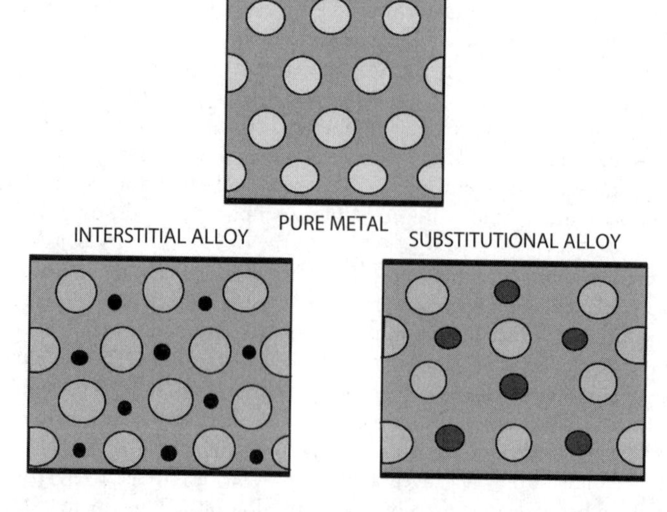

The College Board does not expect you to know all the types and details of metal lattice and alloy structure (covered in Big Idea 2), but they do consider the basic structure of and the distinction between interstitial and substitutional alloys Essential Knowledge.

This question is considered part of Big Idea 1 because the essence of the question is not really about alloys. In order to answer this question correctly, you must understand that **atoms retain their properties when combined with other types of atoms**.

6. (B) Mass spectroscopy (MS, mass spec) **is a technique that allows us to measure the molecular mass of a substance.** Although there are several types of MS, the most common utilizes electron impact and a magnetic sector. The substance is bombarded with a high-energy stream of electrons that dislodges a valence electron from the particles being analyzed and "radicalizes" them. (A radical has an odd number of electrons. Substances with paired electrons are more stable.) Larger particles will often break into fragments and some retain their positive charge. The particles are then subjected to a magnetic field that will deflect them according to their charge-to-mass ratio. Mass spectroscopy is explained in more detail in answer 73.

You should be able to interpret mass spectroscopy data. The mass/charge (m/z) ratio is basically just the mass. This is because, based on the methodology, the charge of the fragment is usually 1. The data show a range of particles between 92 and 100 amu and their abundance in the sample. Tc and Nb may have been candidates for contamination, but notice there are no elements of mass 93 present in the sample (answer choice A provides the necessary information). The percent abundance of naturally occurring Ru isotopes does not match the data pattern (choice C). Contamination would have changed the percent abundance of the 96 and 98 isotopes in the opposite direction of the 100 amu isotope. Although choice D may seem to say the same thing as choice B, many combinations of numbers could produce an average of 95.94, the atomic mass of Mo.

For each chemistry technique you learn in AP Chemistry, understand when the technique is used, what kind of data are produced, and what the data tell us.

7. (B) Infrared radiation is used in **infrared (IR) spectroscopy.** Molecules absorb a particular frequency of IR when it corresponds exactly to the amount needed to augment certain molecular motions. It is common to think of bond lengths as being fixed quantities, but **bonds behave more like springs, stretching and compressing**. This is referred to as the bond's **vibration**. The average of these motions is what we typically think of as the bond length (for example, the C–H bond length is 1.10 Å *on average*). **The bond will absorb energy from incoming radiation if the frequency of the radiation is the same as the vibration.** (In other words, the "spring" of the bond—the alternate compression and stretching—has the same frequency as the radiation – **resonance**!)

IR spectroscopy uses longer wavelengths (λ) than visible light, from about 7.8×10^{-5} cm to 7.8×10^{-2} cm. Frequencies are often expressed in wavenumber, the inverse of the wavelength expressed in centimeters (cm^{-1}) instead of hertz (Hz or sec^{-1}). **IR spectroscopy is particularly useful in determining what functional groups are present in molecules and can be extremely valuable in identifying compounds based on the functional groups they contain.**

8. (A) Ionization energy data provided support for the shell model of electron configuration in the atom, although they hint at the existence of orbitals (see the plot of first ionization energy versus atomic number that precedes question 31 and is part of answer 31). **Ionization energies can be explained qualitatively using Coulomb's law** (and quantitatively, but you won't have to do that; see Coulomb's law below).

The College Board expects you to be able to analyze electron-energy data for patterns and relationships relating the electronic structure of atoms to their properties. You are also expected to understand what ionization energy does and does not tell us about atoms and their electrons.

First ionization energy is related to chemical reactivity. Atoms with a low first ionization energy (like metals) tend to get oxidized during chemical reactions. The lower the first ionization energy, the more reactive the atom (the reaction of sodium versus cesium in water, for example). Atoms with high first ionization energies, like nonmetals, tend to get reduced during chemical reactions. The nonmetals with higher first ionization energies (like fluorine) tend to be more reactive than those with lower first ionization energies (like iodine). **The ionization energy *does not* tell us the energy of a particular electron.** We need photoelectron spectroscopy (PES) for that.

Coulomb's Law

Many atomic properties, trends within the periodic table, and all the forces between charged and polar particles can be understood and explained using Coulomb's law.

$$F_{ES} = \frac{kq_1q_2}{d^2}$$

F_{ES} = electrostatic force

k = Coulomb's constant (8.99×10^9 N • m^2 • C^{-2})

q_1 = magnitude of charge 1

q_2 = magnitude of charge 2

Charges must be multiples of the fundamental charge (the actual, not relative, charge on one proton or one electron): 1.6×10^{-19} C

d = distance between the charges

Coulomb's law states that **the electrostatic force is directly proportional to the magnitudes of the charges and inversely proportional to the square of the distance between them.**

So if the distance between charges is doubled, the force between them decreases to $\frac{1}{4}$.

If the distance between charges is reduced by half, the force between them quadruples!

Remember: The charge on a proton or electron is $\pm 1.602 \times 10^{-19}$ C, not $\pm$ 1 (their relative charges).

Coulomb's law can be used qualitatively to explain many properties of atoms.

Deviations from Coulomb's law are typically attributed to the following confounding effects:

- **Core electrons,** which are typically closer to the nucleus than valence electrons, tend to **shield the valence electrons from the full electrostatic attraction of the nucleus.**
- **Repulsion between electrons** accounts for the differences in energy between electrons in different orbitals of the same shell.

9. (B) Rutherford shot alpha (α) particles (He^{2+}, helium nuclei) at a very thin piece of gold foil. Gold has a very large nucleus (containing 79 protons and an average of 118 neutrons). Neutral gold atoms make up the foil, so the He^{2+} ions are expected to be repelled by the nucleus of the atoms. The question was what the effect of the electrons would be.

Most of the He^{2+} ions went right through the foil without being deflected at all. This suggested that **the atom is mostly empty space**. Some of the He^{2+} particles were slightly deflected from their paths, indicating some He^{2+} ions approached the positive charges enough to be repelled by them, at least a little. But **the smaller number of these deflected particles meant that it was not common for the alpha particles to get close enough to the positive charges in the gold atoms to "feel" the repulsion** (Coulomb's law strikes again; see box following answer 8), indicating that the nucleus of the atom was small and all of the positive charges were packaged together. A rare few of the particles were backscattered, indicating the particle was about to have a head-on collision with the dense bunch of positive charges in the nucleus. **The rarity of this event suggested that the nuclei of the gold foil were greatly spaced apart relative to the size of the nuclei.**

Taken together, the type and frequency of the alpha particle interactions with the gold suggested that the positive charges of an atom were concentrated into a small volume, hence the great repulsion when a He^{2+} particle approached but also the rarity of the close encounter. The atom is also composed of negatively charged electrons that are outside the nucleus. And perhaps surprisingly, the atom is mostly empty space.

This question highlights the importance of having a model in science. **The power of a good model cannot be overstated!** A model provides a framework for asking the right questions—those that will lead to the formation of testable hypotheses. The data provided by these experiments are then interpreted under the guidance of the model and can either lend support to the current model or highlight its weaknesses.

The College Board composed a list of seven science practices it will be evaluating on the AP Chemistry exam. Science practice 1 focuses on models. You must be able to use models and other representations to communicate and solve problems.

Questions 10–13:

$$2\ H_2\ (g) + O_2\ (g) \rightarrow 2\ H_2O\ (l)$$

	Hydrogen	Oxygen
Particles	2	1
Mass	4	32
Volume	2	1

10. (C) The reaction of hydrogen gas and oxygen gas and the particle numbers, masses, and volumes are given in the table above. The number of moles of hydrogen gas must be twice the number of moles of oxygen gas for there to be no limiting or excess reagents. That means **the mass of hydrogen gas relative to oxygen gas is a constant.**

Choice C states that the number of liters of hydrogen or oxygen gas divided by the mass of the particular gas is a constant. This is because **the number of moles per liter is constant**. This is the most direct evidence listed that the volume of a gas is directly proportional to the number of particles (at the same pressure and temperature).

Choice A is true but is more indicative of the stoichiometry of the reaction than the mole-volume relationship between particle number and volume. The mass of oxygen gas is 16 times that of hydrogen gas, but twice as many moles of particles of hydrogen gas (and therefore twice the mass of hydrogen) are needed to react completely, meaning 8 times the mass of oxygen reacts with hydrogen in the reaction.

Choice B is incorrect because the particle-volume relationship is only true for gases, whereas in this reaction, the H_2O that forms is a liquid (notice the units of milliliters, mL, in the table).

Choice D is true, but is evidence of conservation of mass during a chemical reaction, not that the volume of a gas is directly proportional to the number of particles.

11. (B) This question is similar to question 10 but is more specific in its exclusion of mass as a criterion for evidence. Choice C deals only with masses and so it can't provide evidence for the particle-volume relationship according to the criterion of the question. Choice D uses only volumes, but incorrectly. **The particle-volume relationship only holds for gases**, whereas in this reaction, the H_2O that forms is a liquid (notice the units of milliliters, mL, in the table). See the table above answer 10.

12. (B). The masses of hydrogen, oxygen, and water in the table can be used to calculate the percent composition of hydrogen (11%) and oxygen (89%) in water without knowing the formulas or any of the substances. If hydrogen and oxygen were monotomic gases, the mole (and therefore gas volume) ratios required to produce water would be the same as for their diatomic forms: two volumes of hydrogen to one volume of oxygen.

Choice A is incorrect because liquid water is produced in the reaction. Because the H_2O in choice A is a liquid, volume comparisons are not appropriate. The water must be further decomposed into hydrogen and oxygen gases to determine the final masses and gas volumes of the elements involved.

Choice B shows that hydrogen peroxide (H_2O_2) contains equimolar quantities of hydrogen and oxygen. The decomposition of H_2O_2 results in equal volumes of H_2 *(g)* and O_2 *(g)* even though their masses are quite different. See the table above answer 10.

- 1 mol H_2O_2 yields equal volumes of hydrogen and oxygen gas (at STP, 22.4 liters of each gas, 32 grams of oxygen).
- 1 mol H_2O yields twice as much hydrogen as oxygen (at STP, 22.4 liters of hydrogen gas and 11.2 liters of oxygen gas, at 16 grams).

Choice C has the same problem as choice A, it produces a product that is not a gas (carbon is produced as a solid). Without the chemical formula of the products available to early chemists, gases were the most reliable way of comparing the number of particles with their masses.

Choice D contains solids, liquids, gases, and aqueous solutions as well as both elements and compounds. Good luck trying to establish the diatomic nature of oxygen gas with that reaction!

13. No. Because both hydrogen peroxide and water contain two hydrogen atoms, they cannot discriminate between monatomic and diatomic hydrogen, but they can be used to deduce the diatomic nature of oxygen gas.

14. (C) In 1897, J. J. Thomson performed his cathode-ray experiment to determine the charge-to-mass ratio of electrons. He knew **the cathode ray is composed of electrons** (so you can eliminate choice A). In order **for particles to be deflected by an electric or**

magnetic field they must be charged. Neutrons and photons (particles of electromagnetic radiation) are *not* charged and are therefore deflected by an electric or magnetic field (another reason to eliminate choice A).

Electrons, unlike photons, have mass. It is important to remember that **electrons, like all matter, display wavelike properties**, but that is not what is being observed here (eliminates choice B). Choice D is true (alpha particles, which are He nuclei, have twice the charge but over 3,700 × the mass), but with no direct comparison to alpha particles, this cannot be inferred by the data alone.

15. (A) With a charge-to-mass ratio and the charge of an individual electron, we can get the mass of an individual electron. Canceling out units is the easiest way to solve this problem. We need to combine the terms C and $\frac{C}{g}$ to get g, the mass. If we flip $\frac{C}{g}$ to $\frac{g}{C}$, the C unit cancels. (Remember that the fraction as given states that there is a charge of -1.76×10^8 C per 1 gram of electrons, but it can also be stated that 1 gram of electrons has a charge of -1.76×10^8 C.)

$$-1.602 \times 10^{-19}\ \text{C} \times \frac{1\ \text{gram}}{-1.76 \times 10^8\ \text{C}} = 9.10 \times 10^{-28}\ \text{g (usually given as } 9.10 \times 10^{-31}\ \text{kg).}$$

Choices C and D include Avogadro's number. If you multiply your answer (9.1×10^{-28} g) by 6.02×10^{23}, you'll get the mass of one mole of electrons (~5.5×10^{-4} g). If you multiply the charge of a single electron by 6.02×10^{23}, you get Faraday's number (~96,450 C mol^{-1}), the charge on one mole of electrons. (The exact number you get from a calculation depends on how many digits you input.)

16. (B) A particle of high charge-to-mass ratio would experience greater deflection in the same field as a particle of lower charge-to-mass ratio. The field (or force) acts on the particle according to its charge. The force per unit of charge is the same, but a given force can accelerate a small mass with much greater magnitude than a large mass ($F = ma$). Although a real calculation would require charges in units of coulombs and mass in grams, for simplicity we'll use whole number charges and amu. Notice the huge charge on the electron compared to its mass.

Particle	Charge	Mass (amu)	$\frac{\text{Charge}}{\text{Mass}}$	Charge (C)	Mass (kg)	$\frac{\text{Charge}}{\text{Mass}}$ (C kg−1)
β^-, electron	−1	$\frac{1}{1,836}$	−1,836	-1.602×10^{-19}	9.109×10^{-31}	-1.8×10^{11}
H^+, proton	+1	1	+1	1.602×10^{-19}	1.672×10^{-27}	$+9.6 \times 10^7$
α, He^{2+}	+2	4	$+\frac{1}{2}$	3.204×10^{-19}	6.644×10^{-27}	$+4.8 \times 10^7$

17. (C) Two spots would be expected to appear on the screen, one for each of the two directions of spin, $+\frac{1}{2}$ or $-\frac{1}{2}$. The hydrogen atoms with a $+\frac{1}{2}$ spin electron would be deflected away from the centerline in one direction, and the hydrogen atom with a $-\frac{1}{2}$

spin electron would be deflected an equal distance from the centerline but in the opposite direction. If the particles were not deflected by the magnetic field, they would land in the center of the screen, having taken a straight-line path from their source (eliminating choices A and B). The description of the experiment states that a tiny spot develops where an atom strikes the screen. It is highly improbable that the spots would align in a way that creates a wave shape on the detector (choice D). Choice D is an attempt to confuse the property of electron spin with the wavelike behavior of electrons.

For each chemistry technique you learn in AP Chemistry, understand when the technique is used, what kind of data it produces, and what the data tell us.

18. (D) Emission lines represent all the possible energy transitions of the electrons in an element. Emission (and absorption) spectra are **"elemental fingerprints"** and are used for elemental identification. Emission spectra can be observed when a gaseous form of the element is electrified. The light emitted is polarized and dispersed. If a magnetic field is applied to the atoms in the gas, it will not change the magnitude of the energy transitions but it *will* cause the electrons to behave as magnets within the field. Because electrons of opposite spins are like opposite poles of a magnet, the field will affect them in opposite ways, literally—as in direction—splitting the emission lines they produce.

Choices A, B, and C are all true but they don't answer the question. Although the experiment *assumes* electrons have particle properties, it does not specifically investigate their wave versus particle nature.

For each chemistry technique you learn in AP Chemistry, understand when the technique is used, what kind of data it produces, and what the data tell us.

19. (D) The experiment was based on the model of the electron as a spinning charge. The electron spin is not literally clockwise or counterclockwise, but you can assume it is to simplify the interpretation of the experiment (the benefits of a model!).

A spinning charge creates a magnetic field. If the charge spins clockwise, the field it generates is repulsive to other clockwise spins of the same charge and attractive to counterclockwise spins of the same charge. The magnetic field can act on all the charges, positive protons and negative electrons, but the field strength in the experiment was adjusted so that it could not deflect a heavy proton with the same quantity of charge (although opposite) as the electron.

Choice A refers to "one kind of charge." Electrons are negatively charged. All magnetic or electric fields will attract one kind of charge and deflect the other. The field in this experiment was constructed to affect the magnetic fields created by the electrons. Choice B implies that the mass of the proton would actually impede the deflection of a hydrogen atom. Choice C is true, but moving charges don't have to be spinning to be deflected, they need only be moving. **The two types of spin can be differentiated by the direction of the hydrogen atom's deflection by the magnetic field**, which is the essence of this experiment.

Protons are also considered spinning charges and the fields they create are the basis of **nuclear magnetic resonance (NMR) spectroscopy** (see answer #84).

For each chemistry technique you learn in AP Chemistry, understand when the technique is used, what kind of data it produces, and what the data tell us.

20. (B) Electron spin is either $+\frac{1}{2}$ or $-\frac{1}{2}$. Choices A, C, and D are all true statements based on **the quantum mechanical model of the atom** which **provides the theoretical framework for this experiment**, but none of those statements were the conclusion of the experiment. This experiment, done by Otto Stern and Walther Gerlach in 1924, gave electrons their fourth quantum number.

21. (C) The property of interference is a characteristic of *all waves.* The ability to be polarized is characteristic *only* of electromagnetic waves. Through the lens of quantum mechanics nothing is either a wave or a particle; **matter and electromagnetic radiation** (which is pretty much everything!) **have both wave properties and particle properties**. So C is the only correct statement.

For all but the smallest of size scales, the wave properties of matter can be safely ignored. Choice D is the opposite of what the quantum mechanical model states. **Energy is quantized**, so there is a smallest "packet" of energy and a finite difference between energy "packets" of different frequencies. Energy may appear continuous to our limited resolution but it is considered discrete.

Charges are typically associated with particles, but remember that electrons (and objects with mass) have wave properties, as well.

22. (D) Differential solubilities are commonly exploited for separating ions and other substances in solution by precipitation. The addition of any solution will increase the final volume of the solution, and this typically needs to be considered in a dilution or solubility problem. This question is qualitative. All of the solutions are dilute, so we only need to consider the ions in the solutions.

Solutions of HCl and NaCl consist of only soluble ions that do not form precipitates with Ba^{2+}, Fe^{3+}, or Zn^{2+}. Lithium sulfate (Li_2SO_4) is a source of sulfate ions. According to the table, only Ba^{2+} ions form a precipitate when in solution with SO_4^{2-}, which would allow for separation. Sodium hydroxide (NaOH) contributes OH^- ions, which form precipitates with both Fe^{3+} and Zn^{2+} ions but not Ba^{2+} ions, which would also allow for the separation of Ba^{2+} ions. The correct choice is D because Li_2SO_4 can be used to precipitate out the Ba^{2+} ions *or* NaOH can be used to precipitate out the Fe^{3+} and Zn^{2+} ions, leaving the Ba^{2+} ions behind.

You don't have to memorize the solubility guidelines and their exceptions for the AP Chemistry exam, but you are expected to know that ***all sodium, potassium, ammonium, and nitrate salts are soluble in water***.

23. (A) Electrons get excited (energized, raised to an orbital of higher energy) when the atom or molecule of which it is a part *absorbs* energy. The excited electrons are not stable at the higher energy levels, so their time in the excited state is limited. When they drop back down to ground state, they emit the same quantity of energy that was absorbed to excite them. The specific amount of energy depends on the electron's transition. (There are a limited number of transitions and each is associated with a particular value of energy. Most, if not all, of these are known).

Choice B is incorrect because there are no "photons of electricity." Choice C may seem true, but it is not an interpretation of the hydrogen spectrum. Cathode rays are streams of electrons that can be produced in vacuum tubes by application of an electric potential, but that does not explain the hydrogen spectrum. (The two phenomena are related.) D is simply not true.

24. (B) Of the four bands, the one with the longest wavelength has the lowest frequency and therefore the lowest energy. The table shows that the 656 nm band has a frequency of $4.6 \times 10^{14}\ sec^{-1}$.

$$\mathbf{E = h\nu} \quad \therefore E = (4.6 \times 10^{14})\ (6.63 \times 10^{-34}) = 3.0 \times 10^{-19}\ J$$

Remember that **energy is proportional to frequency** and inversely proportional to wavelength.

25. (D) The minimum energy needed to ionize the metal can be found by converting the lowest frequency that can ionize the metal into energy using **E** = $h\nu$. The input must be in sec^{-1} and J because those are the units of the value of Planck's constant you are given.

$$\mathbf{E} = h\nu$$
$$E = (6.63 \times 10^{-34})(1.6 \times 10^{15})$$
$$E = 1.1 \times 10^{-18}\ J$$

The question provides us with the conversion factor you need for the energy in joules that you calculated.

$$1.1 \times 10^{-18}\ J\left(\frac{6.2 \times 10^{18}\ eV}{1J}\right) = 6.8\ eV$$

The work function of platinum is 6.4 eV; close enough! It's better that the work function be smaller than the energy input because **the work function represents a minimum energy**. If the energy of the photon was less than the minimum required (the work function of the metal), it wouldn't have enough energy to ionize the metal.

The work function of a metal is the minimum amount of work needed to remove an electron from the metal, but just enough to get it out of the metal, not enough to "launch it." If higher frequencies of radiation are used, the electrons are not only dislodged from the metal but use the extra energy provided by the photon (energy of the photon – work function of the metal = "launch" energy) to gain kinetic energy. In other words, they are quite literally launched from the metal. The greater the frequency of the photons, the higher the kinetic energy of the electrons (the consistency you would expect from conservation of energy).

You don't have to remember all these details to answer the question, which is actually pretty straightforward if you can handle the data and conversions. However, **understanding Einstein's interpretation of the photoelectric effect (what the experiment described) is important.** At the time, the evidence seemed to point to the wave model of light that could not explain the photoelectric effect. Einstein interpreted the photoelectric effect using a particle model of light behavior (or more generally, electromagnetic radiation). **The importance of a model strikes again!**

The College Board expects an AP Chemistry student to understand how scientific models are used to design experiments and interpret data.

When photons of a high enough frequency strike a metal plate, electrons are dislodged from the plate, much like throwing a ball at a Frisbee that got stuck in a tree. Waves do not demonstrate this behavior. It is important to remember that this analogy is imperfect for several reasons. First, all photons travel at (just about) the same speed in the same medium (but not exactly, which explains the dispersion of light in a prism, for example). Second, photons are massless and their energy is determined *only* by their frequency.

26. (C) Each line in a spectrum corresponds to an electronic transition (energy change) of an electron. Any particular electron can produce several lines according to the energy transitions it undergoes. Although elements with a large number of electrons typically produce highly complex spectrums, the number of lines is not *directly proportional* to electron number.

27. (B) Diffraction is the spreading out of waves when passed through small openings. **Diffraction is a property of waves**, and to occur, **the wavelengths of the waves must be comparable in length to the size of the opening through which they are being passed.** X-rays have wavelengths comparable to the spaces between the ions in an ionic crystal lattice and will produce a diffraction pattern that can be used to analyze the lattice structure of the crystal.

Choice A is incorrect because the X-rays are not reflected off the substance (the information in the question states they are "passed through" the sample). Choices C, and D are similarly incorrect. The question clearly states that "X-rays can be passed through a crystal of a pure substance to produce a diffraction pattern," the X-rays do not destroy the sample (choice C). It is the X-rays emerging from the crystal that create the diffraction pattern, not the actual atoms or electrons (choice D) of the crystal.

28. (B) The molarity of the acid based on the *actual* equivalence point:

$$\mathbf{M_A\,V_A = M_B\,V_B}$$
$$(M_A)(0.100) = (0.100)(0.050)$$
$$\mathbf{M_A = 0.050\ M}$$

However, the student will have passed the actual equivalence point by the time the indicator changes color, so she will end up adding more base than was needed to neutralize the acid present. *Overestimating the amount of base added would result in overestimating the amount of acid present.*

29. (A) The **atomic mass is the weighted average** of all the naturally occurring isotopes of a particular element. The atomic mass of bromine is *almost* 80. The two isotopes are of mass numbers 79 and 81. The nonweighted average of these two numbers is 80, which implies that the natural occurrence of the two isotopes is close to equal. When using mass spectrometry, an atom of Br will show peaks at an m/z of 79 and at an m/z of 81—with both peaks of almost equal height, showing that they are in almost equal abundance.

Different elements have different numbers and percentages of naturally occurring isotopes, which are accounted for in the atomic mass calculation. **The difference between the atomic mass and the mass number of a particular isotope suggests something about the isotope's abundance.** For example, from the atomic mass of carbon (12.011), you can infer that the percent abundance of ^{13}C is quite low (1.1%).

Additionally, subtracting the atomic number from the atomic mass gives you the weighted average of the number of neutrons in the naturally occurring isotopes. For example, $12.011 - 6 = 6.011$, the weighted average of the number of neutrons in the isotopes of carbon. The number of neutrons alone is not particularly useful but can be used to make an educated guess about the abundance of a particular isotope. Clearly, ^{12}C, with 6 neutrons, is much more abundant than ^{13}C with 7 neutrons.

30. (D) The difference between the atomic mass of an element and the mass number of a particular isotope of that element suggests something about the isotope's abundance. For example, from the atomic mass of carbon (12.011) you can be infer that the percent abundance of ^{13}C is quite low (1.1%). Since the weighted average of the isotopes of copper is closer to 63 than 65, ^{63}Cu must be the more abundant isotope. The mass of the electron is so tiny that the value of 0.546 amu overestimates it by an order of magnitude. One electron has a mass of $\frac{1}{1{,}836\text{ amu}}$, so the mass of 29 of them is only 0.016 amu.

31. (B) Elements of atomic numbers 2, 10, and 18 are He, Ne, and Ar, respectively. These noble gases have the highest first ionization energy of the elements. The large drops in ionization energy occur mainly because the elements of atomic numbers 3 (Li), 11 (Na), and 19 (K) have one valence electrons that is farther from the nucleus compared to the valence electrons of elements 2, 10, and 18. It is both this large radius and the lower effective nuclear charge (which is one of the reasons for the large radius) that makes the energy requirements for removal of this electron so low (why choices C and D are incorrect).

Coulomb's law (see the box following answer 8) states that the force of the attraction or repulsion between charges is inversely proportional to the square of the distance between them, so increasing the distance between charges means greatly decreasing the force between them. **Generally, the size of the atom indicates the strength of the nucleus's pull on the electrons.** Choice B is incorrect because in a given period, a smaller atomic radius corresponds to a greater nuclear charge.

Selected First Ionization Energy and Atomic Radius Data

Element	Atomic radius (pm, 10^{-12}m)	Valence electron configuration	1st E_I (kJ/mol)
Na	186	$3s^1$	496
Mg	160	$3s^2$	738
Al	143	$3p_x^1$	577
Si	118	$3p_y^2$	786
P	**110**	$\mathbf{3p_z^3}$	**1,060**
S	**103**	$\mathbf{3p_x^4}$	**1,000**
Cl	199	$3p_y^5$	1,256
Ar	No data	$3p_z^6$	1,520

The data show that, generally, **a larger atomic radius corresponds to a lower first ionization energy** (P and Si are notable exceptions). There is no data for d and f electrons, although the Si/P data do indicate that the orbital filling (or removing, as the case may be) has an effect on the ionization energy.

The fluctuations in the plot of the ionization energy versus atomic radius that follows reveals that **both the ionization energy and the atomic radius are the result of multiple factors.**

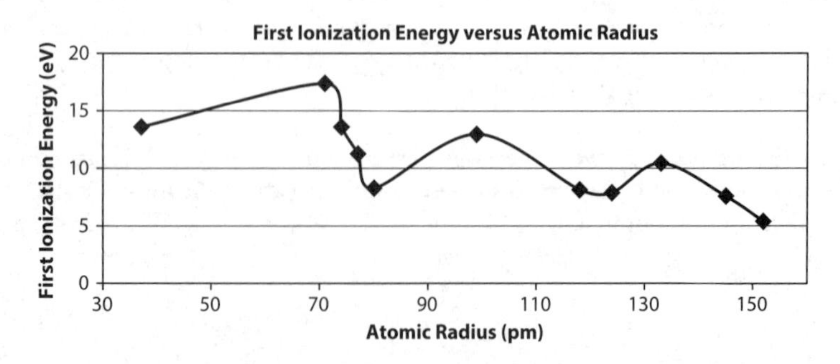

The College Board expects you to be able to qualitatively explain the properties of ionization energies and atomic size based on Coulomb's law (see the box following answer 8).

32. (D) Remember that we are looking for the statement that is *not* a logical explanation. The repulsion between p^1 and p^4 electrons (which are of opposite spins in the same p orbital) can be inferred from the slight drop in ionization energies between N and O (atomic numbers 7 and 8) and P and S (atomic numbers 15 and 16). Answer choice A accounts for these two decreases. The shielding of p electrons by the s electrons is noticeable in the drop in ionization energies between Be and B (atomic numbers 4 and 5) and Mg and Al (atomic numbers 12 and 13).

Choice B states the electrons in a filled s orbital are more effective at shielding the electrons in the p orbitals of the same energy level than each other. Remember that greater shielding means a lower effective nuclear charge and a lower first ionization energy, so we would expect effective shielding of p electrons to reduce ionization energy. This helps explain the drop in ionization energies between Be and B as well as Mg and Al. Choice C helps explain why the first ionization energies of atoms with small radii are larger than the energies of atoms with larger radii.

Atomic radius is determined by the combined effect of all the forces of attraction and repulsion in the atom, but **very generally, atoms with a smaller radius will have electrons closer to the nucleus that will require more energy to remove**. This effect is seen in the drop between He and Li (atomic numbers 2 and 3), Ne and Na (atomic numbers 10 and 11), and Ar and K (atomic numbers 18 and 19).

The information you need to answer the question is all in the graph even if you don't know *why*. Remember that *all the statements are true and what you need to determine is which of the choices provides the least valuable explanation of the data.* If you compare the ionization energy with the electron configuration using our periodic table, the trend becomes clear. Choice D provides a foundation for the limits of the explanatory scope of ionization energies, but it does not explain why the first ionization energies would increase and decrease with increasing atomic number.

Questions asked in the negative are not common but do occur on the AP Chemistry exam, and they can be tricky. When answering a question asked with a "not," try circling the negative word in the question (to remind you that you're looking for the opposite) and then mark each answer choice as true or false (or yes or no). When you're done with the four choices, the one that's false (or no) is the correct answer.

33. (C) Ionization energy is an indicator of effective nuclear charge (and, therefore, atomic radius; see answers 8 and 31 for more on ionization energies). Choice A is incorrect because the standard state of the element has little to do with effective nuclear charge. Choices B and D are false. Choice D states the *opposite* effect of shielding on ionization energy. **Valence electrons that experience less shielding have greater ionization energies.** This is because the effective nuclear charge is larger with less shielding, so the electron "feels" the pull of the positively charged nucleus more strongly, evidenced by a greater amount required to remove the electron. It's important to remember that **the valence electrons of elements in the same period basically experience the same amount of shielding,** but the greater nuclear charge of the rightmost elements makes for a greater effective nuclear charge, attracting the valence electrons and reducing atomic radius.

34. (C) Each peak represents one or more electrons of the same energy. The two peaks of Be are of equal height, so they have the same number of electrons. In this diagram, each of the vertical lines represents a peak height of 2 (the next line up is 4, then 6). The leftmost peak represents the two 1s electrons, each of which requires 11.5 MJ to remove. The rightmost peak represents the two 2s electrons, each of which requires 0.90 MJ to remove.

Choice A is incorrect because the peak is shifted to the left, which means the electrons it represents require more energy than the electron in hydrogen to remove (not less). Choice B is incorrect because you do not split the value of energy of the peak by the number of electrons it represents. Each electron in lithium's 1s orbital requires 6.26 MJ to remove. Choice D is incorrect because the energy to remove the 1s electrons is greater than the energy to remove the 2s or 2p electrons.

PES data can be confusing because, **according to the Aufbau principle, the energy of the 2p electrons is greater than the energy of the 1s electrons.** The Aufbau principle states that electrons fill lower energy orbitals first. (You can remember the energy of these orbitals by referring to the order of elements on the periodic table, for example, 4s electrons "have less energy" than electrons in 3d orbitals.) However, that is *not* what you are measuring with PES. **PES measures the amount of energy required to remove a particular electron according to Coulomb's law** (see answer #8). **The closer an electron is to the nucleus, the greater the electrostatic force between the electron and the positive charges in the nucleus and therefore the greater the energy needed to remove it.**

PES versus Ionization Energy Data

Light exhibits some properties of particles. Particles of light (or more accurately, the particle behaviors of light) are called **photons**. According to an equation you should be familiar with

$$E = h\nu$$

where E = energy, h = Planck's constant, and ν = frequency
each frequency of light has a particular energy associated with it. As you have seen in the photoelectric effect (discovered in the mid- to late 1800s), in order to dislodge an electron from the surface of a metal, the photons bombarding it must have sufficient energy (see answer 25). **Photoelectron spectroscopy (PES) is an application of the photoelectric effect**, the observation that metals emit electrons when particular frequencies of light are shined on them. PES is a general term for a variety of techniques that are used to measure the energy of electrons ejected from solids, liquids, or gases when electromagnetic radiation is applied to them. These techniques were developed in the mid-1900s, after **Einstein interpreted the photoelectric effect as a demonstration of the particulate property of light.**

X-ray and UV PES are used collectively to study the energies of all the electrons in an atom. **The benefit of PES compared to ionization energy is that the PES technique allows *any* electron to be removed from the atom regardless of which orbital it inhabits.** Successive ionization energies only pull off the outer electrons, one by one. In other words, getting to the 1s electron of sodium requires the removal of all the electrons outside of it. This makes the calculation of the individual electron energy difficult because the relative charge of the nucleus to the electrons increases with each electron that is removed. PES can remove one electron at a time and is not confined to only the outermost electrons, allowing the energy of individual energy levels to be measured independently of the others. **The intensity of the photoelectron signal at a given energy is a measure of the number of electrons at that energy level.**

Perhaps most important, the measurement of electron energies by this technique provided a method to deduce the electron shell structure of the atom.

How to Interpret PES Data

1. The x-axis is ENERGY. Notice that energy typically DECREASES from left to right—although there is no law that says the data must be presented this way, so check!
2. The y-axis is the INTENSITY of the signal, which represents the total number of electrons with that energy (i.e., the number of electrons that required that particular frequency of energy to be removed).

Each peak represents a subshell.
Each block represents an energy level (*n*, principle quantum number).

Block I: The 1s electrons will always require the most energy to remove since they are closest to the nucleus. This data shows two 1s electrons (a peak height of one electron is shown in the leftmost peak of block V).

Block II: Left to right, peaks represent two 2s and six 2p electrons.

Block III: Left to right, peaks represent two 3s and six 3p electrons.

Block IV: Left to right, peaks represent ten 3d, two 4s, and six 4p electrons.

Blocks I to IV show the progression of PES data from helium to krypton. **Adding the peak heights of all of them ($2 + 2 + 6 + 2 + 6 + 10 + 2 + 6 = 36$) gives you the total number of electrons** in krypton.

Block V: *Real PES data would not look like this.* This block is to show what peak heights of 1–6 would look like (from left to right). The relative size of the peaks is probably the only way you'll have to determine the number of electrons at a particular energy level. It is unlikely you will have the benefit of the horizontal gray lines on your AP exam.

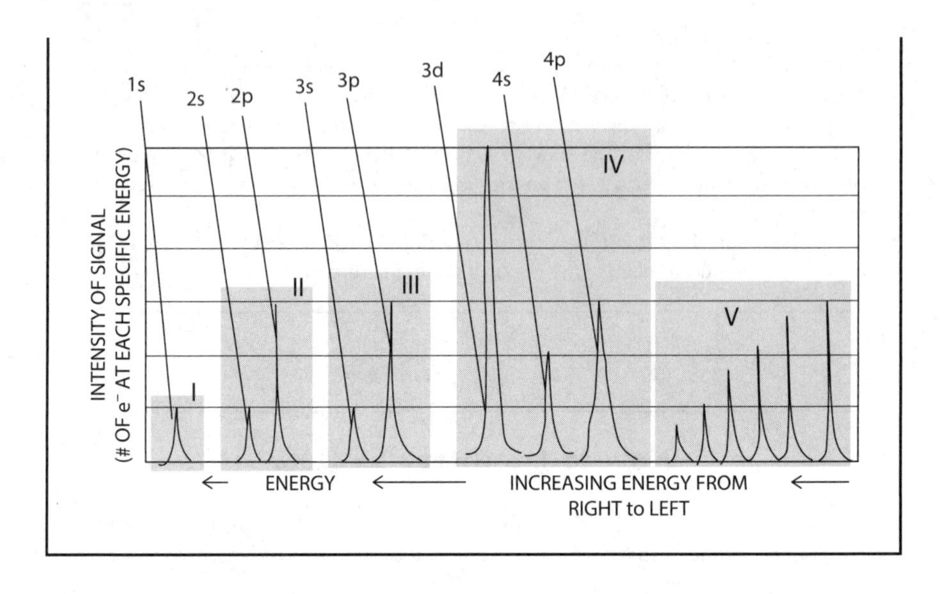

The College Board expects you to understand the development of the shell and orbital models of the atom. Memorization of the quantum numbers does not lead to greater understanding of how the model works (which is why it is not a topic assessed on the exam), but knowing how the model was developed should help you understand the model.

35. (C) PES spectra for hydrogen and helium show one peak, which would correspond to the first energy shell. PES spectra for lithium and beryllium show two peaks, indicating two shells. Choice A is incorrect because there is no discrepancy between the PES data and the shell model. Choice B is incorrect because the peak heights for the second peaks of lithium and beryllium should not be the same; beryllium has one more electron than lithium in the 2s subshell and so the second peak in the PES data should be higher (twice as high, since Be has 2 electrons in the 2s subshell and Li has one).

Boron, carbon, oxygen, and neon have three peaks, which would seem to indicate three shells. However, we know from the current model that they only have two shells. *This is the discrepancy between the data and the model that required a refinement of the model.* This refinement led to the development of the orbital model, which identifies orbitals of different energies within shells.

Choice D is alluring because it does indicate that carbon has three peaks, which is one of the discrepancies between the shell-only model and the PES data, but it's not the peak heights that are the basis of the discrepancy, it is the number of them. In addition, the data for boron, carbon, oxygen, and neon are collectively a much greater discrepancy than that of just one element.

36. (D) Photoelectron spectroscopy is a technique to determine the energy of the electrons in an element. See "How to Interpret PES Data" after answer 34. Refer to the question and answer for more on determining the spin on an electron.

37. (C) Each peak represents a subshell (1s, 2p, 3d, etc.). See "How to Interpret PES data" after answer 34.

38. (B) The peak height represents the number of electrons in a particular subshell. The location of the peak on the *x*-axis represents the frequency of energy required to remove those electrons from the subshell. There can be either one or two electrons in a an s orbital (because there is only one s orbital per energy level the s orbital is also an s subshell) and so there are only 2 possible peak heights for (valence) s subshells. However, there can be anywhere from 1 to 6 electrons in a p subshell and so there are 6 possible peak heights of peak representing a (valence) p subshell. See "How to Interpret PES data" after answer 34.

39. (B) Atomic spectra, also called emission spectra (but can also refer to absorption spectra), have **discrete lines, each of which represents the energy of a particular electronic transition.** There are a large but finite number of these transitions that produce the characteristic series of lines with dark spaces in between. The lines of atomic spectra are clean and thin because they represent single frequencies that correspond to a discrete amount of energy. (Fuzzy lines are due to the technical difficulties in producing spectra.) **The dark spaces indicate the range of frequencies, and therefore energies, in which electronic transitions do *not* occur for that element.**

The spectrum for each element is unique and is one way to identify a particular element. For this reason, atomic spectra are often compared to fingerprints.

Spectroscopy is a broad term describing the study of the interaction between matter and electromagnetic radiation. The figure below shows how advances in spectroscopy were critical in the creation and refinement of the quantum mechanical model of the atom.

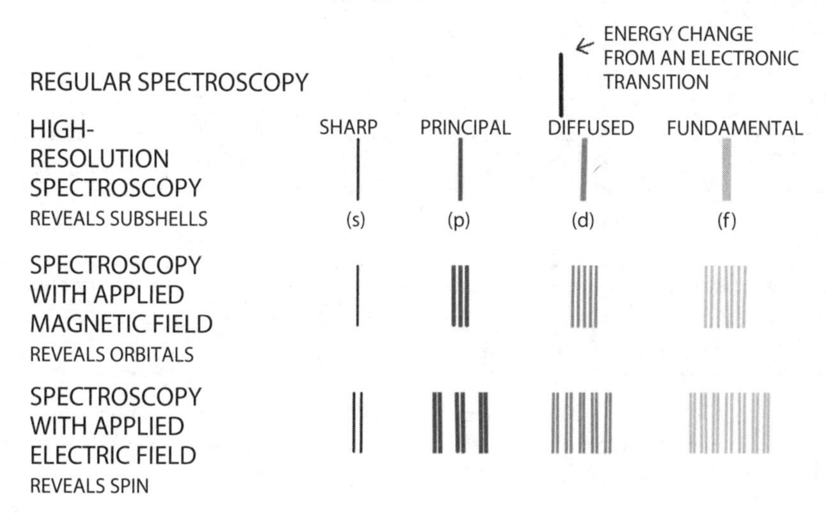

40. (D) The first ionization energy of B is less than that of Be, which is surprising. The increase in ionization energy between hydrogen and helium makes sense because the two electrons are in the same shell. The electron in helium requires more energy to remove because the nuclear charge holding it in the shell is twice that of hydrogen. The same is true for lithium and beryllium. The two electrons are in the same shell, but the shell they are in ($n = 2$) is farther from the nucleus than the first shell, so the drop in first ionization energy is expected from helium to lithium. However, all the valence electrons of the elements in period 2 are in the second shell. That means the trend in first ionization energies would continually increase across the period unless there were another factor that influenced the ionization energy: subshells. See answer 31 for a detailed explanation of ionization energies.

41. (C) Coulomb's law (see the box following answer 8) shows the relationships between the electrostatic force, quantity of charge, and distance between charges. **There is an inverse square relationship between force and distance;** in other words, halving the distance increases the force by a factor of four whereas doubling the distance decreases the force to one-fourth. **There is a direct relationship between the electrostatic force and the quantity of charge.**

42. (C) This is a stoichiometry question. The oxidation state of the lead ion, +2, indicates the number of chloride ions that can bond to it. The formula for lead (II) chloride is $PbCl_2$. The drawings of the original solutions show that the chloride ion is larger (and in the figure, lighter colored) than the lead, so only choices C and D show the correct product of the reaction, $PbCl_2$. Choice A incorrectly shows precipitation $PbCl^-$ and choice B incorrectly shows Pb_2Cl^+.

Choice D is incorrect because it is not in accordance with conservation of atoms. The original solutions were of the same concentration, so if twice as many chloride ions as lead are needed to react, there should be excess lead in the solution, which is what choice C shows.

43. (D) Tables 1 and 2 below list elements by what are now called groups. Although lead and osmium have a common difference of 4, they are not in the same group as the other elements in their respective lists.

Table 1

Simple substance	Weight	Simple substance	Weight	Common difference
Group 17 Halogens		Group 15		5

Table 2

Simple substance	Weight	Simple substance	Weight	Common difference
Group 2 Alkali earth except lead		Group 15 except osmium		4

44. (A) All the spectral lines from hydrogen and helium are present in spectrum 1, and there are no lines in spectrum 1 that are not accounted for in the spectra of hydrogen and helium. See the figure under answer 38 for more on spectral lines.

45. (D) The three spectral lines of sodium are present in spectrum 2. There is another element present, mercury. See the figure under answer 38 for more on spectral lines.

46. (B) Choice A is incorrect because the actual number of oil drops was not what allowed the calculation of the charge on a single electron. Besides, a mole of oil drops is *a lot* of oil drops (but he did analyze thousands of drops, which is impressive when you realize he did this in 1909, before modern computers and handheld calculators)! Choice C is incorrect because the drops held various charges, so dividing the charge on each drop by the charge-to-mass ratio of the electron would produce several different values for the charge. Choice D is incorrect because the frequency of the charges was not necessary to interpret the results of the experiment. **Quantization (and conservation) of charge** had been known for at least two centuries prior to this experiment and was part of the underlying principle: **The values for all of the charges on the drops will be whole-number multiples of the elementary charge.**

47. (A) Mass spectra can often (but not always) give the molecular mass of a substance, which can then be used to narrow down the possibilities for the molecular formula. For example, if the molar mass was 110, possible formulas include (but are not limited to) C_8H_{14} and $C_6H_{10}N_2$.

Mass spectroscopy is explained in more detail in answer 73. Photoelectron spectroscopy (choice B) is used to determine the energy of the electrons in an element (described in detail in answer 34). Although some forms of electron microscopy can visualize atomic surfaces, the technique has not been refined enough to determine the molecular formula of a substance (choice C). Density determination (choice D) can be used to determine the molar mass of a compound but only in conjunction with several other techniques.

48. (C) Many early scientists attempted to organize the elements into a table. Chemists were aware that certain elements shared chemical and physical properties, but once the masses of enough elements were known and they were placed in order of increasing mass, their periodicity became clear.

49. (B) Location II on the periodic table represents the group 1 and 2 metals (s block). The reaction of water with the alkali metals (group 1) is legendary. The reaction is highly exothermic, partly because the reaction produces a strong base whose total dissociation in water is highly exothermic. The reaction also produces energy in the form of light and H_2 *(g)*.

$$2\ Na\ (s) + 2\ H_2O\ (l) \rightarrow 2\ NaOH\ (aq) + H_2\ (g)$$

$$Ca\ (s) + 2\ H_2O\ (l) \rightarrow Ca(OH)_2\ (s) + H_2\ (g)$$

When added to water, the interior of a chunk of sodium metal, for example, will melt before it is consumed due to the high temperature produced by the reaction. (The melting point of Na is ~800° C.) In addition, the high heat ignites the flammable H_2 *(g)* that the reaction produces. The alkali earth metals (group 2) behave similarly. Generally, the metals at the bottom of the groups (with the largest atomic radii and the lowest first ionization energies) react more vigorously than those at the top because of the ease of oxidation of these elements.

Metal hydroxides thermally decompose at high temperatures, however, so it is possible that the hydroxides produced may end up as the metal oxide and water.

$$Ca(OH)_2\ (aq) \rightarrow CaO\ (s) + H_2O\ (l)$$

50. (C) Location V on the periodic table represents the noble gases. In general, first ionization energies are low for metals and high for nonmetals. **The noble gases, on average, have the highest ionization energies of all the elements.** See answer 31 for a detailed explanation of ionization energies.

51. (C) Location IV on the table contains the halogens and all the nonmetals with exception of hydrogen and the noble gases. Fluorine is the element with the highest electronegativity. **Electronegativity is a measure of an atom's ability to attract electrons to itself while in a bond.** Because the atom must be in a bond to measure its electronegativity, the noble gases He, Ne, and Ar do not have measured values for electronegativity (the reason choice D is incorrect). The first ionization energies of Kr, Xe, and Rn, however, are sufficiently low to allow these noble gases to form covalent bonds with other atoms (mostly those with a high electronegativity, like F) and therefore have had their electronegativity values measured. Location IV on the table also contains the metalloids and some metals as well.

52. (D) With the exception of technetium ($Z = 43$) and promethium ($Z = 61$), only elements of $Z = 84$ (polonium) and higher are radioactive. That makes radon ($Z = 86$), a noble gas in location V, radioactive. Location VII contains both the lanthanides and the actinides. All of the actinides are radioactive but the only radioactive lanthanide is promethium.

53. (A) Location II represents the alkali and alkali earth metals. They have, on average, the lowest ionization energies of the elements. Metallic character is not a measured quality of an element. It is a set of properties ascribed to metals, but underlying all the properties of metals is the "one property to rule them all": **their readiness to lose electrons.** This is due to **metallic bonding**, which can be thought of as the most "sharing" form of bonding. Metals are **lattices of positively charged ions** surrounded by a **"sea of electrons."** These electrons are **highly mobile** and account for just about all the properties of metals, especially their ability to conduct heat and electricity. The lack of tight pulling on the electrons by the nuclei allows them to be pulled off easily, resulting in the tendency of metals to take on (almost) exclusively **positive oxidation states.** Metals with the highest metallic character can be considered as those having the lowest ionization energies (though this is a bit of a simplification).

54. (C) The element represented is silicon (left to right from the PES data: $1s^2\ 2s^2\ 2p^6\ 3s^2\ 3p^2$). To get the electron configuration for an element from the spectrum, start at the left end. The first peak represents the 1s electrons, and the second peak represents the 2s electrons. They are the same height and are one-third the height of the third peak, so you can infer that they both have two electrons and the third peak, which represents the 2p electrons, has six electrons. The fourth peak represents the 3s electrons, and the fifth peak (all the way at the right) represents the 3p electrons. These peaks are the same size as each other and as peaks 1 and 2, so we can infer they each have two electrons. You can also add the number of electrons to get the total number of electrons and therefore the atomic number (PES data will most likely be for neutral atoms, not ions). From left to right, $2 + 2 + 6 + 2 + 2 = 14$. See answer 34 for a detailed explanation of the PES technique and spectrum analysis.

55. (C) The actual percent occurrences of the isotopes are 10% ^{86}Sr, 7% ^{87}Sr, and 83% ^{88}Sr. If they occurred in equal numbers, the atomic mass would be the average of their mass numbers, 87. **Since the mass number is *greater* than the average, the isotopes of higher mass must be present in greater quantities.** Since the atomic mass is closest to 88, the most abundant isotopes should be 88. The higher the percent occurrence of isotopes 86 and 87, the closer the atomic mass would be to 87 or 86. See answers 29 and 30 for a more detailed explanation.

56. (B) In gravimetric analysis, a substance is added to solution in which another substance is dissolved. The purpose is to precipitate out the substance in solution. The mass of the precipitated solid is used to infer the amount of the dissolved substance (called the analyte) in the original solution.

In this question, two solutions were analyzed, Ag^+ (*aq*) Pb^{2+} (*aq*). The purpose of adding NaCl (*aq*) was to provide an excess of chloride ions to precipitate AgCl (*s*) from solution 1 and $PbCl_2$ (*s*) from solution 2. As long as excess chloride ions were present in both solutions, the actual amount of chloride added is not important. All you need to do is calculate the molar masses. The molar mass of AgCl is 143.4 g/mol. The molar mass of $PbCl_2$ is 278.2 g/mol, almost twice the mass of AgCl. This makes choice A inviting, but the mass difference alone doesn't explain why the masses of the precipitates were equal.

Choice C is incorrect because $PbCl_2$ is more massive than AgCl, so if twice the amount of $PbCl_2$ was precipitated, the mass of the precipitate of solution 2 would be twice as great as that of solution 1. Choice D, like choice A, is true, but by itself is not enough to explain the equal masses of the precipitates.

Recognizing the limits of our measurements is crucial to truly understanding what our measurements tell us, how much confidence we can have in what they tell us, and what they don't tell us.

Keep asking yourself, your teacher, the Internet, everyone, "How do we know that?"

57. (B) Isotopic labeling is a technique in which, during a chemical synthesis, a specific atom is replaced with a traceable isotope. Common isotopes used for labeling are ^{13}C, ^{18}O, and ^{2}H. By labeling one or more of the atoms in a substance, the fate of that atom can be traced, or followed, through the reaction. Typically, two or more isotopes will be used in separate reactions. For example, labeling the oxygen in the alcohol would show that it ends up in the ester. In a different reaction, labeling the oxygen in the hydroxyl group of the acid would show that it ends up in water. If the oxygen of both the alcohol and the acid were labeled in the same reaction, the result would be that one showed up in water and the other in the ester—but it wouldn't differentiate the origin of either.

Choice C tries to tempt you with the word *acid*, but an acid-base titration is a technique typically used to determine the concentration of an acid or base in solution. Choice D tries to tempt you with *hydrolysis*, the type of reaction that breaks down the ester into an acid and an alcohol. For a detailed explanation of mass spectroscopy, see answer 73.

For each chemistry technique you learn in AP Chemistry, understand when the technique is used, what kind of data it produces, and what the data tell us.

58. (D) The first ionization energies of Kr, Xe, and Rn are sufficiently low to allow these noble gases to form covalent bonds with other atoms. Krypton difluoride, KrF_2, was the first compound of krypton discovered. Xenon can form compounds with oxygen (XeO_3 and XeO_4) and fluorine (XeF_4 and XeF_6). Radon can form compounds with fluorine (RnF_2). Oxygen and fluorine are highly electronegative atoms and it is their strong electron-attracting abilities that "persuade" Kr, Xe, and Rn atoms to share their electrons with them.

59. (C) The table lists **successive ionization energies, the minimum energy requirements for the further ionization of an element by removal of successive electrons.** From the table we see that 786 kJ of energy per mole of silicon is required to remove the first electron (a p^2 electron), leaving Si^+. Removing *another* electron (the p^1) requires an additional 1,577 kJ for per mole of Si^+.

The trick to answering this kind of question is to find a very large "jump" in ionization energies. For silicon, it's between the fourth and fifth ionizations. This indicates that the fifth ionization energy is "digging into" the core electrons with noble gas configuration because all the valence electrons have been removed. In this example, we're looking for an element with four valence electrons. The other elements in group 14 show a similar trend in successive ionization energies (C, Ge, Sn, Pb). See answer 8 for a detailed explanation of first and successive ionization energies.

60. (B) The alkali metals form *strong bases* when they react with water, not strong acids (hence the name *alkali* metals). See answer 49 for the details of the reaction between the alkali metals and water.

61. (C) Photoelectron spectroscopy is a technique to determine the "binding energy" of electrons in an atom or molecule. The binding energy refers to the amount of work that must be done to remove the electron from the substance. See answer 84 for a description of NMR spectroscopy, answer 83 for a description of IR spectroscopy, answer 7 for a description of spectrophotometry, and answer 34 for a detailed description of the PES technique and spectrum analysis. The table corresponding to questions 83-85 compare NMR, IR, and UV-visible spectroscopy.

For each chemistry technique you learn in AP Chemistry, understand when the technique is used, what kind of data it produces and what the data tell us.

Chapter 1: Free-Response Answers

62. Your chart should look like this:

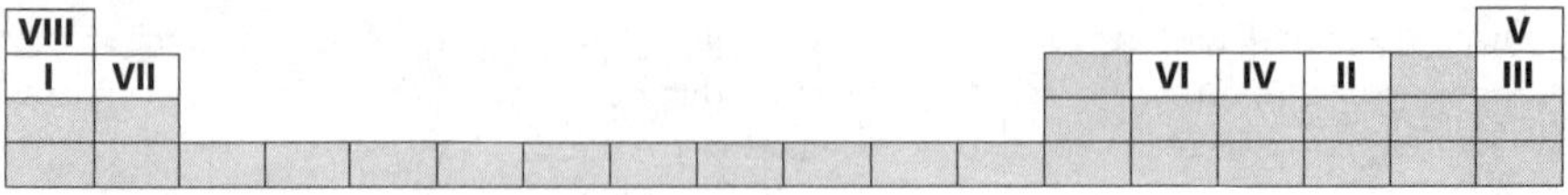

VIII																	V
I	VII												VI	IV	II		III

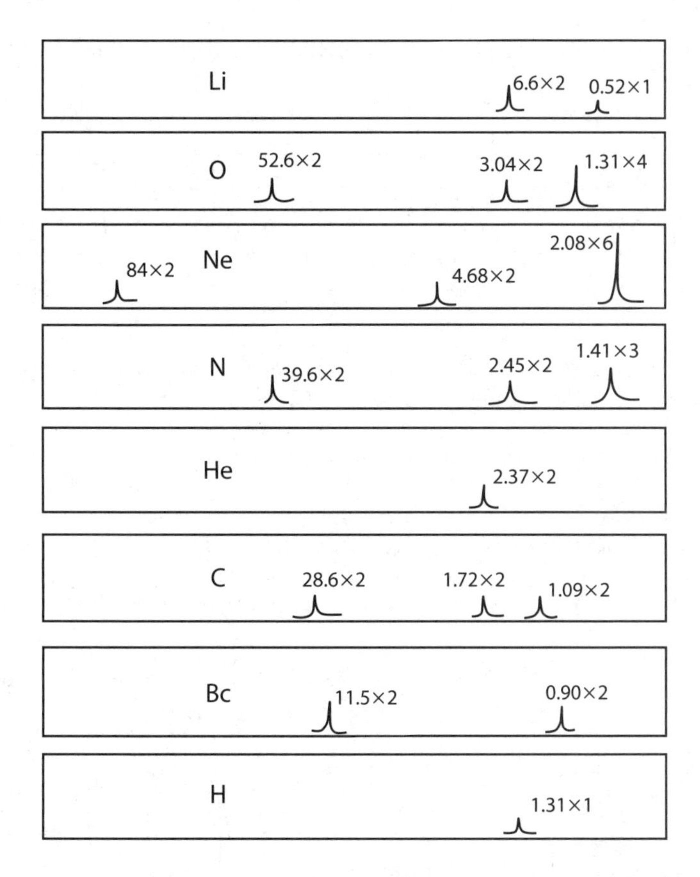

See answer 34 for a strategy for interpreting photoelectron spectra. Your justification should be based on the identity of the element, electron configuration, or number of electrons.

63. Carbon. Three peaks of identical heights means that three subshells each have the same number of electrons. Since there are only three peaks, you can assume the peak height is two. The leftmost peak is the 1s subshell, so the following peaks represent the 2s and 2p subshells. The electron configuration would be $1s^2\ 2s^2\ 2p^2$.

You can ignore the exceptions to the Aufbau principle for the purposes of the AP Chemistry exam (for example, the predicted electron configuration of Cr is [Ar] $4s^2 3d^4$, but the actual configuration is [Ar] $4s^1 3d^5$). The rationale is that the rote memorization of the exceptions does not match the goals of the curriculum revision. However, you may be given an exception and asked to provide possible reasons based on the current model (see the box below answer 8, specifically, the two items listed under Coulomb's law).

64. Your diagram should look something like the following. Be sure to label the peaks.

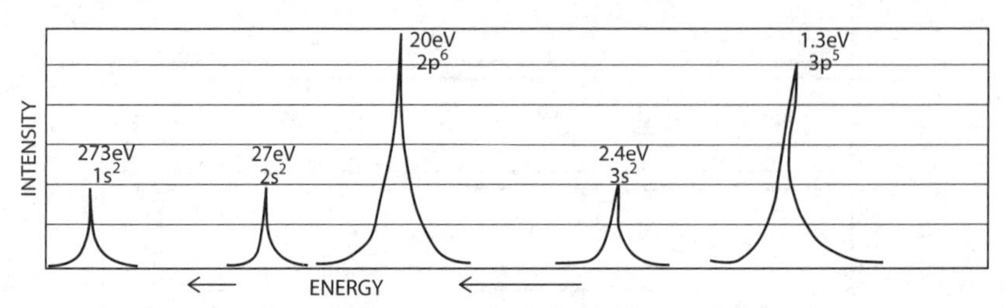

65. Scandium: $1s^2\ 2s^2\ 2p^6\ 3s^2\ 3p^6\ 3d^1\ 4s^2$. As a shortcut (or a check), you could identify the element by adding the number of electrons represented by each peak. In this case, from left to right, $2 + 2 + 6 + 2 + 6 + 1 + 2 = 21$ (the atomic number of scandium). See answer 34 for a detailed explanation of photoelectron spectrum analysis.

66. Sodium: $1s^2\ 2s^2\ 2p^6\ 3s^1$. The large jump between the values of the first and second ionization energies suggests this is a group 1 element, or an element with one valence electron.

67. Your answer should look like the following. Be sure to include labels.

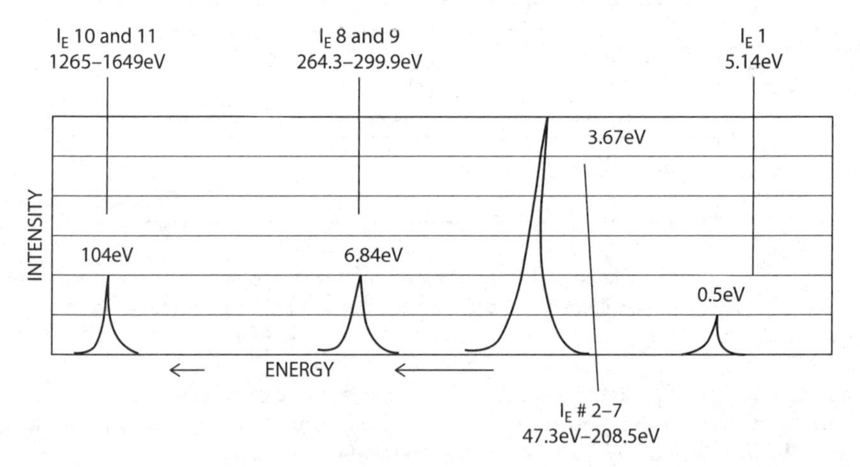

68. Photoelectron spectroscopy (PES) measures the amount of energy required to remove a particular electron from an atom or molecule. The data, like ionization energy, are consistent with Coulomb's law (see box under answer 8). The closer an electron is to the nucleus, the greater the electrostatic force between the electron and the positive charges in the nucleus and therefore the greater the energy needed to remove it. Both data sets show that electrons closer to the nucleus require more energy to remove.

The difference in the values obtained are a function of the techniques used to measure them. Successive ionization energy data are produced by removing the outer electrons, one by one. In other words, to get to the 1s electron of sodium requires the removal of the 10 electrons outside of it. This makes the calculation of the individual electron energy difficult because the relative charge of the nucleus to the electrons increases with each electron that is removed. **The benefit of the PES technique is that it can remove any one electron at a time; it is not limited to removing only the outermost electrons.** This allows the energy of individual energy levels to be measured independently of the others. See answers 8 and 34 for detailed explanations of ionization energy and photoelectron spectroscopy, respectively.

69. Electrons in the same shell will have similar energies because they are located at similar distances from the nucleus (Coulomb's law; see box below answer 8):

Electron(s)	Ionization energies (eV)	Energies measured by PES (eV)
1s	1,265 and 1,649	104.00
2s	264.3 and 299.9	6.84
2p	47.3–208.5	3.67
3s	5.14	0.50

Notice the PES data for sodium show a large difference in energies between the first, second, and third shells. An electron in the first shell requires more than 100 eV to remove, an electron in the second shell requires an average of about 5 eV, and an electron in the third shell requires only $\frac{1}{2}$ eV! Evidence for the existence of subshells comes from the energy discrepancy between the 2s and the 2p electrons.

The ionization energies agree, but with less resolution. Electrons in the first shell have ionization energies of more than 1,000 eV, electrons in the second shell have ionization energies in the 10s to 100s range, and electrons in the third shell have an ionization energy in the 1s range.

70. The range of electron energies is from 0.5 eV (to remove the 3s electron) to 104.0 eV (to remove the 1s electrons).

Convert eV to joules.

$$0.5 \text{ eV} \times \frac{6.2 \times 10^{-18}\text{ J}}{\text{eV}} = 3.1 \times 10^{-18}\text{ J}$$

$$104 \text{ eV} \times \frac{6.2 \times 10^{-18}\text{ J}}{\text{eV}} = 6.4 \times 10^{-16}\text{ J}$$

Use E = *hv* to solve for *v* (frequency).

$$v = \frac{E}{h} \text{ for } \mathbf{3s\ electron}$$

$$= \frac{3.1 \times 10^{-18}\text{ J}}{6.63 \times 10^{-34}\text{ J sec}}$$

$$= \mathbf{4.7 \times 10^{15}\ sec^{-1}}$$

$$v = \frac{E}{h} \text{ for } \mathbf{1s\ electron}$$

$$= \frac{6.4 \times 10^{-16}\text{ J}}{6.63 \times 10^{-34}\text{ J sec}}$$

$$= \mathbf{9.7 \times 10^{17}\ sec^{-1}}$$

Use *c* = λ*v* to convert frequencies to wavelength.

$$\lambda = \frac{3.0 \times 10^{8}}{v} \text{ for } \mathbf{3s\ electron}$$

$$= \frac{3.0 \times 10^{8}}{4.7 \times 10^{15}\text{ sec}^{-1}}$$

$$= \mathbf{6.4 \times 10^{-8}\ m}$$

$$\lambda = \frac{3.0 \times 10^8}{v} \text{ for } \mathbf{1s\ electron}$$

$$= \frac{3.0 \times 10^8}{9.7 \times 10^{17}\ \text{sec}^{-1}}$$

$$= \mathbf{3.1 \times 10^{-10}\ m}$$

Identify region of electromagnetic spectrum.
Mostly X-ray (there may be some overlap with ultraviolet region).

Infrared (IR) Spectroscopy

Covalent bonds are not static entities and the lengths that have been recorded for them are actually averages, not rigid distances between atoms (an example of a limitation the models impose on our imagination). Real bonds between atoms are stretched and compressed, like springs (an example of the refinement of the bonding model).

The energy of all electromagnetic radiation, including infrared, is a function of its frequency. If the frequency of a vibrating bond "spring" (the alternating stretching and compression) corresponds to the frequency of electromagnetic radiation, the vibrating bond will absorb it, producing a peak on the spectrum. The frequency of a spring is the number of times it stretches and compresses per second (sec^{-1}, Hz).

71. Empirical and molecular formula: $C_5H_9NO_4$ (glutamic acid, one of the 20 amino acids that makes up proteins).

Determine the number of moles of each element. When percent mass data are given, assume a 100-gram sample so that each percent represents the mass in grams.

$$40.80\ \text{g}\ \mathbf{C} \times \frac{1\ \text{mol}}{12.01\ \text{g}} = \mathbf{3.40\ mol\ C}$$

$$6.16\ \text{g}\ \mathbf{H} \times \frac{1\ \text{mol}}{1.01\ \text{g}} = \mathbf{6.10\ mol\ H}$$

$$43.50\ \text{g}\ \mathbf{O} \times \frac{1\ \text{mol}}{16.00\ \text{g}} = \mathbf{2.72\ mol\ O}$$

$$9.52\ \text{g}\ \mathbf{N} \times \frac{1\ \text{mol}}{14.01\ \text{g}} = \mathbf{0.680\ mol\ N}$$

Calculate the mole ratio. Identify the element with the smallest number of moles (in this case, nitrogen, 0.680 moles) and divide the number of moles of the other elements by the number of moles of the element present in the least quantity (0.680), making the number of moles of that element equal to 1.

$$\mathbf{C}: \frac{3.40}{0.680} = \mathbf{5}$$

$$\mathbf{H}: \frac{6.10}{0.680} = \mathbf{\sim 9}$$

$$\mathbf{O}: \frac{2.72}{0.680} = \mathbf{4}$$

$$\mathbf{N}: \frac{0.680}{0.680} = \mathbf{1}$$

The empirical formula is the lowest, whole number ratio of the elements in the compound: $C_5H_9O_4N$.

Calculate the "molar mass" of the empirical formula to determine if it is also the molecular formula (147 g mol^{-1}). Because the molar mass of the empirical formula is equal to the known molecular mass, the empirical formula *is* the molecular formula.

If it were *not* the same, the molecular formula would be a whole-number multiple of the empirical formula. In that case, divide the molecular mass by the mass of the empirical formula to obtain the whole-number multiple, then multiply the number of moles of each element present in the empirical formula by that number to get the molecular formula.

72. The molecular mass of a molecule can be determined in several ways, but the one you're most likely to learn about in AP Chemistry is **mass spectrometry** (also called mass spectroscopy). Although mass spectrometry is quite complicated, it can be understood in simple terms.

A good answer includes the name of the specific technique, a brief description of how it works, and how it could be used to determine molar mass. The following paragraph represents an ideal answer. Italics are used to indicate information that is useful but not necessary for a good answer. There are other methods of molar mass determination, like osmometry, that could be used to determine molecular mass.

> A small amount of a pure sample is introduced into the mass spectrometer and then bombarded by a stream of high-energy electrons, which often dislodge a valence electron from the substance to produce cations *(and often radicals—atoms or molecules with an odd number of electrons)*. The cations *(or radicals)* usually fragment after being ionized. Some of those fragments will retain a positive charge while others will be neutral. **The particles are then passed through a strong magnetic field where their mass/charge ratio can be calculated by their deflection.** Only the positively charged fragments will make it to the detector. **The data are recorded as *m/z* ratios (mass/charge).** Since *z* (the charge) is usually 1, the ratio is usually the mass of the fragment.

Interpreting mass spectrum data is fairly complex and beyond the scope of the AP Chemistry exam; however, you should be aware that the AP Chemistry exam does attempt to assess your ability to interpret data from sources you've had no previous experience with. If they show you a mass spectrum, they will supply you with a legend and the information needed for you to figure out how to deal with the data.

73. Titrations are most commonly used to determine the concentration of a substance (the analyte) in a solution. In this technique, a substance that reacts with the analyte is added to the solution. This substance is called the titrant, and the quantity of titrant added is used to determine the concentration of the analyte. The stoichiometry of the reaction between the titrant and analyte must be known to perform this technique quantitatively. The **equivalence point** of the reaction indicates the point at which the amount of titrant added has completely reacted with the analyte in solution. In effect, the titration has achieved its goal. The equivalence point is indicated by a change in the solution's property, like its color. The event that signals that equivalence has been reached is called the **end point**.

The problem facing this student is that she or he does not know the reaction stoichiometry. If the base is a group 1 hydroxide, the number moles of HCl used to reach equivalence will be equal to that of the base. If the base is a group 2 hydroxide, 2 mol of HCl will be required to neutralize each mole of base.

Hypotheses I and II are both incorrect. The solutions do not need to be composed of the same base, nor do they need to be of the same concentration. However, both solutions have the concentration of hydroxide ions (so one can argue that hypothesis II is somewhat correct).

If one of the solutions contains a group 2 hydroxide and the other contains a group 1 hydroxide with twice the concentration, equal concentrations of hydroxide ions will be present. For example, if solution A is a 0.1 M $Sr(OH)_2$ solution, solution B may be a 0.2 M NaOH solution. Either way, both solutions contain 0.01 mol of hydroxide.

74.

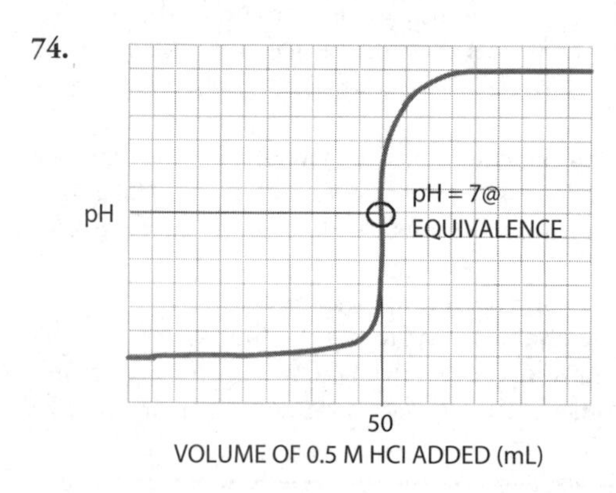

The pH of equivalence for the titration of a strong base with a strong acid (or vice versa) is always 7. The volume of HCl added was taken from the data in the table. The *x*-axis is always the volume of titrant added (make sure you include units) and the *y*-axis of an acid-base titration is always pH.

75.

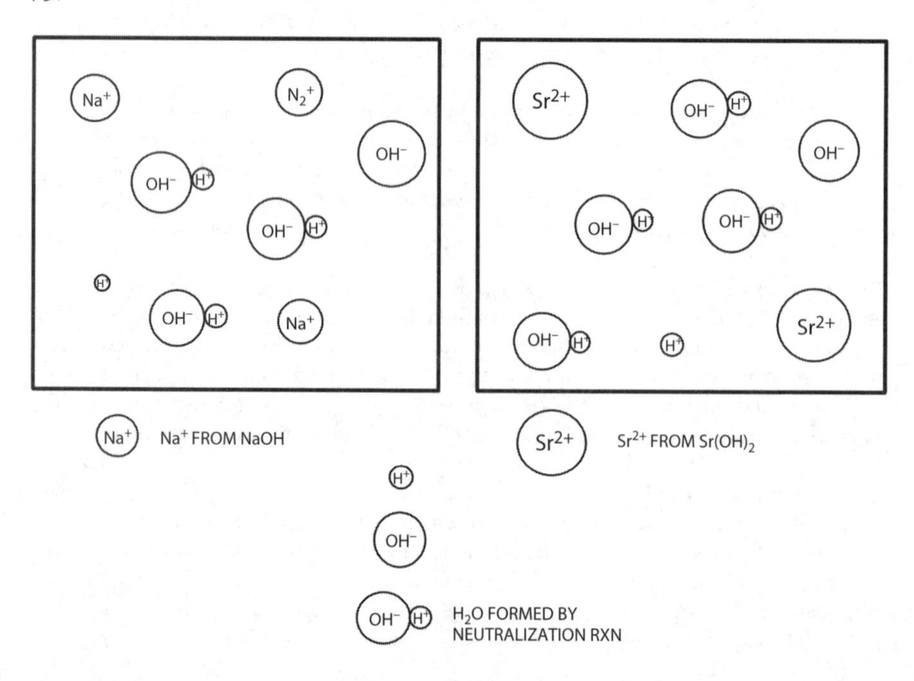

Only one of the two drawings was needed, either for the group 1 or group 2 hydroxide. Note that for NaOH, there are 3 Na^+ ions for each molecule of water that formed. That's because each NaOH provided one OH^- ion for each H^+ ion added. Na^+ is a spectator ion in the neutralization reaction. If you decided to use a group 2 hydroxide, you'll need twice as many water molecules to be stoichiometrically correct. Each $Sr(OH)_2$ provides 2 OH^- ions to the solution, so 2 H^+ ions are needed for neutralization. That's why 2 Sr^{2+} ions and 4 water molecules are shown.

A neutral solution (pH 7 at 25°C), contains an equal concentration of H^+ and OH^- ions, which is indicated by one H^+ and one OH^- ion in each answer.

76. The substance was **not pure.**

Use the formula for the standard curve of absorbance versus concentration curve to calculate the concentration from the absorbance.

$$y = 16x$$

$$x = \frac{y}{16} = \frac{0.547}{16} = \mathbf{0.0342\ M}$$

Make sure the answer makes sense by graphical interpolation:

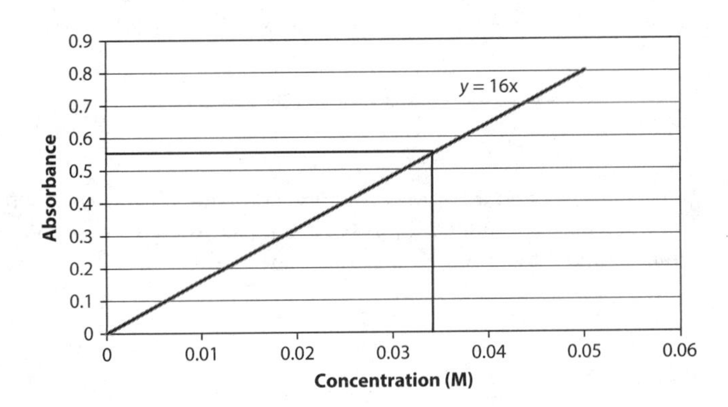

Calculate the number of moles of the substance and the molarity of the 10-mL sample.

$$1.152\ \text{g} \times \frac{1\ \text{mol}}{166\ \text{g}} = 0.00694\ \text{moles}$$

$$0.00694 \times \frac{\text{moles}}{0.2\ \text{L}} = \mathbf{0.0347\ M}$$

The conclusion that the substance was impure was based on the difference in concentrations between the absorbance data and the mass data. Recall from the question that you were to assume that impurities do not absorb at the same wavelengths as the purified substance, so the spectrophotometer only "sees" the substance of interest. The balance, however, does not discriminate.

77. $\mathbf{C_{15}H_{10}O_7}$ (quercitin, a plant pigment).

Calculate the number of moles of carbon in the sample by calculating the number of moles of carbon dioxide produced, then calculate the mass of the carbon (you'll need it later).

$$6.60 \text{ g } CO_2 \times \frac{1 \text{ mol}}{44.01 \text{ g}} = \frac{0.150 \text{ moles } CO_2}{\mathbf{0.150 \text{ moles C}}}$$

$$0.150 \text{ moles C} \times \frac{12.01 \text{ g}}{1 \text{ mole}} = \mathbf{1.80 \text{ g carbon}}$$

Calculate the number of moles of hydrogen in the sample by calculating the number of moles of water produced and multiply by 2 (since there are two moles of hydrogen for each mole of water). Then calculate the mass of hydrogen (you'll need it later).

$$0.901 \text{ g } H_2O \times \frac{1 \text{ mol}}{18.02 \text{ g}} = 0.050 \text{ moles } H_2O\ (\times\ 2) = \mathbf{0.100 \text{ moles H}}$$

$$0.100 \text{ moles H} \times \frac{1.01 \text{ g}}{1 \text{ mole}} = \mathbf{0.101 \text{ g H}}$$

Subtract the mass of carbon and hydrogen from the mass of the original sample to determine the mass of oxygen, then **calculate the number of moles of oxygen.**

$$3.02 \text{ g substance} - (1.80 \text{ g carbon} + 0.101 \text{ g H}) = \mathbf{1.12 \text{ g O}}$$

$$1.12 \text{ g O} \times \frac{1 \text{ mole}}{16.0 \text{ g}} = \mathbf{0.0700 \text{ mole O}}$$

Calculate the mole ratios of C, H, and O (see answer 72, step 2, for details).

C: 0.1500
H: 0.1000
O: 0.0700

Dividing each number of moles by the least number of moles (oxygen, 0.0700) is more difficult than multiplying each number by 100. **As long as you keep the mole ratio the same, you can manipulate the numbers any way you wish.**

C: $0.1500 \times 100 = \mathbf{C_{15}}$
H: $0.1000 \times 100 = \mathbf{H_{10}}$
O: $0.0700 \times 100 = \mathbf{O_7}$

78. 1.5 mg/ml

The relationship between percent absorbance and concentration is linear (unlike percent transmittance and concentration, so absorbance values will always be easier to handle mathematically). You can use two values of absorbance and concentration to calculate a slope and use $y = mx$ to calculate the concentration *or* you can use a ratio. These techniques only work with absorbance because of its linear relationship with concentration.

Slope method:

x = concentration
y = % absorbance

$$\text{slope} = \frac{\Delta y}{\Delta x} = \frac{(0.602 - 0.301)}{(2 - 1)} = 0.301$$

$$y = mx$$

$$0.450 = (0.301)x$$

$$\mathbf{x = 1.5 \text{ mg mL}^{-1}}$$

Ratio:

$$\frac{0.301}{1} = \frac{0.450}{x}$$

$$x = \mathbf{1.5\ mg\ mL^{-1}}$$

79. The emission spectrum of each element is unique. All the naturally occurring, stable elements have been discovered, so it is no longer used to identify new elements. Rather, it serves as an elemental "fingerprint."

80. The extra mass in the "atmospheric nitrogen" is argon. The method of removing oxygen assumes that only nitrogen and oxygen are present, while argon composes about 1% of the partial pressure of the atmosphere and has a mass of 40 g mol^{-1}, significantly heavier than N_2 gas. Because argon is relatively unreactive, the methods of removing other substances from the air by chemical reaction would not have been effective at removing argon. The N_2 prepared chemically is pure N_2.

81. The positively charged alpha particles would not have passed through the gold foil. They would have been repelled by the positively charged "pudding" and backscattered.

82. Emission spectroscopy can be used to identify the elements present. The greatest percentage of atoms are hydrogen (91.2%) and helium (8.7%).

83. Infrared spectroscopy is used to identify functional groups in a molecule. Different functional groups show peaks at different frequency ranges regardless of the molecular details of the rest of the molecule. The frequency ranges of these peaks correspond to energies of approximately 2–12 kcal/mol, the amount of energy needed to stretch a bond but not enough to break it (most bond energies are in the range of hundreds of kJ/mol (about 25–200 kcal/mol). Although the details of spectroscopy are complex, the energy range provides clues as to what each particular type of spectroscopy is best suited for. Generally, **low energy detects nuclear "shifts and flips," intermediate energy detects bond stretching and molecular vibrations,** and **high energy detects electronic transitions.** The box below answer 70 also explains IR spectroscopy.

84. Nuclei in atoms behave as tiny magnets. Certain nuclei, particularly those of hydrogen, behave as if they were spinning. Spinning magnets create a magnetic field. By applying an external magnetic field to hydrogen-containing compounds, the nuclei of the compound can either line up parallel (with) or antiparallel to (against) the field. Nuclei aligned with the field have a lower energy than those aligned against it. Increasing the intensity of the external field allows nuclei to "flip" between lower and higher spin states. This "flip" creates peaks and shifts in an NMR spectrum that allows the identification of the chemical environment of the hydrogen atom; in other words, it allows some structural determination of the compound. Although the details of spectroscopy are complex, the energy range provides clues as to what each particular type of spectroscopy is best suited for. Generally, **low energy detects nuclear "shifts and flips," intermediate energy detects bond stretching and molecular vibrations, and high energy detects electronic transitions.** See answer 7 for more information about IR spectroscopy.

85. UV-visible spectroscopy is used to detect conjugation. Generally, molecules with only one or two double bonds do not absorb in the visible-ultraviolet region (200–800 nm), but molecules with conjugated double bonds absorb in this range. Although the details of spectroscopy are complex, the energy range provides clues as to what each particular type of spectroscopy is best suited for. Generally, **low energy detects nuclear "shifts and flips," intermediate energy detects bond stretching and molecular vibrations, and high energy detects electronic transitions.**

For each chemistry technique you learn in AP Chemistry, understand when the technique is used, what kind of data it produces, and what the data tell us.

Chapter 2: Structure: Property Relations

86. (B) If all the gases were at the same temperature and pressure, the gas of highest molar mass would have the greatest density. However, the data show that the pressure in container C (F_2) is one-half the pressure of the other two containers. The density of each gas can be determined by dividing the mass of the gas by the volume. If the mass is not known, use the ideal gas law: $\boldsymbol{PV = nRT}$. (It may be helpful to remember the name "**pivnert.**")

R is a constant (0.0821 L mol^{-1} K^{-1}) and V and T are the same for all containers (1 L and 295 K). Since you only need to compare relative densities, you may want to round the gas constant down to 0.08 or up to 0.1 and round the temperature up to 300 K to make the calculations easier. Rounding won't affect the answer because you don't need the actual density, and it will affect the answers "equally enough" therefore, $P(1) = n\,(0.1)\,(300)$ and solving for $n = \frac{P}{30}$. (Without rounding you'd get $n = \frac{P}{24.4}$).

For N_2 and O_2, $P = 4$ atm $\therefore n = 0.133$ (or 0.163 without rounding).
Because O_2 has a greater mass than N_2, the O_2 will have a greater density, so you can skip that calculation.

$$\textbf{Density of } \mathbf{O_2} \textbf{ gas} = 0.165 \text{ mol } O_2 \times \frac{32\text{ g}}{\text{mol}} = \mathbf{4.27\ g/L} \text{ (or 5.22 g/L)}$$

The pressure of F_2 is one-half the pressure of O_2 and N_2 (2 atm), so the number of moles is also half $\therefore n = 0.0667$ (or 0.0816).

$$\textbf{Density of } \mathbf{F_2} \textbf{ gas} = 0.0667 \text{ mol } F_2 \times \frac{38\text{ g}}{\text{mol}} = \mathbf{2.53\ g/L} \text{ (or 2.61 g/L)}$$

87. (C) The purpose of this question is to assess if you can recognize that since all of the gases are nonpolar (they have no permanent dipoles) they will condense due to weak London dispersion forces. **The attractive strength of London dispersion forces increases with increasing molar mass.** Although **the strength of London dispersion forces actually depends on the number of electrons in a substance**, the number of electrons tends to be proportional to its molar mass. At a constant temperature (above the vaporization/condensation temperature), the gas of greatest mass (and the greatest number of electrons) is fluorine, so it will condense at a lower pressure than nitrogen and oxygen gas. The lower "initial" pressure of F_2 can be ignored because the question asks which pressure will be lowest at the point of condensation.

88. (B) The polarity and therefore the water solubility of a molecule is largely determined by its molecular geometry. The shape of NH_3 is trigonal pyramidal. NH_3 has a lone pair of electrons, but boron trihydride, BH_3, does not (choice D), which makes the shape of NH_3 pyramidal (but does *not* give nitrogen a negative charge, as choice C states). **All molecules with a trigonal pyramidal shape are polar, regardless of the polarity of the bonds (the same holds true for bent molecules).** BH_3 has a trigonal planar shape. (Recall that boron is an important exception to the octet rule, "B is stable with 3" bonds but it *can* form 4.) The small difference in electronegativity between boron and hydrogen makes the B–H bond mostly nonpolar (choice A), and the trigonal planar shape gives BH_3 a symmetrical shape and nonpolar behavior.

89. (B) A nonpolar molecule, by definition, has no net dipole. Choice A is incorrect because although the C–Cl bond is polar (3.16 − 2.55 = 0.61), the molecular shape is tetrahedral, so all the dipoles cancel each other out, leaving no net dipole. You can imagine that in a molecule where all the electrons are evenly distributed around it, there is no place that "feels" more negative or positive than anywhere else in the cloud of electrons surrounding the molecule. There are no lone pairs of electrons on either molecule. Lone pairs of electrons on the central atom tend to (but not always!) make a molecule polar by producing an asymmetrical shape. For example, the S–O is weakly polar but SO_2 is polar due to one lone pair of electrons on the sulfur atom, causing the molecule to take on a bent shape.

90. (A) All four of the substances are nonpolar, so the only interparticle forces of attraction operating are London dispersion forces. The strength of attraction of London dispersion forces is proportional to the molar mass of the substance. Fluorine has the lowest mass and therefore the weakest dispersion forces of the elements listed. (See answer 98 for more on London dispersion forces.)

If the relationship between the substances were not so clear (i.e., there were other forces besides London dispersion at work), the boiling point, melting point, and vapor pressure could all be used. **The substance with the lowest boiling point (and/or melting point) and the highest vapor pressure has the weakest interparticle forces of attraction.** The vapor pressure data for fluorine and chlorine are not included because they are gases at standard temperatures and pressures. Generally, the term *vapor* applies to the gaseous state of substances that are liquids or solids (i.e., not gases) under standard conditions.

91. (B) This is an easy but important question. The state of matter of a substance is typically considered under "standard conditions," which is *not* the same as STP (standard temperatures and pressures). **Standard conditions are 25°C (approximately room temperature) and 1 atm of pressure. STP conditions are 0°C and 1 atm of pressure.** STP is typically only used when dealing with gases. Finally, since the standard condition is the condition you're probably pretty comfortable with (physiologically), you can recognize it easily. The melting point of I_2 is way higher than room temperature, so it's a solid under standard conditions. The melting point of Br_2 is much lower than room temperature, so we know it's either a liquid or a gas. The boiling point is higher than room temperature, so it must be a liquid. **"Normal" means 1 atm of pressure.**

92. (C) The boiling and melting points show that I_2 is a solid under standard conditions, but there is a measurable vapor pressure showing that some of the solid I_2 has been converted to a gas. Because the melting and boiling points are much greater than 25°C, it is unlikely that the I_2 melted and then evaporated.

93. (B) The forward reaction is clearly favored ($K_P = 2.63$) but the reaction does not go to the completion under the conditions in which it is occurring. The actual pressures are not important because no calculations need to be done; only the original stoichiometry matters.

Three particles of gas ($2\ SO_2 + 1\ O_2$) are consumed to produce *two* particles of SO_3 gas. **The ideal gas law ($PV = nRT$) expresses the direct relationship between the number of particles of a gas and the pressure.**

Choice A may be tempting because all else being equal, gas particles of lower speed will collide less forcefully with the walls of the container, producing less pressure. **The absolute temperature of a substance is directly proportional to the average kinetic energy of the particles in the substance.** If the temperature is constant, the average kinetic energy of the particles can also be considered constant. The formula for kinetic energy ($KE = \frac{1}{2}mv^2$) states that **the mass is inversely proportional to the square of the velocity** ($m \propto \frac{1}{v^2}$ or $v \propto \sqrt{m}$). Exactly how the force of the collision relates to kinetic energy is important, but unnecessary to answer this question. You only need to consult the ideal gas law: $PV = nRT$. The temperature and volume are constant (and so is R, the gas constant), so let's take them out of the equation $\therefore P \propto n$.

In reality, things are a little more complicated. The ideal gas law is an outstanding approximation, but it's not perfect. Real gases do not behave ideally, mainly because there are typically forces of attraction between the particles, however weak. This is where choice C provides a tempting explanation. Sulfur di- and trioxide are both polar gases, and the conversion of at least some O_2, which is nonpolar, would create **a mixture of gases that experiences increased interparticle forces of attraction and would be expected, under certain conditions, to produce lower pressures than predicted by the ideal gas law.** (The van der Waals equation deals with deviations from the ideal gas law.) *But*, the effect is very small (not the 25% decrease in pressure observed here) *and* the temperature is so high that interparticle forces of attraction are not experienced by the particles. (Although the pressure is high, too, it's not high enough to overcome the effect of the temperature.)

An important aspect of Kinetic Molecular Theory (a model of the behavior of gases), which is the basis for this question, is that **the pressure is a macroscopic behavior produced by an enormous number of tiny particles that are not directly observable.** Only their collective behavior is observable, which is why we use the tools of statistical mechanics to deduce what is happening on the microscopic level. **The pressure of a gas is defined as the pressure it exerts on the walls of the container**, not on the collisions between the particles (why choice D is incorrect). Of course it's all related, but it is critical to understand and remember that **every measuring tool and technique has its limits and if we do not acknowledge and understand their limits, we will never truly understand what our measurements mean.**

The Power of a Good Model Cannot Be Overstated!

A model provides a framework for asking the right questions—those that will lead to the formation of testable hypotheses—and the ability to make predictions about the behavior of a system. **The Kinetic Molecular Theory (KMT) of Gases** is an exceptional example of a scientific model. It supported atomism at a time when scientists were divided as to whether or not atoms existed and provided a way to imagine how an incredibly large number of microscopic particles created observable, macroscopic effects by their collective behavior. **KMT** and the **Collision Model** of chemical reactions can explain a vast number of chemical phenomena.

The College Board composed a list of seven science practices it evaluates on the AP Chemistry exam. Science Practice 1 focuses on models. You must be able to use models and other representations to communicate and solve problems.

94. (A) We know from the question that the pressure of N_2 is 4 atm. We can deduce from the diagram that each particle represents $\frac{1}{2}$ atm of pressure or that two particles represent 1 atm of pressure. From that information, we can conclude that the four particles of oxygen represent 2 atm of pressure (half the number of particles exerts half the amount of pressure at the same temperature and volume) and the two particles of argon represent 1 atm. If all the gases were combined in a 1 L container, the total pressure would be $4 + 2 + 1 = 7$ atm (Dalton's law of partial pressures), but combined in a 2 L container means the pressure would be half of that, or 3.5 atm.

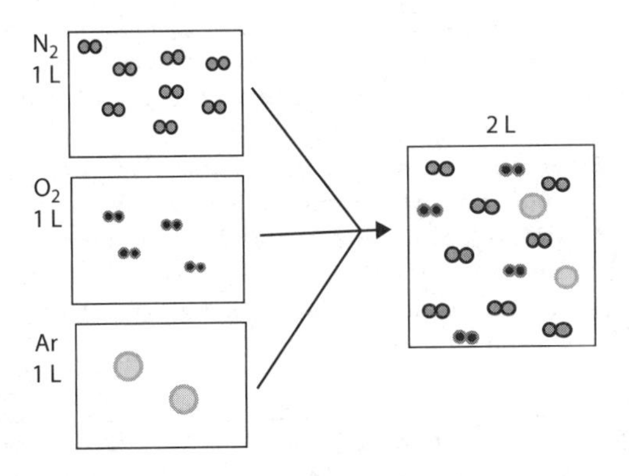

The purpose of this question is to assess your understanding of the relationship between the volume, pressure, and number of particles of gas at a particular temperature and your ability to work with particulate representations of atoms and molecules.

The College Board expects you to be able to understand, work with, and produce particulate representations of substances and processes.

95. (A) Diethyl ether. There are two factors that determine the phase of a substance at a particular pressure. **The motion of the particles, determined by their kinetic energy (the average of all the particles' kinetic energy is directly proportional to the absolute temperature). You can think of a particle's kinetic energy as its intensity of motion, like speed, and fast, high-intensity motion tends to separate particles. The interparticle forces of attraction tend to hold them together.** The greater the force of attraction between particles, the higher the temperature needed to separate them.

The vapor pressure of a substance is an indicator of the interparticle forces of attraction. **A high vapor pressure means the particles of the liquid are not strongly attracted to each other** (at that temperature and pressure) and can readily escape the liquid phase to evaporate.

↑ vapor pressure = ↓ boiling point = ↓ (weak) interparticle forces of attraction

96. (D) Ethanol. Only substance D can form hydrogen bonds (because of its O–H groups). **Hydrogen bonds are the strongest of the interparticle forces of attraction, resulting in the low vapor pressure of ethanol** compared with the other compounds listed. (See answer 95 for an explanation of vapor pressure.) Remember that **when a hydroxyl group is attached to a metal, the compound is a base**, but **when the hydroxyl group is bonded to a carbon, the compound is an alcohol.**

97. (C) The nonpolar compounds in the table are A (diethyl ether), B (carbon disulfide), and C (carbon tetrachloride). **A volatile compound is one that readily evaporates, has a high vapor pressure, and has weak interparticle forces of attraction.** Carbon tetrachloride has the lowest vapor pressure of the three and thus the lowest volatility.

98. (A) The monatomic nature of the gases means they have no permanent dipoles, so the forces of attraction between the gas particles are very, very weak. Therefore, all of the group 18 elements exist as gases at STP (0°C, 1 atm). London dispersion forces are weak interparticle forces of attraction that, although present between all particles, are only significant between particles that have no permanent dipoles (nonpolar molecules or the noble gases).

The strength of dispersion forces is proportional to the number of electrons surrounding the particle. **Since the number of electrons in a particle tends to correlate with its molar mass, it is convenient to think of the strength of dispersion forces as proportional to molar mass.** Trends are not laws! The shape of the molecule makes a big difference. For example, the boiling points of the different isomers of a particular hydrocarbon can vary greatly based on their branching.

Group 18 element	Normal boiling point (K)	Molar mass (g mol^{-1})
He	4.2	4.00
Ne	27.3	20.18
Ar	87.5	39.95
Kr	120.9	83.80
Xe	166.1	131.30

Although choice B is technically true, the distance between gas particles is typically so large that the size of the gas particles is irrelevant except at extremely high pressures. (Remember that in our typical application of Kinetic Molecular Theory, the volume of the actual gas particles can be ignored.) It is important to know that "normal" refers to 1 atm of pressure when used to describe temperatures such as boiling and melting point.

Choice C is a true statement, which implies, but does not directly explain, why boiling point increases down group 18. The implication is that a higher temperature is required to get more massive particles moving fast enough to overcome the interparticle forces of attraction that hold them together in a liquid (or solid).

Choice D confuses the relationships among kinetic energy, speed and mass.

$$\textbf{Kinetic Energy } (KE) = \frac{1}{2} mv^2.$$

The temperature of a substance is proportional to the average kinetic energy of the particles in the substance. At the same temperature, more massive particles move more slowly on average than less massive particles, Substances at the same temperature have, on average, the same kinetic energy, however, the speed at which the particles move at that temperature varies inversely with mass, See the graphs in answer 382 for a comparison of temperature, mass, and particle speed.

99. (C) The formation of ionic bonds is a chemical change. A new substance is formed by a chemical reaction. **Phase changes are physical changes.** Phase changes only involve the formation of or disruptions to the forces of attraction between particles.

Water, unlike most substances, is *less dense* as a solid because there are four hydrogen bonds per water molecule in ice (compared to the two to three hydrogen bonds per water molecule in liquid water), and the crystalline arrangement of them causes the water molecules to spread farther apart from each other. Water has its greatest density at 3.8° C, a temperature at which it is a liquid. Because fewer molecules of water are present per volume of water in ice, the mass and therefore the density is lower. (You've probably noticed that **ice floats.**) Although amorphous ice exists, it does not form at 1° C and 1 atm.

The College Board would not expect you to know that specific fact about amorphous ice but it would be reasonable for them to ask you to propose a hypothesis to explain the connection between the phase change conditions and the structural differences between crystalline and amorphous ice given the appropriate data.

100. (D) Atomic and molecular vibrations occur at temperatures above 0 K. Higher temperatures increase speed (or intensity) of the vibrations, which can either prevent the particles from forming interparticle forces of attraction or can disrupt the interparticle forces of attraction that were present, which is why high temperatures favor liquid and gas phases and why phase changes in this direction are endothermic.

You should know that dry ice (solid carbon dioxide) is *not* a network solid (choice A); carbon dioxide molecules exist as individual molecules regardless of their state of matter. Solidification of CO_2 is a phase change, not a chemical change, so only interparticle forces of attraction are being formed. Choice C is an attempt to confuse solid CO_2 structure with metallic bonding. Choice B is incorrect because dispersion forces do not produce permanent dipoles, by definition dispersion forces are the result of weak, temporary, instantaneous dipoles. They are most significant in particles *without* permanent dipoles.

The following phase diagram of carbon dioxide shows that at normal pressure (1 atm), solid carbon dioxide only form at very low temperatures (<–79°C). This is because the linear, symmetrical carbon dioxide molecules must be moving very slowly in order for weak dispersion forces to allow them to adhere to one another. Carbon dioxide is never a liquid at normal (1 atm) pressure (note the logarithmic scale on this axis).

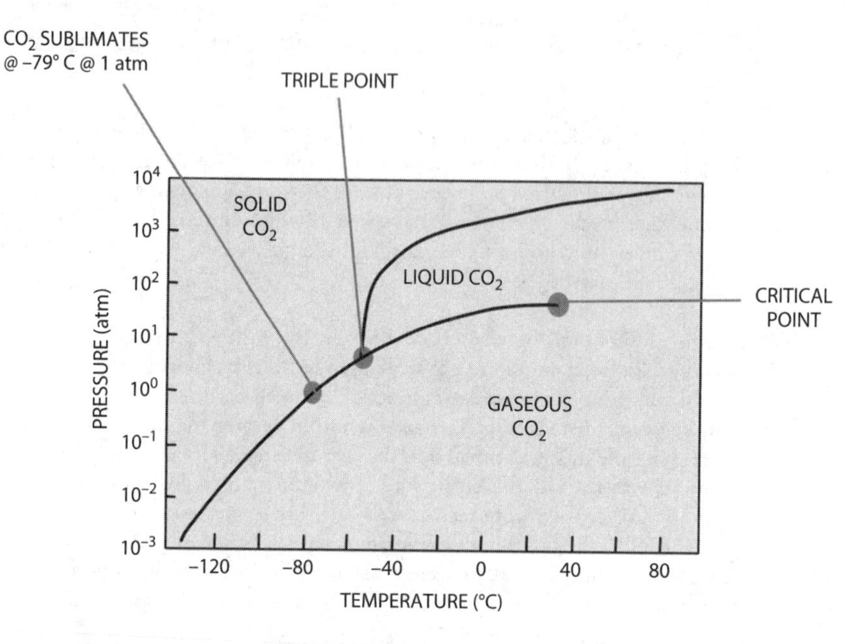

101. (D) Bond polarity is quantified by the electronegativity difference between atoms. The most polar bond will contain the two atoms with the greatest difference in electronegativities, fluorine and iodine (provided it is a covalent bond; ionic bonds form between atoms whose electronegativity difference is very large). It is important to remember that **the polarity of bonds falls along a continuum**. There is no abrupt switch between ionic and covalent bond formation at a particular value of electronegativity difference, rather, bonds can be described as having a particular percent of ionic character (see the following table).

Bond	Electronegativity difference	Percent ionic character
H–N	0.84	16
H–O	1.24	32
F–O	0.54	7
F–I	**1.32**	35

The greater the bond polarity, the greater the partial-charge difference between the atoms and the greater the potential strength of the dipole attraction it may form (if the molecule of which it is a part is polar). **The smaller the electronegativity difference between the two atoms in the bond, the more uniformly distributed the shared electrons between them and the weaker the potential dipole.**

102. (A) A meniscus forms due to surface tension. A high surface tension causes the surface of a liquid to appear to have a film. In reality, the tension, or firmness, on the surface of the liquid is caused by the strong attraction of the particles at the surface for each other. The unit of surface tension is a pressure (force per area, or milliNewtons per square meter). This is the force required to break the surface tension over a particular area.

These attractive forces between particles tend to minimize the surface area at the interfacing surfaces, creating shapes like the droplets of water on a leaf surface. (The hydrophobic leaf surface further encourages the minimization of droplet surface area.) This tendency promotes steeper meniscuses to form on liquids of high surface tension. Whether the meniscus of a liquid within a cylinder is concave or convex depends on the attractive forces (or lack thereof) between the liquid and the cylinder surface. A concave meniscus increases the area of contact between the liquid and the cylinder surface whereas a convex meniscus minimizes it.

103. (D) The viscosity of a liquid is its resistance to flow. A liquid of high viscosity, like honey, resists flowing.

104. (D) The properties of viscosity and surface tension are both attributable to the strength of a substance's interparticle forces of attraction. The high viscosity of glycerol indicates the presence of strong interparticle forces of attraction. Its surface tension is comparable to, though less than that of water, another indication of strong interparticle forces of attraction. **The great difference in viscosity relative to the small difference in surface tension suggests that the influence of the interparticle forces of attraction on glycerol is greater than that of water.** The IUPAC name, propane-1,2,3-triol, indicates the present of three hydroxyl groups, which means that relatively strong hydrogen bonds are one of the interparticle forces of attraction that form between these molecules.

105. (D) *Miscible* means that two or more liquids are soluble in each other. Upon mixing, they will form a homogeneous solution. *Miscible* sounds like *mixable* and you can think of them as synonyms. Water and glycerol are both very polar and can both form hydrogen bonds. (You should already know that about water. As for glycerol, both the -ol in glycerol and the IUPAC name "propane-1,2,3-triol" should be clues to the presence of hydroxyl groups.) **The high surface tensions of both compounds as well as the high viscosity of glycerol are further suggestive that they form strong interparticle forces of attraction.** Olive oil and castor oil are also miscible but that pair was not presented as a choice.

106. (C) Surface tension is measured by the force per area (pressure) required to break it. A Newton (N) is a unit of force. The magnitude of the force required to break the surfaces of these liquids falls in the milli-Newton (mN) range. The surface tension of castor oil is expected to be lower than water and glycerol since the only interparticle forces of attraction in oils are weak London dispersion forces. We would expect the surface tension to be similar to that of olive oil (castor oil and olive oil are composed mainly of triglycerides, or triacylglycerols). They have high but variable molar masses (in the 600s to 700s) because their fatty acid composition is not constant.

107. (B) There are generally **two competing forces** that determine the state of matter a substance will be in at a particular pressure. **The kinetic energy of the particles (the average speed of their motion, which is a function of the temperature), which tends to push them apart,** and the **interparticle (or interparticle) forces of attraction, which tend to hold them together.** Choice A is a consequence of choice B. The decrease in density (choice A) as a function of temperature is a direct consequence of increased kinetic energy, which will continue to diminish the Interparticle forces of attraction, lowering the viscosity and surface tension until evaporation occurs.

108. (C) Larger charges on an ion create a greater force of attraction, increasing the melting point. The force of the electric field produced by a charged particle is inversely related to the square of the distance from that particle, so ions with smaller radii will be closer to the ions they are attracting and will thus exert a greater force on them, thereby increasing the melting point.

Choice A is incorrect because NaCl and BeO are both ionic compounds (the electronegativity difference between Na and Cl is 2.1 and the difference between Be and O is 2, giving NaCl and BeO ionic characters of 67% and 63%, respectively).

Choice B is incorrect. Be^{2+} and O^{2-} are smaller ions than Na^{+} and Cl^{-}, respectively. The relationship between the melting point of an ionic compound and the distance between its ions can be explained through Coulomb's law (see box below answer 8). Larger ionic radii tend to decrease melting points because there is a greater distance, and thus a smaller force of attraction, between the ions.

Choice D is incorrect because atomic number alone does not predict anything about the strength of ionic bonds or the melting point of ionic compounds.

109. (A) All of the elements are nonpolar, so the only interparticle force of attraction to consider is dispersion. Br_2 has the highest molar mass (and the greatest number of electrons) of the choices listed, and therefore has the strongest dispersion forces. The strongest evidence, however, is that Br_2 is also the only liquid in the list (under standard conditions). That fact alone would indicate the highest boiling point. (Because O_2, N_2, and Cl_2 are gases under standard conditions, their boiling points must be below 25° C.) Remember that although it is commonly stated that a larger molar mass corresponds with stronger dispersion forces, the strength of dispersion forces is positively related to the number of electrons, which usually corresponds with molar mass.

110. (D) Phase changes of pure substances are <u>not</u> accompanied by temperature changes, which are measures of changes in the average kinetic energy of all the particles in the substance. The density of a liquid is usually less than that of a solid, but water is an important exception (choice C does not fulfill the requirement of <u>*must*</u> be true). **If the interparticle force holding the solid together decreased to zero, evaporation and/or sublimation would occur, not melting.** Even so, however small those forces may be, all gases, even nonpolar ones, experience interparticle forces of attraction that are greater than zero, evidenced by the fact that all gases can be liquefied.

111. (B) The solubility of gases in solution is greatest under high pressures and low temperatures. High pressures force particles closer together, favoring the formation of interparticle forces of attraction. Low temperatures favor slower, less intense particle motion, also favoring the formation of interparticle forces of attraction.

112. (C) Salts dissolve by dissociating so their ions form ion-dipole interactions with polar solvent molecules. Water-soluble salts are *not* soluble in nonpolar solvents like CCl_4. A sodium ion (Na^+), for example, will be attracted to the partially negatively charged oxygen atom in an O–H bond of water, while the chloride ion (Cl^-) will be attracted to the partially positively charged hydrogen atom of the O–H bond.

Choice A is ethanoic acid. Both the hydroxyl and carbonyl groups of the carboxyl group on organic acids are polar, as is the alcohol (hydroxyl) group on methanol (choice B). Hydrobromic acid is a strong, polar acid, and an aqueous solution of HBr would easily dissolve NaCl mostly because of the water in the solution (choice D).

113. **(A)** This question is asking us to identify the gas that is *least soluble in water*. The substance least soluble in water, if collected above water, will produce the highest yield because the least amount of it will be dissolved into the water.

114. **(A)** Different proportions of solute and solvent can produce different enthalpy changes, but the solvation of ethanol and water is unusual in that it starts out exothermic at low concentrations of water, changes to endothermic in the midrange (when the amounts of ethanol and water are closer to equal), and then reverts back to exothermic at high concentrations of water. Solutions are complex, but we can still arrive at a fairly simple but logical interpretation of this observation. **If a particular solvation process is exothermic, then the effect of the exothermic formation of interparticle forces of attraction is greater than the endothermic "breaking" of interparticle forces of attraction**.

115. **(C)** The formation of hydrogen bonds (and all interparticle forces of attraction) is exothermic. "Breaking" interparticle forces of attraction is endothermic (choices A, B, and D).

116. **(D)** **Any particle with electrons will experience dispersion forces**, however small, so our answer must contain dispersion forces. The structures of ethanol and water contain O–H bonds, so they are both capable of **hydrogen bonding, the strongest of the interparticle forces of attraction.** Choice B is not true. Although the oxygen and hydrogen atoms in the O–H bond carry partial negative and positive charges, respectively, they are both neutral molecules.

117. **(D)** **Diamond is a network solid composed exclusively of carbon atoms.** In diamond (as opposed to graphite) each carbon atom is covalently bonded to four other carbon atoms (making choice A is incorrect). Individual atoms are not hard, soft, or even dense. Nor does the stability of an atom in a substance account for the hardness of that substance. Only **the arrangement of atoms (or particles) in a substance determines the hardness of the substance.** Graphite, another allotrope of carbon, is quite soft, for example. Although diamonds are formed under extreme high heat and pressure, that fact alone does not sufficiently explain the hardness of diamonds. (See answer 118 for a comparison of the structures of diamond and graphite.)

118. **(B)** **Diamond and graphite are allotropes of carbon (see diagram).** Allotropes are two or more forms of the same element whose structural differences account for their different properties. In order for diamond to become graphite, covalent bonds must be broken.

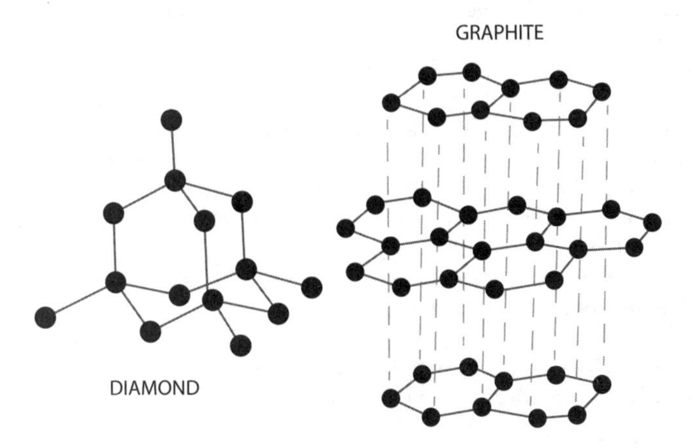

Graphite is a network solid composed of sheets of carbon stacked on top of one another. Within each sheet (called graphene), the carbon atoms are covalently bonded to three other carbon atoms. The sheets are *not* covalently bonded to each other, however. Because each carbon atom is covalently bonded to three (instead of four) other carbon atoms, the hybridization of carbon in graphite is sp^2. The electrons in the unhybridized p orbitals are *delocalized*, so they "spread out" over several carbon atoms and can exert fairly strong dispersion forces, allowing the individual sheets to stick together. Graphite is soft because even though the dispersion forces created by the pi electrons are strong, they are very weak relative to covalent and ionic bonds. **Graphite is remarkable in that it is a nonmetal solid that conducts electricity (due to those delocalized electrons—they're mobile!).** Like a metal, graphite has luster, but like a nonmetal, it is not malleable. It is soft and flexible, but inelastic (it doesn't re-form after being de-formed).

Choice A represents the sublimation of naphthalene. Phase changes do not involve the breaking or formation of covalent bonds. Choice C represents melting of ionic compounds, which breaks ionic bonds. Choice D represents the dissolution of ammonium nitrate, an ionic compound, in water.

119. (A) The arrangement of particles ultimately determines the physical properties of the substance (choice C). **The shape of the particle determines its polarity and which forces of attraction it will exert** (choices B and D). Although the elemental composition does determine the physical properties of a pure element, it is really the forces of attraction between them that is responsible for the bulk, physical properties. For compounds, the relationship between the elements (for example, the electronegativity difference or the size and charge of ions) determines the relationship they have and ultimately the physical properties. Two compounds with very different elemental composition can have similar properties because the arrangement of matter within them is similar.

Remember that **the bulk properties of a substance are typically unpredictable based on the elemental composition of a substance.** Sodium is a highly reactive metal that explodes on contact with water whereas chlorine is a toxic gas. Combine them and you get sodium chloride, or table salt, something we eat every day!

When answering a question that asks you to find the *least*, circle the word *least* (so you remember what you're looking for) and then evaluate each answer choice in the positive, or which contributes the most, or a lot. You can even rank the answers. When you're done, evaluate all four answers to identify which ranked lowest as a contributor.

120. (D) An ideal gas is an imaginary gas whose behavior with regards to temperature, pressure, and volume is completely described by the ideal gas law, $PV = nRT$. These ideal gases are assumed to experience no interparticle forces of attraction and to have completely elastic collisions (the kinetic energy is conserved in collisions). Actual measurements of T, P, and V of real gases vary (very) slightly from the predictions made by the ideal gas law and are correctable using the van der Waals equation. **Real gases deviate from ideal behavior mainly because of the forces of attraction and repulsion between the particles.** *Low pressures* maintain a low density of the gas, so particles are far enough apart so that attractive and repulsive forces between them aren't experienced by the particles. Conversely, *high temperatures* overcome forces of attraction and repulsion. With these forces of attraction and repulsion minimized, real gases behave quite ideally!

121. (B) Kinetic Molecular Theory assumes that the volume of the gas particles is so small relative to the distance between them (and the size of the container) that it can be considered negligible, and this is usually a reasonable assumption. However, **under high pressures, the particle size becomes nonnegligible as the distance between them decreases,** particularly if the particles have any polarity, permanent or induced. Under these circumstances, interparticle forces of attraction and repulsion become significant.

122. (C) Liquids can only form between particles that experience forces of attraction between them, so the forces that exist between gas particles are real. This is conclusive proof that real gases *do not* behave completely ideally, even the noble ones!

123. (C) The particular type of steel referred to in this question is an example of **an interstitial alloy** (see answer 5 for an explanation of the two types of alloys). In order to answer this question, the relative size of the carbon and iron atoms must be taken into account. The carbon atoms, which are much smaller than the iron atoms, fit in between the iron atoms (shown below).

Choice A is incorrect because adjectives like *hard* do not accurately describe the properties of individual atoms. More importantly, **the bulk properties of a substance are typically unpredictable based on the elemental composition of a substance.** Choices B and D are incorrect because there is no covalent or ionic bonding, respectively, in metal alloys

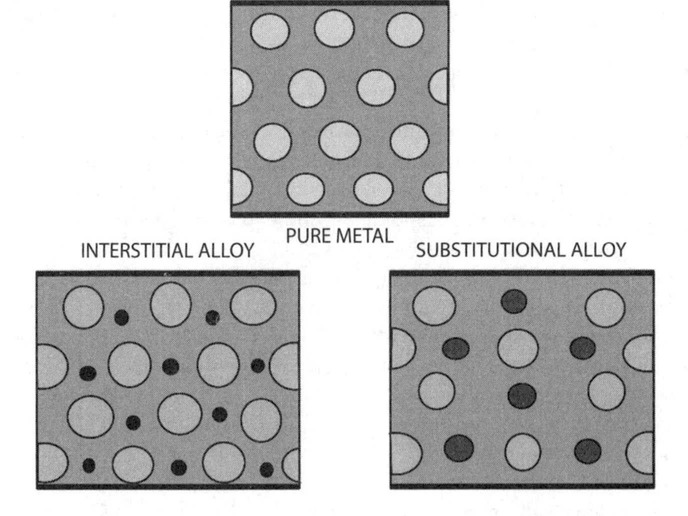

The College Board does not expect you to know all the types and details of metal lattice and alloy structure, but they do consider the basic structure of and the distinction between interstitial and substitutional alloys Essential Knowledge.

124. (A) The arrangement of particles ultimately determines the physical properties of the substance. Although the elemental composition determines the physical properties of a pure element, it is the forces of attraction between the particles that is responsible for the macroscopic (bulk) physical properties. For compounds, the relationship between the elements (for example, the electronegativity difference or the size and charge of ions) determines the type and strength of their attractions and ultimately, the physical properties of the substance they create. Two compounds with very different elemental composition can have similar properties because the arrangement of matter within them is similar. **The temperature and pressure affect physical properties but only by affecting the attraction between the particles.**

125. (C) There are only five bonds listed (all single), so the conclusions that can be drawn from the data are very limited. Choice A incorrectly states the relationship between bond length and strength (energy). Although for the bonds listed, increased length correlates with decreased bond energy; the relationship is not inverse, which would mean that bond length $\propto \frac{1}{\text{bond energy}}$, and that is not true. Choices A and B both overgeneralize the relationship based on very limited data. Choice D is not true because the bonds shown are all covalent. Hydrogen bonds are not the covalent bonds that form between hydrogen and another atom, but are the weak interparticle forces that involve attractions between the O–H, N–H, and F–H groups on different molecules.

126. (B) The dispersion forces increase for larger molecules, and that accounts for the increased boiling point. (Remember that the relationship between dispersion forces and molar mass is due to the fact that substances with higher molar masses typically have more electrons than substances with lower molar masses).

Although choice A is true, all the compounds listed contain a single alcohol (hydroxyl, O–H) group, so they should all be able to form the same number of hydrogen bonds. In reality, the longer carbon chains on the larger molecules can pose spatial problems in the pure liquid that limit their hydrogen bonding, but that doesn't answer the question in this case (see answer 127).

Choice C is does not answer the question directly. Generally, water soluble means polar, and polar compounds typically have higher boiling points than nonpolar compounds of similar mass. Choice D is incorrect. Substances with high polarity typically form stronger interparticle forces of attraction (dipole-dipole and hydrogen bonds) and would therefore tend to have higher boiling points.

127. (A) Each of the alcohols listed has one hydroxyl (O–H) group. This is the part of the alcohol that can form hydrogen bonds with water molecules. Because there is only one hydroxyl group per molecule, each alcohol listed can form the same number of hydrogen bonds. However, the portion of the molecule that is nonpolar increases as you go down the list. The six-carbon alcohol at the bottom is mostly hydrocarbon and only "a little" alcohol, resulting in mostly nonpolar behavior.

128. (D) The boiling point of decane is the highest on the graph (174° C), but there is no data that can distinguish the state of matter of decane at that temperature. It likely exists as a solid at the lowest temperatures, but with no melting point data, there's no way to definitively predict the temperature at which the transition occurs.

129. The expected boiling points for an unbranched, five-carbon hydrocarbon is 36° C. Due to the nonpolar nature of the hydrocarbons, weak dispersion forces are the only interparticle forces that hold the particles together in a liquid. The boiling point of 2-methylbutane is 28° C, higher than unbranched butane (0° C) but less than unbranched pentane, suggesting that **the branch on the hydrocarbon reduces its boiling point.** The boiling point of 2,2-dimethylpentane is even lower, so the additional branch appears to further lower the boiling point.

With this limited data it is reasonable to propose that increased branching decreases the boiling point by reducing the surface area of the molecule and thereby reducing the strength of the dispersion forces between them.

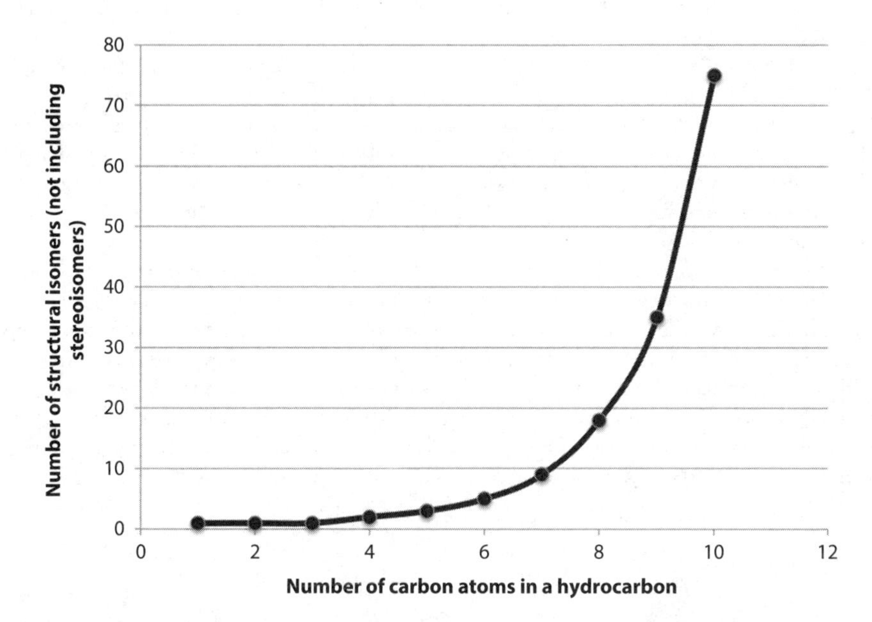

130. (C) In VSEPR theory, electron domains for nonbonding electron pairs exert a greater force on neighboring electron domains than bonded electron domains. CCl_4 has both tetrahedral electron geometry and molecular geometry. When the molecular and electron geometries are the same, it means that each electron domain is bonded to an atom. The 109.5° bond angle agrees perfectly with the angles representing the four corners of a tetrahedron.

The PCl_3 molecule has a tetrahedral electron geometry but a trigonal pyramidal molecular geometry because there is one pair of unbonded electrons on the central phosphorus atom. The bond angles are less than they are for a tetrahedral—approximately 107°—although some have reported bond angles of around 100°, probably due to the further repulsion between the unbonded pair and the chlorine atoms.

Water has tetrahedral electron geometry but is a bent molecule because there are two unbonded electron pairs. The bond angles in the bent water molecule are 104.5°.

131. (C) Molecules with dipoles experience coulombic (electrostatic) interactions that result in a net attraction between them. Hydrogen fluoride (HF) has the largest dipole moment of the molecules listed because the electronegativity difference between two atoms (H and F) is the largest. PH_3 is polar, but its dipole moment is less (0.58 μ, or debye, a coulomb meter), whereas the dipole moment of HF is a whopping 1.91 μ. **The dipole moment is calculated as the product of the magnitude of the charges (or partial charges) and the distance between them (values range from about 0 to 11 μ).** You don't need to calculate the dipole moments of the molecules to answer the question. The symmetrical shapes of O_2 and CO_2 result in molecules with no net dipole. PH_3 is trigonal pyramidal, so there is a net dipole on this molecule, but the electronegativity difference between P–H is much less than that between F–H.

132. (A) Carbon dioxide, CO_2, has two double bonds. Double and triple bonds contain pi (π) bonds and therefore pi electrons.

Pi (π) bonds are covalent bonds involving the overlap of the two lobes of an unhybridized p orbital between the two atoms that share them. The electron overlap occurs above and below the plane of the nuclei of the two atoms involved, but does not occur between the two nuclei (as in a sigma, σ, bond). **Only double and triple bonds involve pi bonding.** All bond orders contain a sigma bond. A single bond is simply one sigma bond; a double bond consists one sigma bond and one pi bond; and a triple bond consists of one sigma and two pi bonds.

Pi electrons are often delocalized because they are not usually confined to the orbital overlap between the two atoms they join. They seem to "spread out" over multiple adjacent atoms. **These delocalized electrons are responsible for the phenomenon of resonance.** Because of the relative promiscuity of pi electrons, pi bonding makes it easier to induce a dipole in a molecule (by dispersion), which is how the sheets of graphite stick together.

Choices B and C contain only single (sigma) bonds. Choice D is incorrect because although O_2 contains a pi bond (and thus pi electrons), the double bond is a function of the valence of oxygen, not resonance, and there is no other Lewis structure that can describe it.

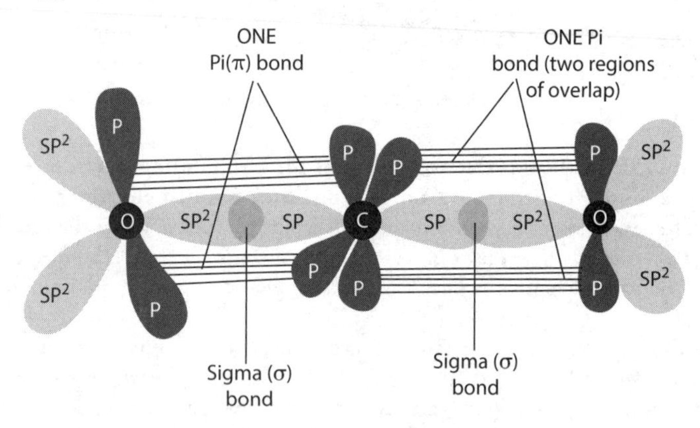

133. (D) A combustion reaction is a self-propagating exothermic reaction that combines oxygen with a substance to produce an oxide of the element (or elements) of the substance being combusted. All combustion reactions are oxidation-reduction (redox) reactions. **The element (or elements) of the substance combusted gets oxidized while oxygen gets reduced.** Combustion often, but not always, refers to the oxidation of a carbon-containing fuel.

134. (B) This molecule is phosphorus trihydride (also called phosphane and phosphine). All single bonds are sigma (σ) bonds, so PH_3 has three of them. CO_2 has two double bonds; each double bond has one sigma and one pi bond, so there are two sigma bonds in CO_2. HF and O_2 both have one sigma bond each.

135. (B) Ions of a lower ionic radius will have a greater electrostatic attraction to the partial positive charges on the water molecule because of the smaller distance between the particles.

(See the Coulomb's law box following answer 8.) Choices A and D are incorrect because the force of gravity is negligible on individual particles like atoms and monatomic ions. Choice C is incorrect because it confuses ionic radius with mass. **It is important to remember that a single word like *large* is not very descriptive of a particle in terms of chemistry. Does the particle have a large mass or a large radius?** Across a period, atomic and ionic radius tends to decrease even though atomic mass generally increases.

136. (C) Be^{2+} is the cation of smallest radius and it has a 2+ charge. O^{2-} is also has a small radius and it has a 2− charge. **The force of attraction between charges is proportional to the product of the quantity of the charges and inversely proportional to the square of the distance between them.** The larger charges (2+) of the ions coupled with the small distance between them (due to their small radii) would produce the greatest force of attraction. (See the Coulomb's law box after answer 8.)

137. (A) Hydrogen bonds form between the partially positive hydrogen on one molecule (or one part of a very large molecule, like a protein) and the partially negative oxygen, nitrogen, or fluorine on another. Hydrogen bonds are the relatively strong interparticle forces of attraction (except in very large molecules, like proteins). Choice A shows the attraction between the partially positive hydrogen atom in the upper left water molecule with the partially negative oxygen in the lower left water molecule. Choice B incorrectly shows an attraction between two partially positive hydrogen atoms in the lower two water molecules. Choice C is a covalent bond between the oxygen and hydrogen atoms within a single water molecule. Choice D incorrectly shows an attraction between the two partially negative oxygen atoms between the water molecules on the right side of the diagram.

You are expected to know the typical types of hydrogen bonding that occur when hydrogen covalently bonds to a highly electronegative atom such as N, O, or F. Cases of weaker hydrogen bonding are beyond the scope of the AP Chemistry course and exam.

138. (A) Molecule A is propane, a nonpolar hydrocarbon. The C–C bonds are nonpolar and C–H bonds are considered relatively nonpolar. Both of the central carbon atoms have a symmetric, tetrahedral geometry. There is nothing polar about this molecule!

Molecule B is water. The strongly polar O–H bonds and the bent shape make it the most polar.

Molecule C is carbon dioxide. Although the C=O bond is polar, the linear shape causes the two dipoles made by the two C=O bonds to point in opposite directions, canceling each other out. The symmetry of molecule makes it nonpolar despite the bonds being polar. However, the polar bonds make it slightly more polar than propane.

Molecule D is methanal, or formaldehyde. This molecule has a trigonal planar shape, but because the three atoms bound to the central carbon are not the same, the molecule is not symmetrical and is nonpolar simply on the basis of its shape. Further, the C=O bond is polar (but the C–H bonds on the opposite sides are nonpolar). Overall, this molecule is polar, but the dipole created by the C=O bond is not as strong as the dipole in water.

139. (D) CCl_4 has both tetrahedral electron geometry and molecular geometry and therefore bond angles of 109.5°. The PCl_3 molecule has a tetrahedral electron geometry but a trigonal pyramidal molecular geometry because there is one pair of unbonded electrons on the central phosphorous atom. The bond angles are less than they are for a tetrahedral—approximately 107°—although some have reported bond angles of around 100°, probably due to the further repulsion between the unbonded pair of electrons on the phosphorus and the chlorine atoms. SCl_2 has tetrahedral electron geometry but is a bent molecule because there are two unbonded electron pairs. The bond angles in the sulfur dichloride molecule are 103°. The actual bond angles are not needed to answer this question. The bottom line is that an increasing number of unbonded electron pairs on the central atom tends to decrease the bond angles in the molecule.

140. (B) Electron domains for nonbonding electron pairs exert a greater force on neighboring electron domains than bonded electron domains. See answer 139 for the actual bond angles of these molecules.

141. (B) It is important to identify these acids as oxyacids (they all contain oxygen). The strength of oxyacids increases with

- increasing electronegativity of the central atom (the atom the O–H attaches to)
- a greater oxidation number of the central atom
- other atoms of high electronegativity in the compound
- an increasing number of oxygen atoms.

Generally, these four factors allow electron density to be drawn away from the hydrogen in the already polar O-H bond, allowing it to more readily dissociate from the oxygen atom.

H_3PO_4	H_2SO_4	$HClO_4$	↑ **Increasing acid strength**
H_3AsO_4	H_2SeO_4	$HBrO_4$	
H_3SbO_4	H_2TeO_4	HIO_4	

Increasing acid strength →

The pH of a solution depends on the concentration on H^+ in solution and therefore the concentration of the acid in solution, but **the strength of an acid is intrinsic to the acid and does not depend on the concentration.**

142. (B) Chlorine's relatively high electronegativity draws electrons away from the O—H bond, which increases its polarity and stabilizes the conjugate base. This is called the **inductive effect**. The inductive effect refers to the ability of atoms or groups of atoms within a molecule to change the distribution of electrons within the molecule. In this case, it refers specifically to atoms other than the oxygen to which the acidic hydrogen is bound.

See answer 141 for factors that contribute to the strength of oxyacids.

143. (D) An increased number of hydrogens does not make an oxyacid a stronger acid. If an acid has more than one dissociable proton, it is a *polyprotic acid.* The K_A of the first H^+ is the highest (most acidic) and the K_A decreases for each successive H^+. See answer 141 for factors that contribute to the strength of oxyacids.

Questions asked in the negative are not common but do occur on the AP Chemistry exam, and they can be tricky. When answering a question asked as "except," try circling the word *except* in the question (to remind you you're looking for the opposite) and then mark each answer choice as true or false (or yes or no) When you're done with the four choices, the one that's false (or no) is the correct answer.

144. (C) The speed at which a gas diffuses at a given temperature is inversely proportional to the square root of its molar mass. If the two gases released at opposite sides of a tube diffused at the same rate, they would meet in the middle. To diffuse at the same rate requires the two gases be the same mass; the larger mass of HCl causes it to diffuse more slowly than NH_3.

The kinetic energy (KE) of the gases is proportional to the absolute temperature, which is the same for both gases. The kinetic energy is proportional to the mass and the velocity squared, so at the same kinetic energy, the velocity is inversely proportional to the square root of the mass.

$$KE = \frac{1}{2} mv^2$$

$$velocity \propto \sqrt{molar\ mass}$$

Let 1 = HCl, molar mass = 35.5 g mol^{-1}

$$v_1 \propto \sqrt{36.15}$$

Let 2 = NH_3, molar mass = 17 g mol^{-1}

$$v_2 \propto \sqrt{17}$$

$$\frac{v_1}{v_2} \propto \sqrt{\frac{36.5}{17}} = 1.5$$

∴ NH_3 diffuses 1.5 times faster than HCl, so in the same period of time, NH_3 travels 1.5 times the distance.

x = the distance HCl diffuses
$1.5x$ = the distance NH_3 diffuses
$1x + 1.5x = 2.5x$ = 100 cm = total distance
x = 40 cm
$1.5x$ = 60 cm

145. (C) The speed at which a gas diffuses at a given temperature is inversely proportional to the square root of its molar mass. **If the two gases had the same molar mass, they would diffuse at the same rate (at the same temperature).** If the two gases released at opposite sides of a tube diffused at the same rate, they would meet in the middle. The molar mass of HCl is 35.5 g mol^{-1}. NH_4OH is the gas with the mass closest to 35.5 g mol^{-1}.

N_2H_4 = 32 g mol^{-1}
CH_3NH_2 = 31 g mol^{-1}
NH_4OH = 35 g mol¯1
$(CH_3)_2NH$ = 45 g mol^{-1}

146. (C) The particles in compartment C are in either the solid or liquid phase at the bottom, and the fewer particles in the gas phase relative to D indicate a **lower vapor pressure therefore stronger interparticle forces of attraction.**

147. (B) Substance A has a lower vapor pressure than B, indicating that substance A has stronger interparticle forces of attraction. Choice C is incorrect because as long as there is still some liquid remaining (and the amount of gas does not change over time, as indicated by the state of equilibrium having been reached), the vapor pressure is not affected by the number of moles of the liquid. Because of this, the volume is irrelevant as well (which is why choice D is incorrect). For example, if you had two sealed bottles of water at the same temperature, one 75% full and the other only 10% full, the vapor pressure would still be the same, approximately 23.8 torr at 25°C. **The only factors affecting the vapor pressure are the interparticle forces of attraction between the particles and the temperature.**

148. (C) Choice A is incorrect because covalent bonds *can* be broken at 25° C. Many factors besides temperature drive a chemical reaction and therefore the breaking (and/or forming) of covalent bonds. Choice B is incorrect because bond strength is a measure of how much energy is needed to break the bond. **Bonds do not typically form at very high temperatures because the kinetic energy of the atoms or particles exceeds the energy needed to break those bonds.** Choice D is incorrect because melting and boiling of most molecular compounds involves breaking weak forces (Van der Waals and hydrogen bonds). Molecular substance also melt and boil over a large range of temperatures. However, melting ionic compounds does break ionic bonds and typically requires high temperatures.

Van der Waals Versus Other Interparticle Forces of Attraction

The term "Van der Waals forces" applies to several types of interparticle forces of attraction. It includes (but is not limited to) London dispersion forces (the attraction between two induced dipoles) and dipole-dipole attractions (the attraction between two permanent dipoles) but does not usually include hydrogen bonding, a specific case of dipole-dipole attraction.

Chapter 2: Free-Response Answers

149. Coulomb's law (see box following answer 8) relates the electrostatic force between two charged particles (q_1 and q_2) with the quantity of charge and the distance (d) between them:

$$F_{ES} \propto \frac{q_1 q_2}{d^2}$$

The table shows that **the strength of the attractive force is weaker at longer distances**. Coulomb's law states that the electrostatic force is inversely proportional to the square of the distance. The electrostatic force is proportional to the product of the charges. **Ionic bonds have the largest energies (and are therefore the strongest) because they are the result of attractions between charged particles.** The smallest charge on an ion is ± 1 (actual charge would be $\pm 1.6 \times 10^{-19}$ C). Polar covalent bonds are an unequal sharing of electrons, and the atoms that are sharing the electrons have a "partial charge," which means the product of the charges would be less than the weakest ionic bond. Hydrogen bonds and Van der Waals forces are forces that occur between particles, and they are the result of weak partial charges over fairly long distances.

150. Many weak forces can impart stability in large molecules due to their collective action. There are a tremendous number of weak forces acting in a molecule such as a protein, so the cumulative effect can be quite large. Weak forces have a large influence on biological molecules such as proteins and nucleic acids.

151.

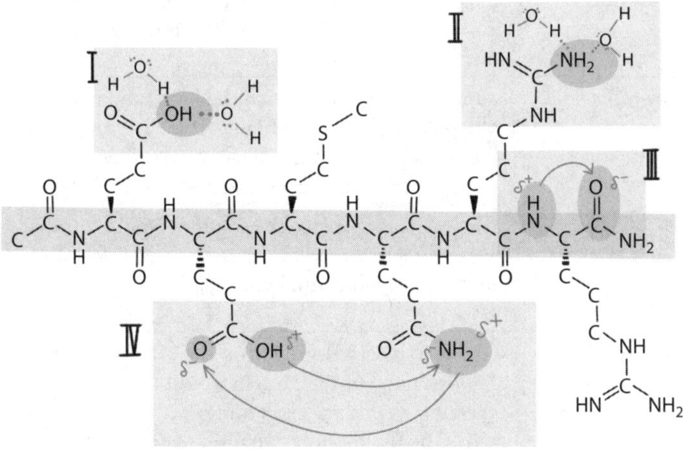

If you are asked for three types of interactions, only provide three. The figure shows 4, and more are possible than what is shown. What is NOT shown is that at physiological pH, the carboxyl groups on amino acids are deprotonated (and therefore ionized) and the amino groups are protonated (and ionized) and the oppositely charged groups can interact ionically.

Box I shows what water can form hydrogen bonds (indicated by three dots) with O–H groups. The O–H groups on water can also form dipole-dipole attractions to carbonyl (C=O) groups (not shown). Box II shows that water can form hydrogen bonds with amino and N–H groups. Box III shows where the N–H group on one amino acid can form a dipole-dipole interaction with the C=O group of another amino acid. In reality, these two groups are too close together in the chain to fold in such a manner, but as the question stated, you could ignore their location in the chain). Box IV shows where amino groups can form hydrogen bonds with O–H groups or dipole-dipole attractions with carbonyl groups.

152. Structural complementarity and **charge distribution complementarity** are the primary factors determining the affinity of the active site for the substrate. **Weak interparticle forces are readily reversible because the kinetic energies of the molecules involved are only a few times weaker than the forces of attraction between them.** A substrate fits into an active site due to complementary shapes but is also held in position by forces of attraction such as hydrogen bonds and Van der Waals forces (and others that may not have been mentioned in your AP Chemistry class). Once the substrate has been converted to product, the shape of the molecule as well as the functional groups will be different. That means the product (or products) will most likely have *less* of an attraction (lower affinity) for the active site. **The thermal motion (kinetic energy) of the particles would then be high enough relative to the affinity to release the product from the enzyme.**

153. For this question, it may be helpful to set up a table.

Proteins in solution	Fold only in solution ∴ hydrophilic R-groups in the protein's exterior face the solution while hydrophobic R-groups face the interior.
Peripheral proteins	One surface interacts with the hydrophilic heads of the phospholipid membrane. These R-groups would form weak, reversible interactions such as dipole-dipole attractions, ionic attractions, and hydrogen bonding.
	The surfaces that face the solution would form weak interactions with water, such as dipole-dipole attractions, ionic attractions, and hydrogen bonding.
Integral proteins	The amino acid R-groups on the exterior of the "midsection" of the protein would face outward, forming weak dispersion forces with the phospholipid tails. The hydrophobic environment would also cause hydrophilic amino acid R-groups to face the interior of the protein.
	The hydrophilic R-groups on the surface of the protein face the aqueous intra- and extracellular solutions.

154. *Structural comparison*: Water has a bent molecular geometry while ammonia has a trigonal pyramidal geometry. **This geometry affects the number of hydrogen bonds per molecule that can form.** Water has two hydrogen atoms available for hydrogen bonding and two lone pairs of electrons. In H_2O *(l)* approximately two hydrogen bonds can form per molecule at any one time. Ammonia has three hydrogen atoms available for hydrogen bonding and one lone pair of electrons but will typically form approximately one hydrogen bond per ammonia molecule at any one time.

A mechanism based on interparticle attractions: **Hydrogen bonds are constantly forming and breaking.** The average lifetime of a hydrogen bond in H_2O *(l)* is on the scale of 10 picoseconds. **The number of hydrogen bonds that form per molecule is dependent on the temperature. Higher temperatures mean higher kinetic energy, which means there is an increased tendency toward molecular dispersal.**

Both ammonia and water are polar. The **oxygen-hydrogen bond in water is more polar** than the nitrogen-hydrogen bond (electronegativity difference of 1.24 versus 0.84), **which makes the hydrogen bond between water molecules stronger** (21 kJ mol^{-1}) than the hydrogen bond between ammonia molecules (13 kJ mol^{-1}).

A thermodynamic comparison: The kinetic energy of the particles tends to disperse them while the interparticle forces of attraction tend to hold them together in a liquid (or solid). As kinetic energy increases, the dispersive forces will overpower the attractive forces and the molecules will break free of all their attractive forces, resulting in evaporation. The temperature (and therefore the average kinetic energy) at which this occurs depends on the strength of the interparticle forces of attraction that hold the particles together in a liquid (or solid).

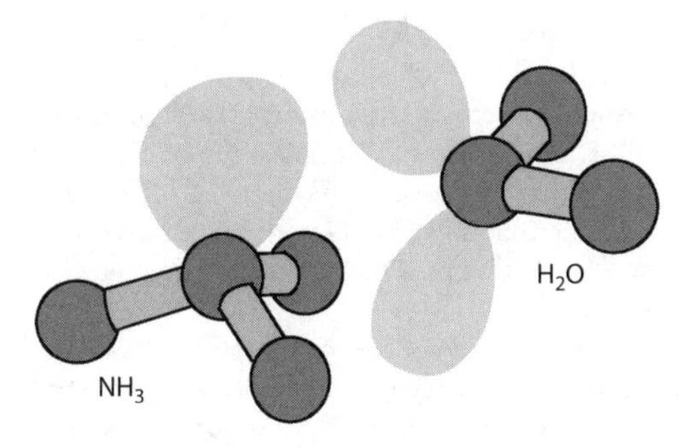

155. **(a)** The ΔHvap is the amount of heat absorbed to vaporize water at 373 K whereas the ΔHsub is for the sublimation of water at 273 K, or 0° C. For 1 kg of water to melt (333.5 kJ), get heated to 100° C $\left(\frac{1 \text{ kg} \times 4.18 \text{ kJ}}{\text{kg °C}} \times 100° \text{ C} = 418 \text{ kJ}\right)$, and vaporize (2,257 kJ) requires 3,009 kJ. **The "missing" enthalpy change is from the heating of water that is necessary to vaporize it at its boiling point.**

(b) The heat of melting ice is about 15% the heat of vaporization and about 11% of the heat of sublimation. In ice, there are approximately four hydrogen bonds per water molecule, and in the vapor phase of water, there are none. Sublimation breaks *all* the hydrogen bonds at once, whereas melting only breaks about 1–1.5 hydrogen bonds per water molecule. Heating reduces the average number of hydrogen bonds in liquid water, and vaporization breaks whatever hydrogen bonds remain.

156. Real gases usually only deviate from ideal behavior by about 5% under standard conditions.

(a) Under conditions of high pressure and low temperature, they can deviate significantly.

(b) The two assumptions of KMT that are not applicable under these conditions are that there are no interparticle forces of attraction or repulsion between the particles of a gas (at low temperatures and high pressures, even weak interparticle forces of attraction become significant) and that the volume occupied by the gas particles is negligible (in a container with a small volume the size of the gas particle becomes significant).

(c) Under high pressure, the volume of a real gas is larger than the volume expected from an ideal gas because volume of the gas particles becomes nonnegligible. At low temperatures, the pressure of a real gas is lower than that expected of an ideal gas because the interparticle forces of attraction between the particles becomes significant.

157. **(a) Real gases are not as compressible as ideal gases because the particles have volume**, however small. The volume of a real gas is therefore larger than that predicted from the ideal gas equation. Constant b depends on the size of the gas particle. Taken alone, **the modified formula $P(V - nb) = nRT$ can correct for particle size.** Notice that the greater the number of moles, the greater the value of nb. When the pressure is low and the volume is large, nb doesn't make much of a difference, but at high pressures and low volumes, nb is no longer negligible.

If, as the ideal gas law assumes, there were no interparticle forces of attraction, gases would never condense to form liquids. The constant a corrects for the lower pressure observed in real gases compared to ideal gases. Notice the value of *a* for NH_3 is very large compared to He due to the polarity of ammonia.

(b) The small size and extremely weak interparticle forces of attraction of He atoms make helium the most ideal of gases. The larger size of carbon dioxide would make the actual volume larger than that predicted by the unmodified ideal gas equation, and the large value of *a* indicates that there are significant interparticle forces of attraction present, so the pressure would be lower than that predicted by the ideal gas law.

158. The C–O–H bond angle is approximately 109.5° because the electron geometry around the oxygen atom is tetrahedral but the molecular geometry is bent (like water; see figure and explanation with answer 159). Because lone pairs of electrons exert a greater repulsive force than bonded atoms, the bond angle is expected to be *less than* 109.5°. For example, the H–O–H bond angle in water is 104.5° (two lone pairs of electrons), and the H–N–H bond angle in ammonia is 107° (one lone pair of electrons).

159. The methyl carbon has four electron domains therefore a tetrahedral geometry (both electron and molecular). The bond angles are 109.5° (see figure).

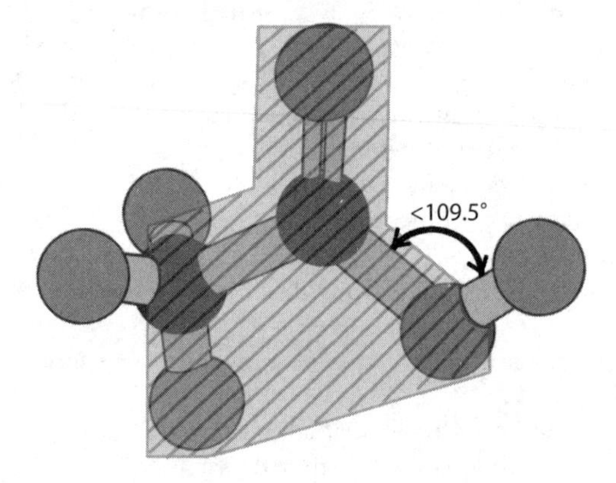

All molecules, even planar ones, exist in three-dimensional space. If the methyl carbon and hydrogen, carboxyl carbon, and the two carboxyl oxygen atoms are viewed in o ne plane (as shown below), then:

- one of the other two methyl hydrogen atoms (top right) is directed into the page and the third is directed out of the page.
- the hydroxyl oxygen (part of the carboxyl group) has a bent molecular geometry (but a tetrahedral electron geometry).
- there are two lone pairs of electrons. One pair is directed out of the page (pointing down, to the left) while the other is directed into the page (pointing down, to the right).

160. Your drawing may vary but should look similar to this:

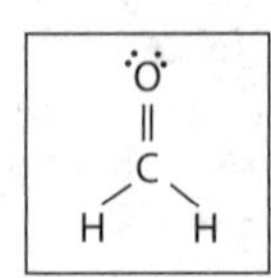

161. Your drawing may vary but should look similar to this:

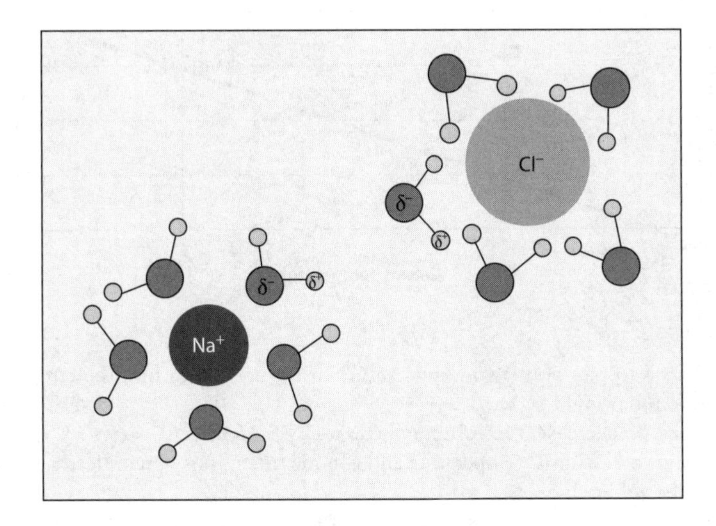

162. Your drawing may vary but should look similar to this:

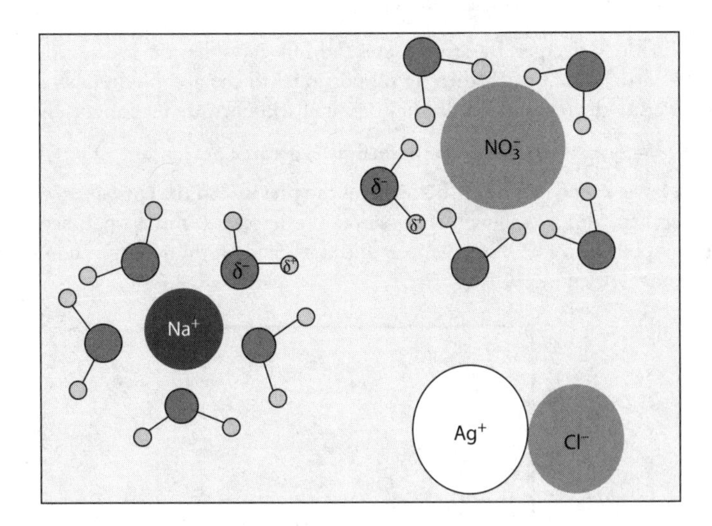

163. The two points you needed to calculate the line was the initial temperature and volume (99.92° C, 2,000 mL) and the final temperature and volume (10.00° C, 1,517.8 mL). Remember that the water that was removed needed to be subtracted from the total volume to calculate the gas volume in the flask (2,000 − 482.1 = 1,517.8 mL).

Temperature should be on the *x*-axis since **the slope should have the units of mL/°C** so that you can determine the change in volume as a function of the change in temperature.

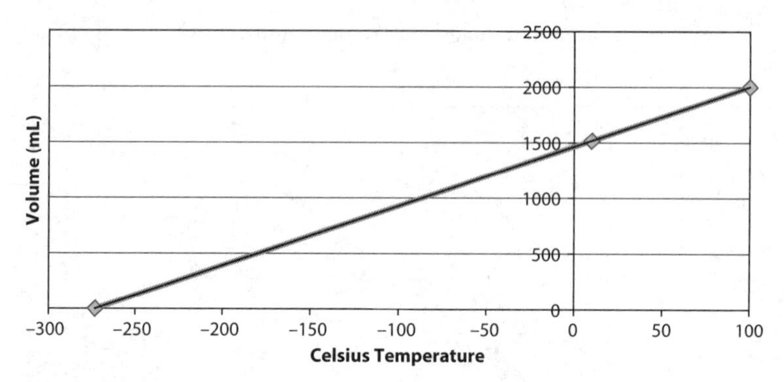

164. Slope $= \frac{\Delta y}{\Delta x} = \frac{482.1}{89.92} =$ **5.36 mL/°C**

You'll need to plot your two points and then extrapolate to find the temperature at which the volume would be zero.

For every °C decrease, the volume decreases by 5.36 mL.

If you start at the initial temperature and volume (the minus sign indicates that you are decreasing the volume):

$$-2{,}000 \text{ mL} \times (°\text{C}/5.36 \text{ mL}) = -373° \text{ C}$$

To decrease the volume by 2,000 mL, a 373° C temperature reduction is needed.

The experiment began at 99.92° C ∴ 99.92° − 373 = **−273.2° C.**

165. The Kinetic Molecular Theory of Gases (KMT) states that the average kinetic energy (KE) of the particles in a gas is directly proportional to the absolute temperature ($KE \propto T_{ABS}$), so lowering the temperature in the container will decrease the kinetic energy and ∴ the speed ($KE = \frac{1}{2}mv^2$). **At lower temperatures, a greater percentage of particles will be moving at a lower speed (see figure below) and the pressure in the container will decrease.** The lower temperatures also allow the formation of interparticle forces of attraction. The gas used in this experiment, H_2O, is very polar and can readily form hydrogen bonds when the temperatures are low enough.

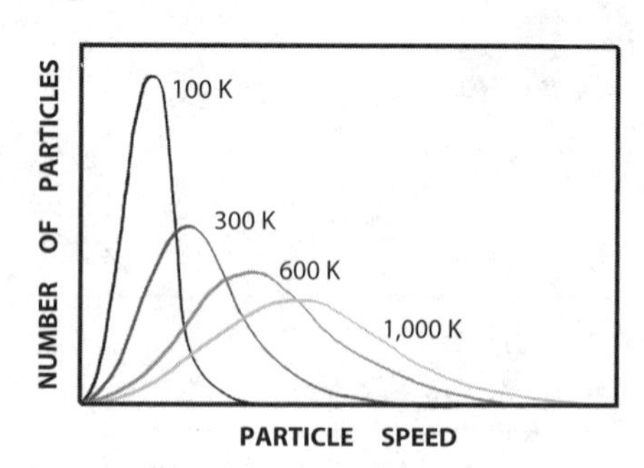

166. Component 7 is the most soluble in acetone because it went the farthest along the chromatography paper. The mixture components that are most soluble in the mobile phase travel the farthest and have the highest R_F values. The abbreviation R_F stands for retention (or retardation) factor. The R_F value describes the ratio of time that a mixture component spends in the mobile phase versus the stationary phase (the paper):

$$\mathbf{R_F = \frac{distance\ the\ substance\ migrated}{distance\ the\ solvent\ front\ migrated}}$$

The higher the R_F value, the more soluble the substance is in that particular solvent.

167. Components #1 and #5 have the lowest R_F values. **The components with the lowest R_F values are the *least* soluble in the solvent.** The R_F value of a substance is a function of the composition of the mobile and stationary phases as well as the substance itself. A substance with a high R_F value with a particular solvent/paper combination indicates that the functional groups and/or overall structure of the substance are similar to that of the solvent to allow them to form more and/or stronger interparticle forces of attraction than the substance can form with the paper.

168. Ethanol is soluble in both polar and nonpolar solvents. The alcohol (O–H) group allows it to hydrogen bond with water. **Dimethyl ether** is nonpolar. It will only dissolve in organic (nonpolar) solvents. Because ethanol molecules can also form hydrogen bonds with other ethanol molecules, ethanol has a higher boiling point than dimethyl ether. (The boiling point of dimethyl ether is about −141° C, so it's actually a gas under standard conditions.)

```
     H   H      ETHANOL
     |   |
  H—C — C — O — H
     |   |
     H   H
                H         H
                |         |
             H—C — O — C — H
                |         |
    DIMETHYL    H         H
     ETHER
```

169. (a) Calculate the volume of oxygen gas needed.

550 L oxygen/day × 10 people × 14 days = **77,000 L O_2 needed**

Convert volume of O_2 gas to number of moles of O_2.

$$PV = nRT$$
$$(1)(77{,}000) = n(0.0821)(298)$$
$$n = \mathbf{3{,}147\ moles\ of\ O_2\ gas}$$

Convert moles of O_2 to mass of O_2.

$$3{,}147 \text{ moles} \times \frac{32 \text{ g}}{\text{mole}} \times \frac{1 \text{ kg}}{1{,}000 \text{ g}} = \mathbf{100.7\ kg\ O_2}$$

(b) For every 1 liter of oxygen gas a person breathes, 0.82 liters of carbon dioxide are released into the environment. **You must use the amount of CO_2 produced for every liter of O_2 gas consumed,** *not* the 2 CO_2:3 O_2 ratio in the balanced equation.

$$77{,}000 \text{ L } O_2 \times \frac{0.82 \text{ L } CO_2}{1 \text{ L } O_2} = \mathbf{63{,}140 \text{ L } CO_2}$$

Convert volume of CO_2 gas to number of moles of CO_2.

$$PV = nRT$$
$$(1)(63{,}140) = n(0.0821)(298)$$
$$n = \mathbf{2{,}581 \text{ moles of } CO_2 \text{ gas}}$$

Convert moles of CO_2 to mass of CO_2.

$$2{,}481 \text{ moles} \times \frac{44 \text{ g}}{\text{mole}} \times \frac{1 \text{ g}}{1{,}000 \text{ g}} = \mathbf{113.6 \text{ kg } CO_2}$$

170. **(a)** Use the coefficients in the balanced equation to **convert moles of CO_2 to moles of KO_2.**

$$2{,}581 \; CO_2 \times \frac{4 \; KO_2}{2 \; CO_2} = \mathbf{5{,}162 \text{ moles } KO_2}$$

Convert moles of KO_2 to mass of KO_2.

$$5{,}162 \text{ moles } KO_2 \times \frac{71 \text{ g}}{\text{mole}} \times \frac{1 \text{ kg}}{1{,}000 \text{ g}} = \mathbf{366.5 \text{ kg } KO_2}$$

(b) For each mole of CO_2 removed from the environment, one mole of K_2CO_3 is formed ∴ 2,581 moles of K_2CO_3 will be produced during the two-week period.

$$2{,}581 \text{ moles } K_2CO_3 \times \frac{138 \text{ g}}{\text{mole}} \times \frac{1 \text{ kg}}{1{,}000 \text{ g}} = \mathbf{356.2 \text{ kg } K_2CO_3}$$

(c) The benefits of removing CO_2 by this method are (1) a solid is formed, so the volume is minimized, and (2) the ratio of O_2 formed and CO_2 removed is close to physiological value (1.5 O_2 to 1 CO_2 in the process versus 1.22 O_2 to 1 CO_2 for breathing).

171. **Answers will vary.** The following table is one of several ways to present your answer. You did not need to set up a table to correctly answer this question. However, your answer should include most of these details.

	Metals	Ionic compounds
Composition	Composed of metallic elements only Pure metal = 1 kind of metal Alloy > 1 kind of metal	At least two different kinds; ions of opposite charge
Melting temperature		Typically higher than metals

Appearance	Shiny, have luster Usually silver but may be gold or copper colored	Can be white or colored, may be reflective
Texture	Ductile, malleable	Hard, brittle
Conductive properties	Conducts heat and electricity	Does NOT conduct heat or electricity as a solid Can conduct electricity in molten form or in solution
Ease of oxidation	Easily oxidized	
Solubility in water	Solid	Solubility varies from very soluble to mostly insoluble

172. Metal lattice structures are composed of **cations in fixed positions** relative to one another. (A body-centered cubic arrangement is shown in the figure below, on the left. It is for illustrative purposes only. *The knowledge of specific crystal structures is beyond the scope of the AP Chemistry exam.*) The cations are composed of the nucleus of the metal and the core electrons. The **valence electrons are mobile**, or **delocalized**. They do not permanently associate with any particular cation. These mobile electrons are referred to as a **"sea of electrons"** and are responsible for metallic bonding. These mobile electrons are also the key structural feature of metals that gives them the property of heat and electricity conduction.

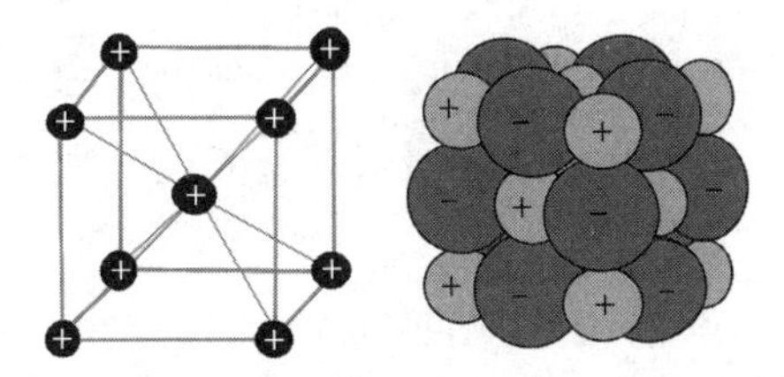

Both **the cations and anions of ionic compounds are fixed** in position. There the cations are arranged to surround (and be surrounded by) anions and the anions are arranged to surround (and be surrounded by) cations (see figure above, right). There are no mobile charges in ionic compounds, which is why they cannot conduct electricity as solids.

173. Metals can conduct heat and electricity but ionic solids cannot.

174. Metal lattice structures are composed of cations in fixed positions surrounded by mobile valence electrons. Both the cations and anions of ionic compounds are fixed in position. There are **no mobile charges in ionic solids.**

175. Sample I is a pure substance. The entire volume of the distillation occurred at one boiling point. Sample II is a mixture of at least four different components, each of which is distinguished by a different boiling point (see figure below).

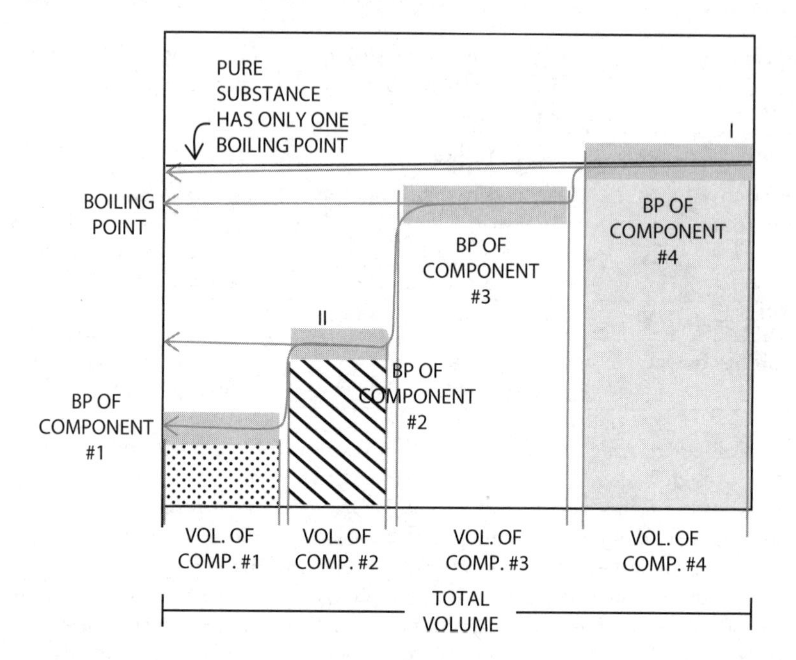

176. There are at least four different pure substances in mixture II (see figure above). Each substance has its own boiling point. The volume of each distillate is on the *x*-axis. Component 3 has the largest relative volume, so it must be present in the largest volume in the mixture. Substance 2 has the lowest volume. The total volume of the mixture will be the sum of the volumes of the four components, so each component can be quantified as a fraction of the total volume.

177. The liquid has no odor and one boiling point so it is not a mixture of two or more alcohols. It is possible that it is water or propanol with a white solid (NaCl or glucose) dissolved in it. The boiling point is only 2° C above boiling, so it is unlikely to be propanol, which would need a very large amount of solute dissolved in it to increase the boiling point so dramatically. There are several methods to determine which compound is dissolved in the water. (Taste is easiest since the two substances are in a normal diet, but never try that in your lab!) A simple one is to determine if the solution conducts electricity. If it does, the solute is NaCl, and if it does not, it is glucose.

Colligative properties are beyond the scope of this course and the AP Chemistry exam, so they will not be directly assessed. However, they are considered prior knowledge, so you are expected to know what they are. You will not have to perform calculations of molality, percent by mass, and percent by volume.

178. Glucose is water soluble and sulfur is not. You can add water to the mixture to solubilize the glucose and filter out the sulfur. You can also solubilize the sulfur with hexanol (or pentanol or even butanol) and filter out the glucose. You'd want to test a small sample of pure sulfur before committing to this test because I assumed that the nonpolar nature of sulfur would allow it to dissolve in a water-insoluble alcohol, but without actual data that is just an assumption.

179. You would need a specific form of chromatography that could separate the charged ions of NaCl from the polar glucose molecules in the mixture. Because both glucose and NaCl are water soluble and their solubility in other solvents is quite similar, they are very difficult to separate from each other. You would need a method that could specifically distinguish between ions and polar substances.

180. Answers will vary but should include the following information:

	Compound	Solubility
A	**Acetone (Ethanone)**	Soluble in water and nonpolar solvents. Any of the solvents would dissolve acetone.
B	**Formaldehyde (Methanal)**	A gas at room temperature. It can dissolve in water and alcohols as well as polar solvents such as benzene.
C	**Benzene**	Benzene is not very soluble in water. The water solubility of the alcohols decreases as the number of carbons in the alcohols increases, so hexanol would be the best alcohol to solubilize benzene in.
D	**Carbon tetrachloride**	Not very water soluble but soluble in alcohols and organic solvents. The water solubility of the alcohols decreases as the number of carbons in the alcohols increases, so alcohols of high carbon number would dissolve a greater amount of CCl_4.

Chapter 3: Transformation

181. (D) In the Brønsted-Lowry definition of acids and bases, **acids are proton donors** and **bases are proton acceptors**. The acid reactant donates its proton to the base, becoming a base itself (it can now, theoretically, accept a proton). The base that the acid donated the proton to becomes an acid because it can now, at least theoretically, donate the proton it just accepted. **For a reaction in equilibrium, this process is a "back and forth" of proton exchange. The terminology of conjugates can help us distinguish between the two acids** (the reactant or the acid formed when the base was protonated) **and the two bases** (the reactant or the base formed by the deprotonation of the acid) **present at equilibrium.** Because the acid loses a proton to become a conjugate base and the base gains a proton to become a conjugate acid, **conjugate pairs only differ by one proton (H^+ ion).** Keep in mind that a strong acid has a weak conjugate base and a strong base has a weak conjugate acid, so conjugates of strong acids and bases may not be very effective bases or acids.

Generally, the **acid and base reactants will be on the left side** of the arrow and **the conjugates will be on the right.**

Choices A and B are incorrect because the pairs do not differ by only one proton (H^+). Choice C is incorrect because CO_2 and HCO_3^- do not appear in the same reaction as they are expressed by equations I, II, and III.

Note that H_2O *(l)* is a reactant in all three equations. Equation I represents a synthesis reaction involving water. In the reaction expressed by equation II, water acts as a base, picking up a proton from carbonic acid (H_2CO_3). The conjugate acid of water is the hydronium ion, H_3O^+ and the conjugate base of carbonic acid is hydrogen carbonate, HCO_3^-. In the reaction expressed by equation III, water again acts as a proton acceptor (the base), its conjugate acid the hydronium ion, H_3O^+ and the conjugate base of hydrogen carbonate is the carbonate ion, CO_3^{2-}.

One of the learning objectives in Big Idea 3 is to be able to identify compounds as Brønsted-Lowry acids, bases, and conjugate pairs. Lewis acid-base concepts have been excluded. However, don't be surprised if the concept comes up as a theoretical question about the construction, application, and refinement of a model. Acid-base theory has gone through an evolution from Arrhenius, Brønsted-Lowry, and Lewis. This would be an application of Science Practice 1: The student can use representations and models to communicate scientific phenomena and solve problems.

182. (C) Conjugate pairs only differ by one proton. The terms *acid* and *base* and *conjugate acid* and *conjugate base* do not describe conjugate pairs (see answer 181 for an explanation of why these terms exist). Conjugate pairs consist of either an acid and its conjugate base (which contains one less proton than its acid) OR a base and its conjugate acid (which contains one more proton than its base).

183. (D) Chlorine has an oxidation state of 0 in Cl_2, –1 in Cl^-, and +5 in ClO_3^- (see figure below).

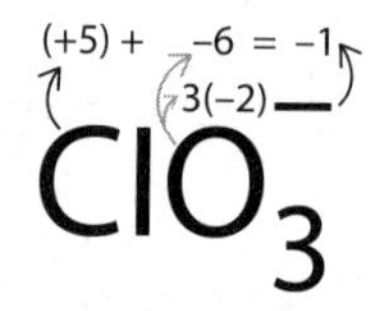

In reaction D, five of the six chlorine atoms in 3 Cl_2 gain one electron (each) to become Cl^- and the sixth chlorine atom loses five electrons (to become Cl^{+5} in ClO_3^-).

Reaction A is an acid-base neutralization, a type of double replacement reaction (aka double displacement). Double replacements are not typically redox reactions, so the oxidation numbers of the elements involved are not expected to change. Reaction B shows the precipitation of silver chloride from the combination of its ions (it could also be the net ionic reaction of a double displacement–precipitation reaction). Again, the oxidation numbers do not change; it is not a redox reaction. Reaction C is a synthesis (and also a combustion, since it uses molecular oxygen as a reactant and produces the oxide of magnesium). Syntheses from elements and combustions are always redox reactions. In this reaction, however, magnesium is oxidized and oxygen is reduced. (See answer 133 for a description of combustion reactions.)

184. (C) Ionic bonds form between oppositely charged ions or elements with large differences in electronegativity. Reactions A, B, and C all produce ionic compounds, but only choice C does it by way of a redox reaction.

185. (C) When you're given the amounts of both (or all) reactants in a problem, check for limiting reactants.

Simple Check to Identify the Limiting Reactant (LR) Using Moles

moles reactant A $\div$ stoichiometric coefficient of A $= P_A$
moles reactant X $\div$ stoichiometric coefficient of X $= P_X$

$\mathbf{P_A < P_X}$ $\therefore$ A is limiting reactant

$\mathbf{P_A > P_X}$ $\therefore$ X is limiting reactant

$P_A = P_X$ $\therefore$ no limiting reagent (only happens in theory)

In this case, 0.400 mol $CS_2 \div 1 = 0.40$ and 1.20 mol $O_2 \div 3 = 0.40$ $\therefore$ **no limiting reactant.**

If there is no limiting reactant, **use the reactant that has the easiest numbers to crunch.** In this case, it is 0.40 mol CS_2 because its stoichiometric coefficient is 1.

$$1 \text{ mol } CS_2 \rightarrow 3 \text{ mol products } (1 \text{ mol } CO_2 \text{ and } 2 \text{ mol } SO_2)$$

$$\therefore 0.40 \text{ mol } CS_2 \times \frac{3 \text{ moles products}}{1 \text{ mol } CS_2} = 1.20 \text{ mol products}$$

186. (C) Three moles of O_2 gas produce three moles of gaseous products $\therefore$ the number of moles of oxygen gas that reacts is equal to the number of moles of gas the reaction produces.

187. (A) At STP, one mole of gas occupies 22.4 L.

$$\therefore 33.6 \text{ L} \times \frac{1 \text{ mole}}{22.4 \text{ L}} = \mathbf{1.5 \text{ moles of gas formed}}$$

One mole of CS_2 reacts for every three moles of gas that form.

$$\therefore 1.5 \text{ moles gas} \times \frac{1 \text{ mole } CS_2}{3 \text{ moles gas}} = \mathbf{0.5 \text{ moles } CS_2}$$

188. (D) There is a one-to-one relationship between the number of moles of SO_2 consumed to the number of moles of SO_3 produced, so **three moles of SO_2 are needed** to form three moles of SO_3. Because O_2 is present in excess, you can assume that all of the SO_2 will react.

Determine the number of moles of O_2 required.

$$3 \text{ moles } SO_3 \times \frac{1 \text{ mole } O_2}{2 \text{ moles } SO_3} = 1.5 \text{ moles } O_2$$

The question asks for 0.5 moles of O_2 to be present in excess, so you'll need 0.5 moles extra $\therefore$ 1.5 to react + 0.5 in excess = **2 moles O_2**.

189. (B) In this reaction, **O_2 is the limiting reactant** (see the box below answer 185 for a quick way to identify the limiting reactant). Therefore, the reaction will only produce as much SO_3 as what one mole of O_2 will produce: two moles (exactly the stoichiometry of the reaction). Two moles of SO_2 are required to **produce two moles of SO_3**, so **two moles of SO_2 will be left** (in excess). Choice B shows an equimolar ratio of sulfur dioxide and sulfur trioxide.

190. (A) A simple way to check for a limiting reactant is shown in the box below answer 185:

$$5 \text{ moles } SO_2 \div 2 = 2.5$$
$$6.5 \text{ moles } O_2 \div 1 = 6.5$$
$$\therefore SO_2 \text{ is the limiting reactant.}$$

Five moles of SO_2 produces 5 moles of SO_3 and consumes 2.5 moles of O_2, leaving 4 moles of O_2 in excess.

191. (D) In the forward reaction, the fluoride ion (F^-) is the base, or proton acceptor, and H_2O is the acid (proton donor). In the reverse reaction, HF is the acid and OH^- is the base. Because **the reaction is at equilibrium, both the forward and reverse reactions are occurring** (and why the terminology for conjugates exists).

See answer 181 for a more detailed explanation of Brønsted-Lowry acids and bases.

192. (D) In the first reaction, H_2O acts as a base, accepting the proton from CH_3COOH. The conjugate acid of water is H_3O^+, the hydronium ion. (Free protons, H^+ ions, don't really exist in solution.) In the second reaction, H_2O acts as an acid, donating a proton to NH_3. The conjugate base of H_2O is OH^-. The reverse reactions have water as a product. In the reverse of reaction 1, water is the conjugate base of the hydronium ion. In the reverse of reaction 2, water is the conjugate base of the hydroxide ion. An important aspect of conjugate acids and bases is that at equilibrium, the acid, the base, and their conjugates all exist in the same solution. The equilibrium arrows indicate that all the species represented in the reaction occur in the same solution at the same time. *The "reaction as written" is a convenience for chemists, not necessarily a physical reality.*

See answer 181 for a more detailed explanation of Brønsted-Lowry acids and bases.

193. (C) Solid copper gets oxidized to produce aqueous Cu^{2+} ions. The nitrogen gets reduced. (See the reaction depicted below.) Hydrogen ions do not get oxidized or reduced (the oxidation state of hydrogen does not change). Remember that in a redox reaction there must be an oxidation *and* a reduction.

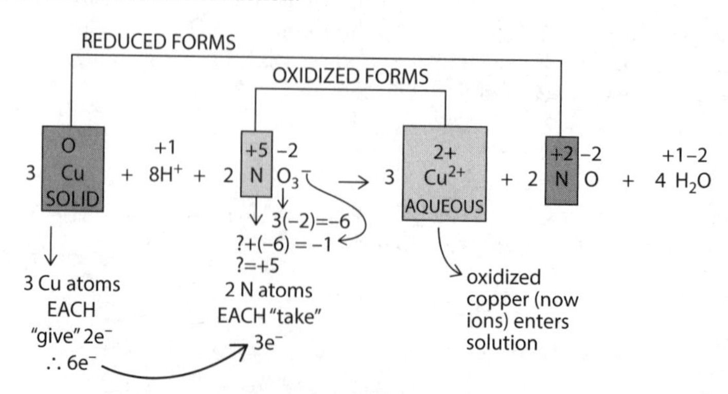

194. (C) Choice C represents **the most important aspect of the reaction: the oxidation and reduction. The quantity of charge being lost and gained and the number and type of atoms they are exchanged between are all balanced.** Choice A is a half-reaction, the oxidation of copper. Oxidations are always accompanied by reductions, so one half-reaction will never accurately represent a redox reaction.

Choice B does not represent anything useful about the reaction. It does not even accurately represent a transfer of electrons. The oxidation state of oxygen is −2 in the actual reaction but has an oxidation state of 0 in choice B. There couldn't be more wrong with choice D! It confuses an ionic dissociation with a redox reaction. Besides not representing the most important part of the reaction being represented—the reduction and oxidation—the dissociation of $CuNO_3$ would not produce NO. In addition, the reaction is not balanced.

195. (B) By convention, the table is set up to show only reductions. The oxidation reaction is the reverse of the reduction (and the sign on the potential is opposite). The left side of the arrow lists the species with the most exergonic reduction on top. This species is **the most easily reduced** (and is therefore the strongest oxidizer; see table below). **The species at the bottom left requires the most energy to reduce.**

The right side of arrow lists the products of the reduction. The bottom of the right side of the table contains a species that has the most exergonic oxidation. You can image the right side as the table of oxidations, but remember that the most easily oxidized species is at the *bottom* of this list.

F_2 *(g)* **Most easily reduced** **Strongest oxidizer**	→	$2\,F^-$ *(aq)*
$2\,H^+ + 2\,e^-$	→	H_2 *(g)*
Li^+ *(aq)*	→	**Li *(s)*** **Most easily oxidized** **Strongest reducer**

The College Board does not use the language of "oxidizing agent" and "reducing agent" on the AP Chemistry exam. However, you do need to understand the concepts of substances that oxidize and substances that reduce.

196. (B) To understand the information in the table, you must understand how the data in the table were obtained. The zero $E°$ value for the reduction of H^+ is a convention based on the **standard hydrogen electrode**, or SHE. Scientists have universally accepted this designation (that the reduction of $2H^+$ to H_2 under standard conditions is assigned a value of 0 V). All of the values in the table are relative to this reduction under standard conditions. Recall that the ° (naught) sign indicates **standard conditions,** 1 M concentrations of solutes and 1 atm gas pressures, which means that you can't assume the $E°$ value is applicable in all situations in which a particular oxidation or reduction is occurring.

Positive $E°$ values represent thermodynamically favorable reductions. Thermodynamically favorable (spontaneous, exergonic) processes can do work. Negative $E°$ values represent thermodynamically favorable oxidations (the reverse reaction). The magnitude is based on measurements taken under standard conditions.

Remember that as a cell operates, the concentrations of the reactants decreases, which changes the $E°$ value.

197. (B) Electrons flow (move in a direction that allows them to do work and does not require work be done on them) because of a difference in potential energy. Electrons in the anode (the substance to be oxidized) have a higher potential energy than those in the cathode. It follows that **electrons will flow spontaneously (exergonically, in the thermodynamically favored direction) from the anode to the cathode, the electrode with the more positive electrical potential.**

The potential difference is the driving, or electromotive force (emf), measured in volts. **A volt the unit of the electric potential** and is defined as 1 joule of work per coulomb of charge transferred, or 1 J/C, or, **the potential difference in which 1 joule of work is needed to move 1 coulomb of charge.** Because the electrical potential measures the potential energy per charge, **the standard reduction potential of a substance is an intensive property**. Increasing the amount of the substance also increases the number of charges. It's like the unit price. If the unit price is $1.00 per item, it doesn't matter whether you buy 10 or 100. Of course the more you buy the more it costs, but the price per unit is the same.

The stoichiometry of the reaction balances the oxidation reaction and the reduction reaction so that the number of electrons transferred are the same. It can also be thought of as the number of times a reaction must occur to satisfy the electron requirements of the reaction. For example, in the reaction involving nickel and silver, the reduction of silver must happen twice for each oxidation of nickel. This obviously depends on the amount of the substances involved and is *not* an intensive property of the substance. **The standard reduction potential is an intensive property because it depends *only* on the potential energy of the electrons involved in the reaction, not on how many times it happens.**

198. (D) The standard reduction potential of each half-cell in a voltaic cell is a measure of the driving force of the reaction. For a voltaic cell operating under standard conditions (25° C, 1 M concentrations for solutes, and 1 atm pressure for gases), the cathode reduction has a more positive $E°$ value that the anode reduction. You can easily remember this because you already know that a positive cell potential means the cell is thermodynamically favorable. **The potential of the cell is calculated from the half-cell potentials by finding the difference between their reduction potentials:**

$$E°_{CELL} = E°_{RED}\,(\text{cathode}) - E°_{RED}\,(\text{anode})$$

The reduction potentials of the cathode reaction are more positive than the reduction potentials of the anode reactions, which makes the $E°_{CELL}$ value positive. Keep in mind that the actual reduction occurs at the cathode and an oxidation occurs at the anode, so the reduction half-reaction from the standard reduction potentials must be read from right to left and the sign of $E°$ must be reversed to describe the reaction as an oxidation. However, when simply comparing reduction potentials, the less positive (or more negative) reduction potential of the (oxidation) reaction that occurs at the anode is more favored than the reduction.

For example, the second column of the chart below shows three reduction potentials for oxidations that occur at the anode (column 1). The most positive reduction potential is for the reduction of I_2, +0.54 V. However, because the reaction that occurs at the anode is the "opposite," the oxidation of 2 I^-, the sign must be reversed and is now −0.54 V, which means the oxidation is *not* thermodynamically favorable. The most negative reduction potential is the reduction of Al^{3+}, which means that the oxidation of Al will be very favorable.

Reaction at the anode	Reduction Potential (V)	E° value for the oxidation (V)
$Zn \rightarrow Zn^{2+}$	**−0.76**	+0.76
$2\ I^- \rightarrow 3\ I_2$	**+0.54**	−0.54
$Al \rightarrow Al^{3+}$	**−1.66**	+1.66

199. (A) Redox reactions are among the most important of the chemical reactions because they are readily exploited to do work. This applies to batteries that power your electronic devices as well as the mitochondria that power your cells. *If the oxidation and reduction reactions were not physically separated from one another, there would be no way to harness the work that could be done by them.* When the two half-cells of a voltaic cell are physically separated except for *one path* through which electrons can flow between them (the wire connecting the anode and the cathode), **the electrons flow** (move in the thermodynamically favorable direction, the direction in which *no work is done on the electrons but work can be done by the electrons*) **from the anode to the cathode producing a current that can do work.**

200. (C) $CaCO_3$ and $Mg(OH)_2$ are both bases, which is why they are often used (with questionable efficacy) as antacids. The basic properties of the compounds are due to the carbonate of $CaCO_3$ and the hydroxides of $Mg(OH)_2$. Because the masses of the two compounds are different and neither of the metal ions are part of the neutralization, an equal number of moles of CO_3^{2-} and OH^- would be the most effective way to compare the efficacy of the compounds. $Mg(OH)_2$ contains 2 hydroxides per mole, however, so doing a "per mole compound" comparison would give $Mg(OH)_2$ an unfair advantage in comparison to the carbonate antacids.

201. (C) The reaction (which goes to completion) produces twice the number of moles of gas, which would double the pressure. The volume of the container is irrelevant to solving this particular problem.

202. (B) Chlorine gas has a mass of 71 g/mole, approximately three times the mass of sodium (23 g/mole). Therefore, equal masses of Na and Cl_2 contain very different numbers of particles. For simplicity, let's use a mass of 71 grams of each substance, but it doesn't matter what mass you choose:

$$71 \text{ g Na} \times \frac{1 \text{ mole Na}}{23 \text{ g Na}} \approx 3 \text{ moles Na}$$

$$71 \text{ g } Cl_2 \times \frac{1 \text{ mole } Cl_2}{71 \text{ g } Cl_2} = 1 \text{ mole } Cl_2$$

According to the reaction, that would leave Na in excess. (See the box after answer 185 for a quick way to identify the limiting reactant.)

$$3 \text{ moles Na} \times \frac{0.5 \text{ mole Cl}_2}{1 \text{ mole Na}} = 1.5 \text{ moles Cl}_2 \text{ needed to completely react with 3 moles Na}$$

or

$$1 \text{ mole Cl}_2 \times \frac{1 \text{ mole Na}}{0.5 \text{ mole Cl}_2} = 2 \text{ moles Na needed to completely react with 1 mole Cl}_2$$

NaCl will be produced; Cl_2, the limiting reactant, will be completely consumed; and 1 mole Na, the reactant present in excess, will remain unreacted.

203. (C) Chlorine gas has a mass of $\frac{71 \text{ g}}{\text{mole}}$, not quite double the mass of potassium $\left(\frac{39 \text{ g}}{\text{mole}}\right)$. The stoichiometry of the reaction requires twice as many potassium atoms as chlorine molecules; however, equal masses of K and Cl_2 will have less than that. If you're not convinced, see the box after answer 185 for a quick way to identify the limiting reactant.

Because this relationship is easier to spot than the one in question 202, you shouldn't need to do the math (but by all means do it if you're anything less than 99% sure!). Because potassium is the limiting reagent this time, KCl will be produced; K, the limiting reactant, will be completely consumed; and some Cl_2, the reactant present in excess, will remain unreacted.

204. (B)

$$2\text{ Al }(s) + 3\text{ Fe}^{2+}(aq) \rightarrow 2\text{ Al}^{3+}(aq) + 3\text{ Fe }(s)$$

$$\text{Al }(s) \rightarrow \text{Al}^{3+}(aq)$$

$$\text{Fe}^{2+} \rightarrow \text{Fe }(s)$$

From cell I: $2\text{ Al }(s) + 3\text{ Cu}^{2+}(aq) \rightarrow 2\text{ Al}^{3+}(aq) + 3\text{ Cu }(s)$ $\quad E°_{CELL} = 2.00\text{ V}$
From cell IV: $\text{Fe }(s) + \text{Cu}^{2+}(aq) \rightarrow \text{Fe}^{2+}(aq) + \text{Cu }(s)$ $\quad E°_{CELL} = 0.78\text{ V}$

If we reverse one of the two reactions and add them, we can eliminate copper. **When you reverse a reaction, reverse the sign of the reaction.** You'll want to choose the reaction you "flip" carefully—the sum of the two reactions should contain the correct oxidation and reduction *and* the sum should add to a positive value for $E°_{CELL}$ (if it doesn't, the cell won't be thermodynamically favorable). **Because the stoichiometry does not affect the cell potential, you can ignore it.**

$$\text{Al }(s) + \cancel{\text{Cu}^{2+}(aq)} \rightarrow \text{Al}^{3+}(aq) + \cancel{\text{Cu }(s)} \qquad E°_{CELL} = 2.00\text{ V}$$
$$\text{Fe}^{2+}(aq) + \cancel{\text{Cu }(s)} \rightarrow \text{Fe }(s) + \cancel{\text{Cu}^{2+}(aq)} \qquad \mathbf{E°_{CELL} = -0.78\text{ V}}$$
$$\mathbf{Al\ (s) + Fe^{2+}(aq) \rightarrow Al^{3+}(aq) + Fe\ (s) \qquad E°_{CELL} = 1.22\text{ V}}$$

205. (C)

$$\text{Zn }(s) + \text{Fe}^{2+}(aq) \rightarrow \text{Zn}^{2+}(aq) + \text{Fe }(s)$$

$$\text{Zn }(s) \rightarrow \text{Zn}^{2+}(aq)$$

$$\text{Fe}^{2+}(aq) \rightarrow \text{Fe }(s)$$

$$
\begin{array}{ll}
Fe^{2+}\ (aq) + \cancel{Cu\ (s)} \rightarrow Fe\ (s) + \cancel{Cu^{2+}\ (aq)} & E^\circ_{CELL} = \mathbf{-0.78\ V} \\
Zn\ (s) + \cancel{Cu^{2+}\ (aq)} \rightarrow Zn^{2+}\ (aq) + \cancel{Cu\ (s)} & E^\circ_{CELL} = 1.10\ V \\
\hline
\mathbf{Zn\ (s) + Fe^{2+}\ (aq) \rightarrow Zn^{2+}\ (aq) + Fe\ (s)} & E^\circ_{CELL} = \mathbf{0.32\ V}
\end{array}
$$

206. (B) Half-cell C contains a strip of Fe *(s)* in 1.00 M $Fe(NO_3)_2$ *(aq)*. In this half-cell, solid iron can be oxidized or iron cations can be reduced. In cell II, Fe^{2+} gets reduced, and in cell IV, it gets oxidized:

Cell II: $2\ Al\ (s) + \mathbf{3\ Fe^{2+}\ (aq)} \rightarrow 2\ Al^{3+}\ (aq) + \mathbf{3\ Fe\ (s)}$
Cell IV: $\mathbf{Fe\ (s)} + Cu^{2+}\ (aq) \rightarrow \mathbf{Fe^{2+}\ (aq)} + Cu\ (s)$

207. (D) Half-cell B contains a strip of Cu *(s)* in 1.00 M $Cu(NO_3)_2$ *(aq)*. In this half-cell, solid copper can be oxidized or copper cations can be reduced. In both cells I and V, Cu^{2+} gets reduced:

Cell I: $2\ Al\ (s) + 3\ \mathbf{Cu^{2+}}\ (aq) \rightarrow 2\ Al^{3+}\ (aq) + 3\ \mathbf{Cu}\ (s)$
Cell V: $Zn\ (s) + \mathbf{Cu^{2+}}\ (aq) \rightarrow Zn^{2+}\ (aq) + \mathbf{Cu}\ (s)$

208. (A) Increasing the concentration of Cu^{2+} will increase voltage in *both* cells because Cu^{2+} is a reactant in both cells:

Cell I: $2\ Al\ (s) + 3\ Cu^{2+}\ (aq) \rightarrow 2\ Al^{3+}\ (aq) + 3\ Cu\ (s)$
Cell IV: $Fe\ (s) + Cu^{2+}\ (aq) \rightarrow Fe^{2+}\ (aq) + Cu\ (s)$

Remember that as a cell functions, the concentrations of reactants decreases and the voltage drops until the concentrations of reactants and products are at equilibrium. The cell is "dead." The Nernst equation allows you to calculate the cell's potential based on concentration. You don't need to know the Nernst equation specifically (see note in italics below), but **it may be helpful to remember the essence of it:**

$$E \propto E^\circ - \log Q$$

E is the potential of a cell under any conditions.
E° is the cell potential under standard conditions.
Q is the reaction quotient.

As the value of *Q* increases, the value of E decreases.

The College Board has excluded calculations involving the Nernst equation from the AP Chemistry exam. However, you do need to be able to qualitatively reason about the effects of concentration on cell potential.

209. (C) Increasing the concentration of Zn^{2+} will increase the voltage in cell W, where Zn^{2+} is a reactant, but decrease the concentration in cell Z, where Zn^{2+} is a product.

Cell W: $2\ Al\ (s) + 3\ Zn^{2+}\ (aq) \rightarrow 2\ Al^{3+}\ (aq) + 3\ Zn\ (s)$
Cell Z: $Zn\ (s) + Fe^{2+}\ (aq) \rightarrow Zn^{2+}\ (aq) + Fe\ (s)$

See answer 208 for an explanation of the effect of concentration on a cell's potential.

210. (D) There is a shortcut to doing calculations involving mass-mass stoichiometry (shown below).

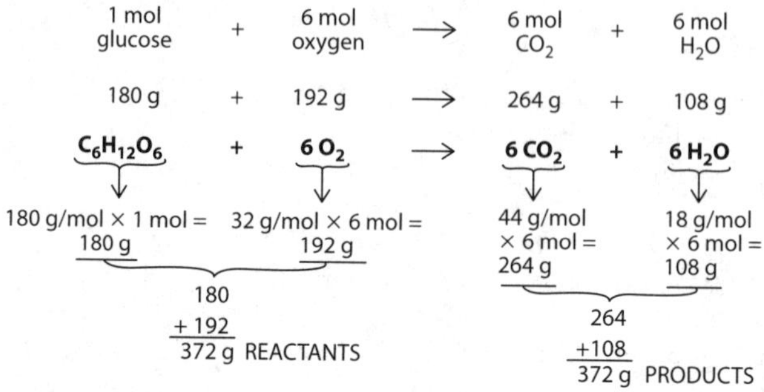

1 mol glucose	+	6 mol oxygen	→	6 mol CO_2	+	6 mol H_2O
180 g	+	192 g	→	264 g	+	108 g
$C_6H_{12}O_6$	+	**$6\ O_2$**	→	**$6\ CO_2$**	+	**$6\ H_2O$**
180 g/mol × 1 mol = 180 g		32 g/mol × 6 mol = 192 g		44 g/mol × 6 mol = 264 g		18 g/mol × 6 mol = 108 g

180 + 192 = 372 g REACTANTS

264 + 108 = 372 g PRODUCTS

Multiply the molar mass of the substances by their stoichiometric coefficients to get the mass ratios of the products and reactants in the reaction, and use the mass ratio instead of the mole ratio in your conversions.

$$1\ \text{g}\ C_6H_{12}O_6 \times \frac{264\ \text{g}\ CO_2}{180\ \text{g}\ C_6H_{12}O_6} = \mathbf{1.47\ g\ CO_2}$$

Choice A is incorrect because the complete oxidation of 180 grams of $C_6H_{12}O_6$ will yield 108 grams of water, but 192 grams of O_2 are required.

Choice B is incorrect because although the mole ratio of oxygen consumed to carbon dioxide produced is the same, the masses of the two substances are different. Carbon dioxide is more massive than oxygen, so 1 gram of carbon dioxide contains fewer molecules than 1 gram of oxygen.

Choice C is almost correct. It appears to represent the conservation of mass in the equation, but without knowing the ratios of the reactants in their combined masses, there's no way to tell how much product will be formed.

For example, if 300 g $C_6H_{12}O_6$ and 72 g O_2 react, then 300 g $C_6H_{12}O_6$ × $\left(\frac{264\ \text{g}\ CO_2}{180\ \text{g}\ C_6H_{12}O_6}\right)$ = 410 g CO_2 should be produced.

That would require:

$$410\ \text{g}\ CO_2 \times \frac{192\ \text{g}\ O_2}{264\ \text{g}\ CO_2} = 298\ \text{g}\ O_2$$

Approaching from the "limited reactant" side:

$$72\ \text{g}\ O_2 \times \frac{264\ \text{g}\ CO_2}{192\ \text{g}\ O_2} = 99\ \text{g}\ CO_2$$

$$72\ \text{g}\ O_2 \times \frac{108\ \text{g}\ H_2O}{192\ \text{g}\ O_2} = 40.5\ \text{g}\ H_2O$$

That's only 140 g of carbon dioxide and water.

There's no need to calculate the product of the mass and molar mass of every substance in the equation if you're looking for only one substance. However, **verifying that the sum of the masses of the reactants is equal to the sum of the masses of the products (conservation of mass) could be a useful check of your math (if you have the time).**

211. **(A)** The amount of ammonia formed will be approximately 20%.

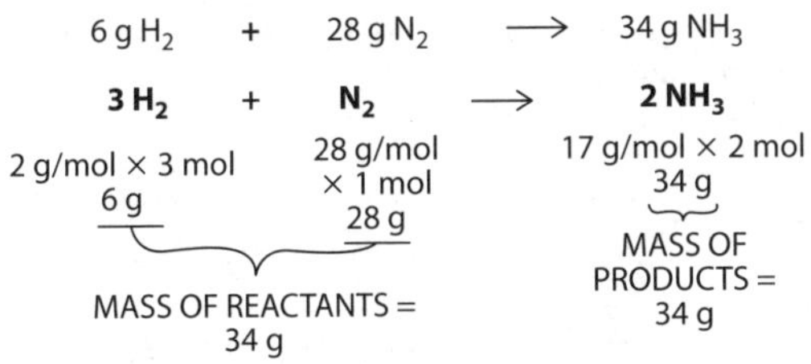

$6\ g\ H_2$	+	$28\ g\ N_2$	⟶	$34\ g\ NH_3$
$3\ H_2$	+	**N_2**	⟶	**$2\ NH_3$**
2 g/mol × 3 mol 6 g		28 g/mol × 1 mol 28 g		17 g/mol × 2 mol 34 g
MASS OF REACTANTS = 34 g				MASS OF PRODUCTS = 34 g

Assume 1 g of H_2 and N_2, but the ratio will be the same no matter what values you use:

$$1\ g\ H_2 \times \frac{34\ g\ NH_3}{6\ g\ H_2} = 5.7\ g\ NH_3$$

$$1\ g\ H_2 \times \frac{34\ g\ NH_3}{28\ g\ N_2} = 1.2\ g\ NH_3$$

1.2 g/5.7 g $\approx$ 21% of the yield if there is no limiting reactant (i.e., the 1 g of H_2 had all the N_2 it needed to react).

An easier way to answer to this question is to start with answer choice B (skip choice A, which clearly requires some time-consuming math). Because you're looking for the answer that is *not* true, you would see that choices B, C, and D are all correct and then you'd know A was incorrect without having to spend the time working it out.

There's also a simple way to consider problems in which only mass comparisons are needed. From the balanced reactions we see that 28 g of nitrogen (1 mole of N_2 @ 28 g mol^{-1}) is needed to react with 6 g of hydrogen (3 moles of H_2 @ 2 g mol^{-1}), so the ratio of N_2 to H_2 is 28:6 or about 4.6:1, so approximately 5× more N_2 (by mass) is needed, meaning that 80% of the reactant's mass should be nitrogen. If nitrogen and hydrogen are provided in equal masses, only 20% of the mass of nitrogen will be consumed and, since it's the limiting reagent, only 20% of the theoretical yield relative to a reaction with no limiting reaction.

212. **(A)** **The synthesis of ammonia involves a change in oxidation state of hydrogen (which gets oxidized) and nitrogen (which gets reduced**; see diagram below). It is not a neutralization reaction (neither acid-base nor charge); however, NH_3 is a weak base, which is how choice B can be misleading.

Choice C is incorrect because the standard enthalpy of formation, $\Delta H_F°$, = −46.11 kJ. The negative value indicates that **the reaction is exothermic**, so the sum of the energy "released" by bond formation exceeds the amount of energy required to break the bonds (remember that **bond formation is always exothermic** and **bond cleavage is always endothermic**).

Choice D is incorrect because there is no hydrogen bonding involved. First of all, the reactants and products are all gases. Ammonia *can* form hydrogen bonds, but in the gaseous state there should theoretically be zero hydrogen bonding (recall Kinetic Molecular Theory). More importantly, hydrogen bonds are weak, interparticle forces of attraction are between molecules, not within small molecules, like hydrogen gas. Hydrogen atoms are held together by covalent bonds in hydrogen molecules.

213. (C) The goal of this problem is to determine which set of reactant combinations has the smallest difference between the excess and limiting reactants.

We know from answer 211 that a 4.6:1 mass ratio of $N_2 : H_2$ will completely react with theoretically no excess reactant (from the balanced equation we know that 28 g of N_2 is needed to completely react with 6 grams of hydrogen gas).

H_2 *(g)* (grams)	**N_2 *(g)* (grams)**	**Mass ratio of N_2 to H_2**
1.0	6.0	**6:1 Too much nitrogen**
2.0	8.0	**4:1 Not enough nitrogen**
3.0	14.0	**4.7:1 Close to the perfect combination!**
4.0	20.0	**5:1 Just a little too much nitrogen**

If you've become accustomed to using the strategy for identifying the limiting reactant as shown in the box below (also included after answer 185), you can use the number of grams given and divide by the amount represented by the stoichiometry in the equation.

Simple Check to Identify the Limiting Reactant (LR) Using Mass

Mass of reactant A ÷ (molar mass of A × stoichiometric coefficient of A) = P_A
Mass of reactant X ÷ (molar mass of X × stoichiometric coefficient of X) = P_X

$P_A < P_X \therefore$ A is limiting reactant

$P_A > P_X \therefore$ X is limiting reactant

$P_A = P_X \therefore$ no limiting reactant (only happens in theory)

H_2 (g, grams)	***Divide by 6 g***	**N_2 (g, grams)**	***Divide by 28 g***	**Divide N_2 (column 4) by H_2 (column 2)**
1.0	***0.17***	6.0	*0.21*	1.2
2.0	*0.33*	**8.0**	***0.29***	0.88
3.0	*0.50*	14.0	*0.50*	1.0
4.0	*0.67*	**20.0**	***0.64***	0.95

The limiting reactants are **bold** (because you divided the nitrogen value by the hydrogen value a fraction less than 1 means that N_2 was limiting and a fraction of greater than 1 shows that H_2 was limiting. The results show that the reactant combinations in choice C, a perfect 1, has no limiting reactant, so that combination of reactant masses will give the greatest yield for the mass of reactants. (The question asked you to determine the most efficient reaction conditions, *not the conditions that would give the greatest amount of product*).

The ratio in the rightmost column tells you just how far from ideal conditions a particular combination of reactants is. Comparisons of choices B and C can illustrate:

Theoretical yields for each reactant for the combination of **2 g H_2 and 8 g N_2**

$$2\text{ g H}_2 \times \frac{34\text{ g NH}_3}{6\text{ g H}_2} = 11.3\text{ g NH}_3$$

$$8\text{ g N}_2 \times \frac{34\text{ g NH}_3}{28\text{ g N}_2} = 9.7\text{ g NH}_3$$

$$11.3 \div 9.7 = \mathbf{1.16}$$

Theoretical yields for each reactant for the combination of **3 g H_2 and 14 g N_2**

$$3\text{ g H}_2 \times \frac{34\text{ g NH}_3}{6\text{ g H}_2} = 17\text{ g NH}_3$$

$$14\text{ g N}_2 \times \frac{34\text{ g NH}_3}{28\text{ g N}_2} = 17\text{ g NH}_3$$

$$17 \div 17 = \mathbf{1}$$

Ratios greater than 1 are useful in this circumstance because there is no need to determine which of the reactants is limiting, so *the higher the number, the greater the disparity between the limiting reactant and the one present in excess*. In calculating the ratios, I divided the larger by the smaller number, regardless of which reactant was represented by that number. However, if the identity of the limiting reactant mattered, the ratio could be used to determine which reactant is limiting and by how much by crunching the numbers in the same order every time.

214. (C) This is a variation of question 213, but in this case, the volumes of gases are used. The first part of the solution is recognizing that **the number of moles of each gas are represented by their volumes**. The stoichiometry of the reaction requires that three moles of hydrogen gas are required for each mole of nitrogen gas, so you need to find which condition has a volume ratio closest to 3:1.

Choices B (2.5:1) and C (3.3:1) are the closest. You can compare how far the deviation is from 3:1 to see which will give you more efficiency ($0.3 < 0.5$, so choice C wins), but let's check to be sure:

Assume the number of liters is equal to the number of moles.

Although this will only be true under certain conditions (see question 215), the assumption is reliable for comparison (only!) purposes.

25 L H_2 and 10 L N_2

$$25\text{ L H}_2 \times \frac{2\text{ L NH}_3}{3\text{ L H}_2} = 16.7\text{ L NH}_3$$

$$10\text{ L N}_2 \times \frac{2\text{ L NH}_3}{1\text{ L N}_2} = 20.0\text{ L NH}_3$$

As expected, hydrogen is the limiting reactant.

50 L H_2 and 15 L N_2

$$50\text{ L H}_2 \times \frac{2\text{ L NH}_3}{3\text{ L H}_2} = 33.3\text{ L NH}_3$$

$$15\ L\ N_2 \times \frac{2\ L\ NH_3}{1\ L\ N_2} = 30.0\ L\ NH_3$$

As expected, nitrogen is the limiting reactant.

215. (B) Use the ideal gas law and substitute 1 for both n and V:

$$PV = nRT$$
$$P(1) = (1)(0.0821)T$$
$$P = 0.0821T$$
$$\therefore \frac{P}{T} = 0.0821$$

You can substitute in the values given, but you can use rounding to limit how many substitutions you need to do.

Round 0.0821 to 0.1 ∴ the P is approximately 10% of the T, eliminating choices A and C. Because you rounded the gas constant up, the real relationship between pressure and temperature is that the P is *less than* 10% of the T, eliminating choice D. Do the math if you're less than 99% confident in your logic:

$$P = 0.0821T$$
$$P = 0.0821(273)$$
$$P = 22.4 \text{ atm}$$

216. (A) The zeroth (0th) law of thermodynamics states that the flow of heat is unidirectional. **Heat moves from a substance of higher temperature to a substance of lower temperature until thermal equilibrium is reached.** In this situation, the student's hand "feels cold" because it is losing heat to the beaker. This may be a tricky thing to imagine because it can appear that the beaker is "getting cold" (remember, **things don't "get cold," they lose heat**). In reality, it is taking heat from the environment to perform chemical and physical processes (dissolution, for example). However, *the sensorial feeling of cold is due to your body losing heat to the beaker and its contents.* Your body lost heat because it was absorbed by the contents of the beaker to perform a reaction in which potential energy of the products is greater than that of the reactants. **An exothermic process loses heat to the environment, so the student's hand would feel warmed by the beaker.**

217. (D) No reaction occurs when NH_4NO_3 *(aq)* is mixed with HCl *(aq)*. **Ammonium nitrate is a completely soluble salt.** The NO_3^- ion is a poor conjugate base (if protonated, it would form HNO_3, a strong acid). The NH_4^+ ion is a weak acid, but there is no good conjugate base present in the solution (Cl^- is a weak conjugate base, like NO_3^-, because their protonated forms are strong acids. Remember, strong acids completely dissociate in solution.

Choice A: HCl *(aq)* and KOH *(aq)* : strong acid + strong base → highly exothermic neutralization ∴ reaction vessel would get warm.
Choice B: $CaCO_3$ *(aq)* and HF *(aq)* : an acid base neutralization that also produces carbon dioxide gas ∴ bubbling would occur.
Choice C: Mg *(s)* and HI *(aq)* : hydrogen gas is produced ∴ bubbling would occur.

When answering a question asked with "not," mark each answer choice as true or false (or yes or no). When you're done with the four choices, the one that's false (or no) is the correct answer.

You don't have to memorize the solubility guidelines and their exceptions for the AP Chemistry exam, but you are expected to know that all sodium, potassium, ammonium, and nitrate salts are soluble in water.

218. (B) The absolute temperature of a substance is proportional to the average kinetic energy (KE) of the particles in the substance. (This relationship is so useful I replace the word temperature with "average kinetic energy" in my mind whenever I am considering problems that involve temperature.) *If the temperature is increasing, the average KE of the particles is increasing.*

During phase changes, the temperature remains constant; therefore, there is no change in KE. Choices A and D are incorrect because **the heat added is used to increase the potential energy of the particles which are changing positions relative to each other during phase changes.** Although lines D and E *could* represent a period of equilibrium if the addition of heat was so infinitesimally small that it could be reversible, the constant addition of heat (as shown during the process continually displaces the solid (or liquid) phase from equilibrium during the melting (or vaporization) process until the transition is complete. You can also reason that **because heat is continually being added, there is net "breakage" of interparticle forces of attraction and an overall increase in entropy of the system,** which can definitely be described as changes.

219. (D) Remember the mnemonic "AN OX! RED CAT," which holds for both voltaic (galvanic) and electrolytic cells.

The anode is made of Zn, so in the process of oxidation, the electrons from oxidized Zn atom move through the wire (diagram 2), leaving the Zn^{2+} ions behind (diagram 4). The positively charged ions repel each other and eventually become dislodged from the solid metal and enter solution (diagram 4). Diagram 1 shows the anode on the left side but a reduction on the right side, both occurring in the same half-cell. The oxidation and reduction reaction must be physically separated in a galvanic cell (see answer 199 for an explanation of why this is true). Diagram 3 shows a Zn anode, but there is no oxidation. There is also a neutral Zn atom being released into solution, which does not occur in the anode (only Zn^{+} ions are released into solution).

The College Board expects you to be able to "zoom in" on processes at the particulate level (where much of chemistry reasoning occurs) while still being able to "zoom out" to the macroscopic level (where we perceive chemistry with our senses).

The College Board will not ask you to label an electrode as positive or negative. In their syllabus they state that "labeling electrodes does not provide a deeper understanding of electrochemistry." Like most of the changes to the AP Chemistry syllabus, this change relieves you of unnecessary rote memorization.

220. (C) The equivalence point is the point at which the number of moles of base (or acid) added equals the number of moles of acid (or base) initially present. It is represented on the titration curve as midway along the steep part of the curve (**the inflection point**). The pH at the equivalence point will be 7 for a strong acid/strong base titration, but will be greater than 7 when titrating a weak acid with a strong base, and less than 7 when titrating a weak base with a strong acid. This is because **the equivalence point *is not* when the number of OH^- ions is equal to the number of H^+ ions in the solution at equivalence;** it is when the number of OH^- ions *added* equals the number of H^+ ions *initially present*, i.e., neutralization. In this case, the acid is weak, so the conjugate base can still be protonated. Recall that the conjugate base of a strong acid is very weak and will not be readily protonated in solution.

Although the addition of a strong base produces OH^- ions in solution that will favor the complete ionization of the acid (Le Chatelier's principle), the anion left behind can still pick up protons from water, decreasing the H^+ (really, H_3O^+) concentration and therefore increasing the pH.

221. (B) The region of the curve between 0 and about 25 mL NaOH added is the **buffer region**. At the lower end of the buffer region (the part where the least volume of base has been added), the amount of weak acid exceeds that of the conjugate base, and toward the end, the conjugate base is present in much greater concentrations than the acid. Midway between them, the concentrations of the weak acid and its conjugate base are about equal.

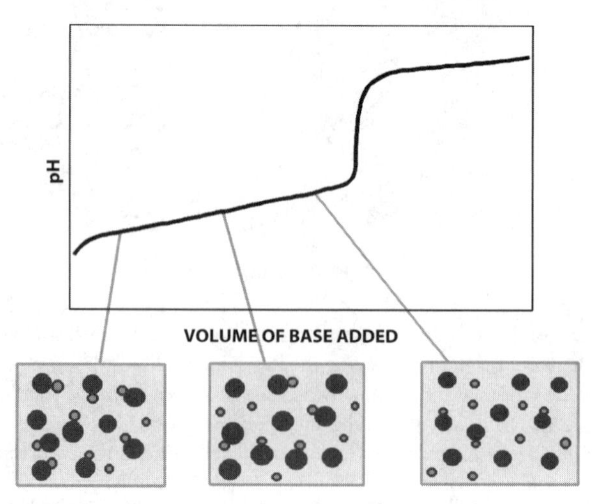

222. (A) A buffer resists changes in pH by acting like a "proton sponge," releasing them into solution when the $[H^+]$ decreases and absorbing them from solution when the $[H^+]$ increases (according to Le Chatelier's principle). A buffer consists of a weak acid or base and the salt of its conjugate base or acid. The region of the curve between 0 and about 25 mL NaOH added is **the buffer region, where the weak acid and its conjugate base are both**

present and very little neutralization has occurred (see diagram above). The weak acid acts as the "releaser" of hydrogen ions and its conjugate base acts as the "absorber."

The term proton sponge *is an example of a simple (and quite crude) model.*

223. (A) Reductions that have a positive $E°$ value are exergonic. Reductions that have a negative E° value have an exergonic reverse reaction (oxidation). The half-reactions at the top of the table of standard reduction reactions, above the reduction of hydrogen, have exergonic reductions. The half reactions at the bottom of the table, below the reduction of hydrogen, have exergonic oxidations (which must be read right to left if considering the oxidation half-reaction). The E° value for the reduction of (1 M) Cl_2 is 1.36 V, while the E° value for the oxidation of (1 M) Br^- is -1.07 V. The difference between them, 0.29 V, is still a positive number, so the redox reaction is favorable ($+\Delta E°$) and reaction can generate electricity.

We can think of this in terms of how much Cl_2 "wants" to get reduced versus how much Br– "doesn't want" to get oxidized (see important note in italics below this answer regarding the use of terms such as want to describe chemical phenomena—they are generally undesirable). The difference in the $\Delta E°$ values quantifies how greatly their "desires" differ and which species will "get what it wants." **Comparing the E° values for the two elements tells us that the ability of Cl_2 to take electrons from Br– is greater than the ability of Br^- to prevent Cl_2 from taking them.**

Imagining and communicating about the behavior of atoms by assigning them "wants" is not particularly helpful to understanding them and should be done sparingly if at all. It relies on a false model with no explanatory or predictive value. Sometimes this makes the language of science seem a little stiff and even cumbersome, but it is usually more important to be accurate. For example, electron affinity can be thought of as how much an atom "wants" to gain an electron, but the real definition is the energy change associated with the addition of an electron to a neutral atom in a gaseous state to form an anion.

224. (B) Reductions occur at the cathode (remember **an**ode = **ox**idation, **red**uction at the **cat**hode, or An Ox, Red Cat; see answer 219 for a visual reminder). The two reactions are written as reductions but only one of them is *actually* occurring as a reduction in the cell. **For a voltaic cell operating under standard conditions** (25° C, 1 M concentrations for solutes and 1 atm pressure for gases), **the cathode reaction has a more positive reduction potential than the anode reaction**. Therefore, Ag^+ *(aq)* + e^- → Ag *(s)* with an E° value of 0.80 V is the reduction reaction. The cathode is made of solid silver, but the half-cell's solution will contain silver ions that pick up electrons from the cathode surface and then adhere to the cathode as solid silver. (This is why the mass of the cathode actually increases as the cell operates.)

225. (C) For a voltaic cell operating under standard conditions, the cathode reaction has a more positive reduction potential than the anode reaction. Therefore, Ag^+ *(aq)* + e^- → Ag *(s)* with an $E°$ value of 0.80 V is the reduction reaction, and Zn *(s)* → Zn^{2+} *(aq)* + 2 e^- is the oxidation reaction (the $E°$ value for the oxidation is +0.76 V). The total cell potential = 1.56 V.

226. (C) Reaction 2, the reduction of Ag^{2+}, has an $E°$ value of +0.80 V. Reaction 1 involves the reduction of Ag^{2+} in the direction written, so you can "remove" reaction 2 from reaction 1, leaving the reaction X *(s)* → X^{2+} *(aq)* and the E° value of (+2.27) − (+0.80) = +1.47 V. However, as written, this is an oxidation. The reaction and the sign of the E° value must be reversed to be written as a standard reduction potential.

227. (D) An ampere (A, amp) is a unit of electric current (*I*). It quantifies the rate at which charge flows through a cross section (or negligible cross-sectional area) of a conductor in coulombs/sec (1.00 ampere = 1 C/sec; this relationship is located in your AP Chemistry reference tables). The charge on *one* proton or electron is $\pm 1.6 \times 10^{-19}$ C (the elementary charge). Note that the values of ±1, 2, 3, etc., that we use for oxidation states and balancing redox reaction refers to *the number of elementary charges* involved, not the actual charge.

A faraday is the magnitude of the electrical charge on one mole of electrons (or protons). The value of Faraday's constant is given in the AP Chemistry reference tables: Faraday's constant = **96,500** coulombs per mole of electrons. If you calculate $1.6 \times 10^{-19} \times 6.02 \times 10^{23}$ you'll get a value of 96,320 because the numbers 1.6 and 6.02 are rounded from 1.602 176 565 and 6.022 141 29, respectively.*

The oxidation state of iron in $FeCl_3$ is +3 so it will require 3 moles of electrons to reduce 1 mole of Fe^{3+}. The charge on 3 moles of electrons:

$$\frac{96{,}500 \text{ C}}{\text{mole of electrons}} \times (3 \text{ moles of electrons}) = 289{,}500 \text{ C required}$$

Delivered at a rate of 1.00 C/sec, that would require 289,500 seconds (more than 80 hours!).

228. (B) The periodic table is your best tool for predicting electron configurations (see note in italics below), but doing an electron count provides a good check and allows you to eliminate some answer choices quickly. The Zn^{2+} contains 28 electrons (not 30, like a neutral Zn atom), eliminating choice C. **Only s and p electrons are considered valence electrons**, so the s and p electrons of the highest *n*-value will be those lost during oxidation processes. In the case of Zn, the $4s^2$ electrons are the "outermost" and will be removed in the formation of the ion.

The College Board does not want you to memorize the exceptions to the Aufbau principle, that is, the electron configurations that do not follow the "expected" trends. If you've never heard of this, don't worry. It may help to be aware that there are several elements whose electron configurations are exceptions to the Aufbau principle, however, as you may be given an exception and asked to provide possible reasons based on the current atomic model (see the box following answer 8, specifically, the two items listed under Coulomb's law).

229. (D) The number of positive charges in the nucleus of Zn^{2+} is still 30, but now those 30 protons are pulling on two fewer electrons (28 instead of 30), which causes them to get pulled in closer to the nucleus. Although you are also "breaking into the noble gas configuration" of electrons, that is not a choice.

*Many of the constants we often take for granted are continuously being refined. To learn more, visit the Fundamental Physical Constants site from the National Institute of Standards and Technology (NIST): http://physics.nist.gov/cuu/Constants/index.html.

230. (A) Oxidations occur at the anode (see answer 219 for a visual reminder of An Ox, Red Cat). The description of the cell states that it is galvanic, so you know the E° value must be positive. Because the reduction of hydrogen, by definition, has an *E*° value equal to zero, Zn must be getting oxidized (so the total cell potential is +0.76). You may have recognized that the reaction $2\ H^+ + 2\ e^- \rightarrow H_2\ (g)$ is the "dividing line" between thermodynamically favorable and non-thermodynamically favorable reductions on the **table of standard reduction potentials**. The *E*° value of 0.00 V indicates that it is the reference electrode (standard hydrogen electrode, or SHE).

Another way to consider this problem is that for galvanic cells, the cathode reduction potential will be more positive than the anode reduction reaction.

231. (C) The **salt bridge** (solution of KNO_3) functions to **provide ions to the half-cells to maintain electrical neutrality**, extending the life of the battery. At the anode, the oxidation of solid Zn produces Zn^{2+} ions in solution. The salt bridge provides NO_3^- ions to allow the neutralization of the excess positive charges. At the cathode, the reduction of the cation in solution to a neutral atom removes positive charges from solution. The salt bridge provides K^+ cations into the cathode half-cell solution to neutralize excess negative charges. This is how cations and anions got their names!

At the cathode, H^+ ions are being reduced to H_2 gas, so the salt bridge provides K^+ ions to replace the lost cations, neutralizing the excess negative charges (Cl^- ions) in the solution.

232. (D) The values on the table of standard reduction potentials are measurements of potentials using the standard hydrogen electrode (SHE) as a reference electrode under standard conditions. Physically, the SHE is a (cathode) compartment of H_2 gas at 1 atm of pressure. The compartment interior is connected to the cell solution (the wet part of the wet cell) by platinum and it is on this platinum surface that the reduction of hydrogen ions to hydrogen gas occurs (see diagram below).

The standard concentration for solutions is 1 M, so this solution must have a pH of zero. This is usually accomplished by using a strong acid. NaCl and $ZnSO_4$ (choices A and C) provide no hydrogen ions, so they are not at all suitable for this cell. A solution of HF could theoretically replace the HCl solution, but a higher concentration of HF must be used to create a 1 M solution of hydrogen ions, so it is not ideal. HNO_3 is a strong acid and NO_3^- ions are commonly used in electrochemical cells. All ionic compounds of nitrate are soluble (one of the few solubility "rules" you should know), so there are no metallic electrodes that would produce precipitates in it.

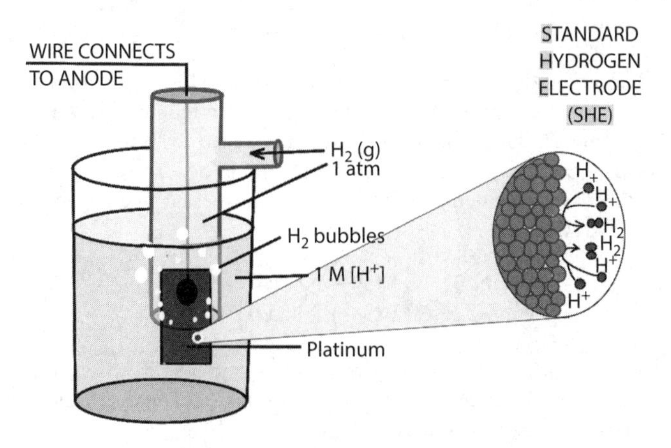

233. (D) The Nernst equation relates half-cell concentrations to the potential of the cell (one form of the equation is $\mathbf{E} = \mathbf{E}° - \left(\frac{\mathbf{0.0592\ V}}{\boldsymbol{n}}\right) \mathbf{log}\ \boldsymbol{Q}$ where n is the number of moles of electrons transferred in the reaction and Q is the reaction quotient (see note in italics below). As the cell operates, the flow of electrons from the anode to the cathode results in the consumption of reactant and the formation of product, which decreases reactant concentration (H^+) and increases product concentration (Zn^{2+}).

Increasing the concentration of $ZnSO_4$ has the effect of increasing product concentration. As the cell approaches equilibrium, Q approaches K_{EQ} and E approaches 0.

The College Board has excluded calculations involving the Nernst equation from the AP Chemistry exam. However, you do need to be able to qualitatively reason about the effects of concentration on cell potential.

234. (A) Zinc is referred to as the sacrificial anode (also called a galvanic anode). The sacrificial anode has a more negative standard reduction potential (therefore a more positive, or thermodynamically favorable, oxidation potential) than the metal it is meant to protect, which allows the sacrificial anode (in this case, Zn) to be oxidized preferentially to the cathode (in this case, Fe).

235. (B) Electrical energy (J) = electrical potential (V) × charge (C). This relationship is located in the AP chemistry reference tables. The sum of the potentials of the two half-cells (or the difference between the standard reduction potentials of the two cells) determines the maximum voltage of the cell, not how much energy the cell can generate (why choice A is incorrect).

You can use dimensional analysis to deduce the relationship or to check that a proposed relationship makes sense. The unit of energy is joules, the unit of potential is V, which is $\frac{J}{C}$, and the unit of charge is C. The product of potential $\left(\frac{J}{C}\right)$ and charge (C) gives the unit of J $\left(\frac{J}{C} \times C = J\right)$. The difference between voltages, for example, can only be a voltage (why choice D is incorrect). Choice C just doesn't make any sense.

236. (C) Na_2CO_3 is a very soluble salt (thanks to the Na^+; see note in italics below) and the CO_3^{2-} it produces in solution is a weak base. It will pick up protons to form HCO_3^- and/or H_2CO_3 (depending on how much Na_2CO_3 was added to the solution). When the CO_3^{2-} solutions is acidified, CO_2 gas is produced by the following chemical equation:

$$H^+\,(aq) + CO_3^{2-}\,(aq) \leftrightharpoons H^+(aq) + HCO_3^-\,(aq) \leftrightharpoons H_2CO_3\,(aq) \leftrightharpoons CO_2\,(g) + H_2O\,(l)$$

All of the compounds listed are white and crystalline. (Most compounds in which the only metal is a group 1 or group 2 metal are white and crystalline.) NaOH is a strong base and would certainly produce a basic solution when added to water, but when an HCl is added, the result is water and a neutral salt; no gas would be formed.

$NaNO_3$ and Na_2SO_4 are neutral salts. To determine the acid and base combination that formed them, "add water":

$$NaNO_3 + H_2O \rightarrow Na\mathbf{OH} + \mathbf{HNO_3}$$
$$Na_2SO_4 + 2\,H_2O \rightarrow 2\,Na\mathbf{OH} + \mathbf{H_2SO_4}$$

The result, in both cases, is a strong base and a strong acid (the conjugates of these are weak). These salts would *not* produce basic solutions when added to water because the NO_3^- and SO_4^{2-} are very weak conjugate bases.

You don't have to memorize the solubility guidelines and their exceptions for the AP Chemistry exam, but you are expected to know that all sodium, potassium, ammonium, and nitrate salts are soluble in water.

237. (C) The graph shows the titration of a weak base with a strong acid (the pH starts at a high pH, indicating that a base is being titrated). HCl is a strong acid, and since the pH at the equivalence point is <7, you know the base being titrated is weak. Methyl red is the best indicator listed for this titration because it **changes color with the pH range of the equivalence point of the titration**, approximately 5. Methyl red undergoes a color change between pH 4.4 and 6.2.

238. (A) The region of the curve between 0 and about 20 mL HCl added (between pH 11 and 8) is the **buffer region** of this titration, where the weak acid and its conjugate base are both present and little neutralization has occurred. At the lower end of the buffer region (the part where the least volume of acid has been added), the amount of weak base exceeds that of the conjugate acid and toward the end, the conjugate acid is present in much greater concentrations. Midway between them, the concentrations of the weak base and its conjugate acid are about equal.

In this titration, phenolphthalein would remain pink in the "lower end" of buffer region and transition to clear by the end of it, when the concentration of conjugate acid has exceeded the concentration of weak base, decreasing the pH of the solution sufficiently for the change in indicator color to occur.

239. (C) The purpose of collecting a gas over water is to keep the gas at a constant pressure as more gas particles are collected. Gases will expand or compress to fit their container, so changes in volume and/or pressure during collection make measurements challenging.

A **eudiometer, a device used to collect gas over water**, works by displacing water with the gas being collected. The apparatus works by equilibrating the gas pressure with the atmospheric pressure (the gas pushing down on the surface of the water in the eudiometer is the same as the pressure of the atmosphere pushing the water up into the eudiometer at equilibrium).

240. (D) HCl, NH_3, and SO_2 are *very* water soluble gases. HCl is a strong acid that completely dissociates in aqueous solution. NH_3 is also very water soluble, because of its ability to hydrogen bond with water and its trigonal pyramidal shape. SO_2 is bent, or V-shaped, and so it is also very water soluble. A large fraction of these gases will dissolve in the water they must pass through to be collected, greatly reducing the yield.

CO_2 is slightly water soluble, and it can react with water to form carbonic acid (H_2CO_3), but only a relatively small fraction will be lost to its dissolving in (and reacting with) water compared with HCl, NH_3, and SO_2.

241. (B) The gas collected in the eudiometer is actually a mixture of two gases—the gas intentionally collected (in this case, H_2) and water vapor that evaporated in the eudiometer. **The partial pressure of the water vapor is determined solely by the temperature.** At 22° C, the temperature at which the experiment was performed, the vapor pressure of water is 19.8 torr. The total pressure of the gases is 770 torr (the atmospheric pressure). Dalton's law of partial pressures states that the total pressure of a mixture of gases is the sum of the partial pressures of each of the gases in the mixture $\therefore 770 - 19.8 =$ **750.2 torr.**

242. (C) The volume of H_2 collected is indicated by the graduations on the eudiometer (100 mL or 0.1 L).

Calculate the number of moles.

$$PV = nRT$$
$$(750 \text{ torr})(0.1 \text{ L}) = (n)(62.4 \text{ L torr mol}^{-1} \text{ K}^{-1})(295)$$
$$n = \mathbf{0.004}, \text{ or } \mathbf{4 \times 10^{-3} \text{ moles}}$$

Calculate the mass of H_2 in grams.

$$4 \times 10^{-3} \text{ mol} \times 2 \text{ g/mol} = \mathbf{8 \times 10^{-3} \text{ g}}$$

Chapter 3: Free-Response Answers

243. Balanced equation: **1** C_5H_{12} *(l)* + **8** O_2 *(g)* → **6** H_2O *(l)* + **5** CO_2 *(g)*

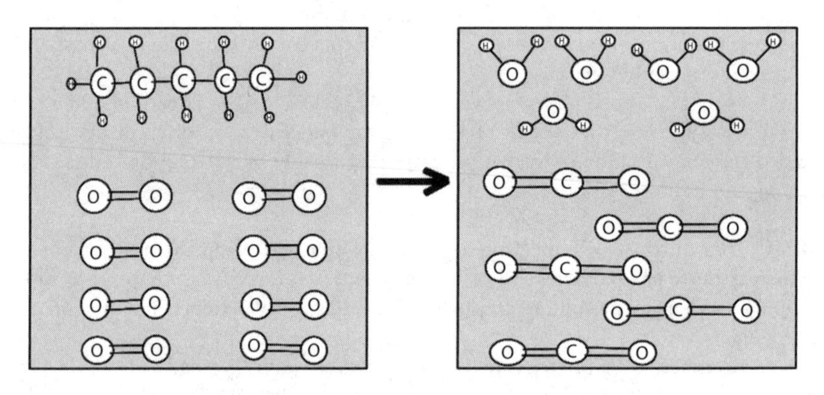

244. Your answer should contain the following information but does not have to be formatted this way.

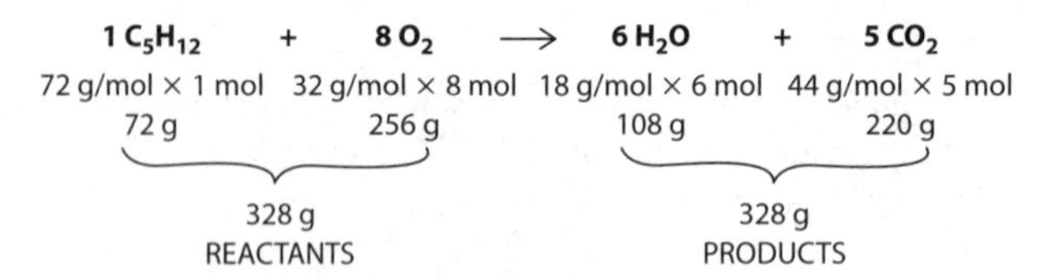

245. There are five significant figures in the values given in the question, so you should have five significant figures in your calculations and answers.

Molar mass of pentane: 72.150 g/mole
Molar mass of carbon dioxide: 44.009 g/mole

Using the **mass-mass method**:

$$100.00 \text{ g } C_5H_{12} \times \frac{220.05 \text{ g } CO_2}{72.150 \text{ g } C_5H_{12}} = \mathbf{304.98 \text{ g } CO_2}$$

or

Using the **mass → mole → mass method**:

$$100.00 \text{ g } C_5H_{12} \times \frac{1 \text{ mole}}{72.150 \text{ g}} = 1.3860 \text{ moles } C_5H_{12}$$

$$1.3860 \text{ moles } C_5H_{12} \times \frac{5 \text{ moles } CO_2}{1 \text{ mole } C_5H_{12}} = 6.9300 \text{ moles } CO_2$$

$$6.9300 \text{ moles } CO_2 \times \frac{44.009 \text{ g}}{1 \text{ mole}} = \mathbf{304.98\ g\ CO_2}$$

246. $$100.0 \text{ g } C_5H_{12} \times \frac{1 \text{ mole}}{72.15 \text{ g}} = 1.386 \text{ moles } C_5H_{12}$$

$$1.386 \text{ moles } C_5H_{12} \times \frac{-3{,}509 \text{ kJ}}{\text{mole}} = \mathbf{-4{,}863\ kJ}$$

247. From answer 245, the mass of CO_2 produced by the combustion of 100.0 grams of pentane with excess oxygen is 305.0 grams (the 100.0 gram masses limits the number of significant digits in our answer to four).

Determine the mass of CO_2 produced by the combustion of pentane in 100.0 grams of O_2 (assume excess pentane).

Using the **mass-mass method**:

$$100.0 \text{ g } O_2 \times \frac{220.0 \text{ g } CO_2}{255.8 \text{ g } O_2} = \mathbf{86.00\ g\ CO_2}$$

or

Using the **mass → mole → mass method**:

$$100.0 \text{ g } O_2 \times \frac{1 \text{ mole}}{31.98 \text{ g}} = \mathbf{3.127\ moles\ O_2}$$

$$3.127 \text{ moles } O_2 \times \frac{5 \text{ moles } CO_2}{8 \text{ moles } O_2} = 1.954 \text{ moles } CO_2$$

$$1.954 \text{ moles } CO_2 \times \frac{44.01 \text{ g}}{1 \text{ mole}} = \mathbf{86.00\ g\ CO_2}$$

248. 70.08 L oxygen consumed − 43.80 L carbon dioxide produced = 26.28 L difference

See answer 247 for calculating the number of moles of oxygen consumed and carbon dioxide produced.

Moles of $\mathbf{O_2}$ consumed: **3.127**

$PV = nRT$

$(1)(V) = (3.127)(0.0821)(273)$

$V =$ **70.08 L**

Moles of $\mathbf{CO_2}$ produced: **1.954**

$PV = nRT$

$(1)(V) = (1.954)(0.0821)(273)$

$V =$ **43.80 L**

249. ***All* combustion reactions are oxidation-reduction (redox) reactions.** Assigning oxidation numbers to the carbons can be tricky because the oxidation number will be different for carbon atoms that have different bonding partners. For example, the 2 carbons on the end each have 3 hydrogen atoms attached, whereas the 3 carbon atoms in the middle have only 2 hydrogen atoms bonded. The carbon atoms with 3 hydrogen atoms bonded to them will have a more negative oxidation state than those having only 2 hydrogen atoms bonded to them. **Do not take oxidation numbers literally!** They are more of an accounting tool for electrons than a representation of physical reality. The illustration below is for reference only; you do not need to include it in your answer.

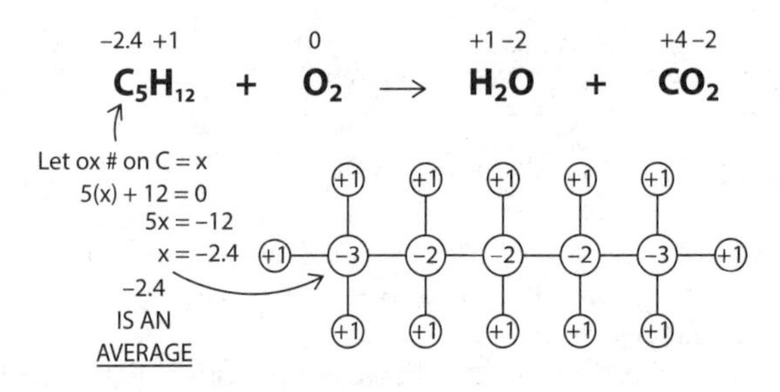

250. The cathode is Ag and the anode is Cu.
From the table of Standard Reduction Potentials:

$Ag^+ + 1\ e^- \rightarrow Ag\ (s) \quad E° = 0.80$ V
$Cu^{2+} + 2\ e^- \rightarrow Cu\ (s) \quad E° = 0.34$ V
(You know it's Cu^{2+} and not Cu^{+1} because the solution contains Cu^{2+}.)

The standard reduction potential is a measure of the driving force for a particular reduction under standard conditions (1M concentrations, 25° C). **The more positive the *E*° value of the reduction potential, the greater the driving force.** For a galvanic cell operation under standard conditions, the reaction at the cathode must have a more positive value of $E°_{RED}$ than the reaction at the anode. **It is the reduction at the cathode that drives the oxidation at the anode.** This fact can be used to **identify the cathode: it will be the more positive of the two *E*° values.**

251. $E°_{CELL} = 0.46$ V
The $E°_{CELL}$ is the difference between the reduction potentials:

$Ag^+ + 1\ e^- \rightarrow Ag\ (s) \qquad E° = 0.80$ V
$Cu^{2+} + 2\ e^- \rightarrow Cu\ (s) \qquad -E° = 0.34$ V

$$\mathbf{E°_{CELL} = 0.46\ V}$$

. . . or the sum of the two half-cell potentials:

Reduction (at the cathode):

$Ag^+ + 1\ e^- \rightarrow Ag\ (s) \qquad E° = 0.80$ V

Oxidation (at the anode):

$Cu\ (s) \rightarrow Cu^{2+} + 2\ e^- \qquad E° = -0.34$ V

$$\mathbf{E°_{CELL} = 0.46\ V}$$

252. The cathode gains mass (another way to identify it) and **the anode loses mass**. You can use the mnemonic "FAT red cat" to remind you that the cathode gains mass. The mass gained by the cathode comes from the Cu^{2+} ions in solution, which, when reduced by the excess electrons on the cathode, are deposited on the cathode. This is sometimes referred to as being "plated out" (and is the basis for electroplating, an electrolytic process).

The anode loses mass as the Ag ions, once oxidized, become positive and repel each other. They become dislodged from the anode and enter the cell solution.

253. Your illustration should look like this. Be sure to use the appropriate labels.

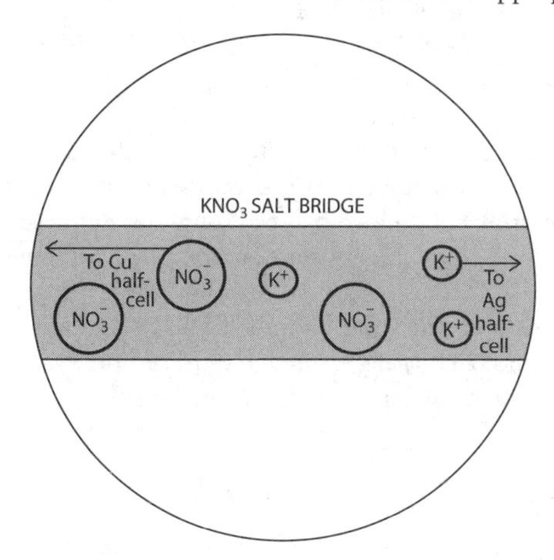

254. For a galvanic (voltaic) cell to work, the two solutions in which the electrodes are bathed must remain electrically neutral. In the case of the Cu anode, Cu^{2+} ions are being released into the solution, so NO_3^- ions from the salt bridge will neutralize the excess positive charges. The Ag cathode is taking Ag^+ ions out of solution (they are being deposited onto the cathode as explained in answer 252), leaving an excess of negative charges that can be neutralized by the K^+ from the salt bridge. **Generally, anions from the salt bridge migrate toward the anode and cations from the salt bridge migrate toward the cathode.**

255. Yes. The ions in the salt bridge must not react with the ions in the electrolytic solutions of the electrodes. **A solution of KCl would release K^+ to the cathode and Cl^- to the anode, and neither would react.** However, if Ag were the anode (not the cathode, as in this cell) the Cl^- ions that would migrate into the cell from the salt bridge would form an insoluble precipitate, AgCl, which would impair the functioning of the cell.

256. The potential would be **equal** because the concentrations of the solutions are equal to each other. We can consult the Nernst equation for some qualitative reasoning on this topic (see note in italics below). The essence of the equation is $E \propto E° - \log Q$. As a voltaic cell is discharged, the reactants are consumed and products are produced, so concentrations in the cell change. The electromotive force (emf) continues to drop until equilibrium is reached and no more work can be done. A simple way to think about it is as follows:

In the (unbalanced) reaction $Cu\ (s) + Ag^+\ (aq) \rightarrow Cu^{2+}\ (aq) + Ag\ (s)$

The **reactants** are Cu *(s)* and **Ag^+** *(aq)*.
The **products** are **Cu^{2+}** *(aq)* and Ag *(s)*.

Remember that for Q and K calculations, pure solids and liquids are not included (or they are equal to 1).

$$Q = \frac{[Cu^{2+}]}{[Ag^+]}$$ **= 1 whether it is 1 M concentration or a 0.6 M concentration.**

Increasing the concentration of reactants or decreasing the concentration of the products would increase the value of Q and result in a higher emf. Generally, **differences in the concentrations between the two half-cells will produce changes in emf.**

The College Board has excluded calculations involving the Nernst equation from the AP Chemistry exam. However, you do need to be able to qualitatively reason about the effects of concentration on cell potential.

257. The standard cell would not exhaust the reactants (reach equilibrium) as quickly as the nonstandard cell with lower concentration, so the standard cell would last longer. If the concentrations of the two half-cell solutions are the same, the same emf will be generated but the cell with the higher concentration will last longer.

258. $Cu\ (s) + 2\ Ag^+\ (aq) \rightarrow Cu^{2+}\ (aq) + 2\ Ag\ (s)$

The reaction is thermodynamically favored in the direction written because the $E°_{CELL}$ is a positive value (0.46 V, calculated in answer 251), making the reaction thermodynamically favorable.

259. 88.8 kJ/mol$_{RXN}$

$\Delta G° = -nFE°$
$\Delta G° = -(2\ \text{mol e}^-/\text{mol}_{RXN})(96{,}485\ \text{C/mol e}^-)\ (0.46\ \text{J/C})$*
$\Delta G° = 88{,}766\ \text{J/mol}_{RXN}$
$\mathbf{\Delta G° = 88.8\ kJ/mol_{RXN}}$

260. No. Although the total reaction would still proceed thermodynamically favorably, there can be no work done by the cell (though it would produce heat). Physical separation of the oxidation and reduction reactions forces the flow of electrons through a wire (or external circuit), which can accomplish electrical work. If there is no separation of the cells, there is no way to harness the work it can do.

Chapter 4: Kinetics

261. (C) Chemical reactions typically occur at surfaces and interfaces. Grinding up the solid increases the surface area of a solid by decreasing the particle size. Increasing the surface area increases the number of surface particles accessible to react. In the case of a tablet, the particles in the center of the tablet are not accessible to react until they have been exposed by the reactions consuming all the particles surrounding them.

262. (A) Collisions are required for chemical reactions to occur (at least according to the collision model of chemical reactions). A **unimolecular** process does not involve a second molecule to undergo a reaction. The particle, in a sense, "rearranges itself" (likely by colliding

*A volt is 1 J/C (joule per coulomb)

with other particles of the same species, but there is *not just one mechanism* to describe all unimolecular reactions). Examples of unimolecular reactions are **radioactive decay, thermal decomposition, racemization, the opening and closing of ring structures, and cis-trans isomerism.** They typically display first-order kinetics which means that the reaction rate is dependent on the concentration.

263. (D) The equilibrium expression uses the stoichiometric coefficients, but the reaction orders of the rate law are experimentally determined. The stoichiometric coefficients can be used to make hypotheses about the molecularity of the elementary steps of the overall reaction, which can then be used to propose a reaction mechanism and even a rate law, but the reaction mechanism and the hypothetical rate law must be experimentally verified.

Reaction mechanisms are difficult to prove; however, **if the mechanism is known, the rate law can be determined from it.**

264. (B) There may be several "elementary" reactions that, if correctly arranged, "add up to" the actual reaction. These elementary reactions are considered the individual steps of a chemical reaction. **To be elementary, the process must occur in a single step.** Choice A is incorrect because if the reaction were to proceed by one elementary step involving two particles (bimolecular), it would have a reaction order of 2 (why choice C is incorrect, too).

Elementary reactions that require two molecules colliding are **bimolecular**. A possible mechanism for the reaction looks like this:

Reaction	Molecularity
$\mathbf{Br_2} \rightarrow Br + Br$	(unimolecular)
$Br + \mathbf{H_2} \rightarrow \mathbf{HBr} + H$	(bimolecular)
$H + Br_2 \rightarrow Br + \mathbf{HBr}$	(bimolecular)
$Br + Br \rightarrow Br_2$	(bimolecular)
$H_2 + Br_2 \rightarrow 2\ HBr$	

The rate law for the reaction $H_2\ (g) + Br_2\ (g) \rightarrow 2\ HBr$ suggests a multistep mechanism for the reaction. The overall order for this reaction is 1.5, according to the student's rate law. Choice D is incorrect because rate laws can have fractional and even negative reaction orders.

The only way a reaction order or rate law can be known with certainty is through experimentation. Only if a reaction mechanism is known *beyond a shadow of a doubt* can the elementary steps provide the molecularity data to construct a rate law. The rate laws written from proposed reaction mechanisms can be used to test the proposed mechanism, but they are not considered true rate laws until the trial data confirm their predictions.

265. (C) The slow step is typically the rate-determining (also called rate-limiting) step of a multistep process. The overall rate of the reaction is usually equal to the rate of the rate-determining step. That makes **identifying the rate-determining step an important strategy for deducing rate laws from reaction mechanisms**. To write a rate law for this reaction, it is important to identify that the $[N_2O_2]$ is directly proportional to $[NO_2]^2$ (choice C, the answer). This allows you to substitute the intermediate (N_2O_2) with reactants or products from the original equation, which is necessary to deduce the rate law.

The concentration of the intermediate (N_2O_2) is *not* directly proportional to $[NO]^2$ because step 1 is in equilibrium, so the concentrations of N_2O_2 and NO are not proportional to each other in a positive way. "Pushing" the equilibrium to the right increases the concentration of N_2O_2 but results in a decreased concentration of NO, for example.

To understand why the concentration of N_2O_2 is proportional to $[NO_2]^2$:

Step 2 is bimolecular, so the proposed rate law is:

$$\textbf{Rate of NO}_2\textbf{ formation} = k_2\,[\mathbf{N_2O_2}][\mathbf{O_2}]$$

For step 1 [the prime (′) indicates the reverse reaction]:

$$\text{Rate}_{\text{forward rxn}} = k_1\,[NO]^2$$
$$\text{Rate}_{\text{reverse rxn}} = k_1'\,[N_2O_2]$$

At equilibrium, the rates of the forward and reverse reactions are equal ∴

$$k_1\,[NO]^2 = k_1'\,[N_2O_2]$$

Solve for $[N_2O_2]$:

$$[N_2O_2] = \left(\frac{k_1}{k_1'}\right)[NO]^2$$

This is the step where you've answered the question, that the **$[N_2O_2]$ is directly proportional to $[NO_2]^2$**.

But you can complete the rate law by substituting the intermediate (N_2O_2) with reactants or products from the original equation into the rate law for step 2:

$$\text{Rate of NO}_2\text{ formation} = k_2\,[N_2O_2][O_2]$$

$$\text{Rate of NO}_2\text{ formation} = k_2\left(\frac{k_1}{k_1'}\right)[NO]^2[O_2]$$

If you were writing the rate law, you'd want to keep all the rate constants, but that is not needed here, so you can simplify:

$$\text{Rate of NO}_2\text{ formation} \propto [NO]^2[O_2]$$

See answer 280 for a procedure to determine the rate law from a reaction mechanism when the intermediate is formed in the slow step.

266. (A) The reaction is third order (see answer 265). The rate law states that the rate of NO_2 formation is proportional to $[NO]^2[O_2]$. Choice B is incorrect because the reactant orders add up to 3, the overall reaction is considered third order. The first elementary step is bimolecular, since 2 molecules of NO are needed to collide to react. (Since 2 particles are required, they both end up in the product, i.e., it's not *just* the collision that matters, it's the collision and the incorporation of both of the colliding particles into the product, unlike the reaction is question 262.)

267. (C) The reaction order is zero. The two graphs collectively show that the concentration of ammonia has no effect on the reaction rate (until it is completely consumed). The rate law is as follows:

$$\text{Rate} = k\,[NH_3]^0 \;\therefore\; k = \text{the rate}$$

The rate of the appearance of products is twice as fast as the disappearance of the reactants as given by the stoichiometry of the reaction: 2 moles of reactants → 4 moles of products.

268. (D) Generally, increasing the surface area of a solid will increase its reaction rate. Chemical reactions typically occur at surfaces and interfaces, so increasing the surface area of a solid increases the number of particles accessible to react. Breaking up a solid doesn't affect its concentration (choice C), although concentration isn't a term typically applied to solids. Choice A may address the surface-area issue (by using the term *exposure*), but the iron needs to react with *sulfur*, not oxygen, and the oxidation state of iron should be 0 to react with sulfur, since it will lose electrons to reduce sulfur. If the iron were partially oxidized, it would be much less effective at reducing sulfur. Choice B is incorrect because pure, solid iron is typically in the reduced state (Fe, not Fe^{2+}) whether it is in the form of filings or a large block (or any shape).

269. (B) Reactant orders can be positive, negative, or fractional. A negative reactant order indicates that the reaction rate decreases with an increased concentration of a particular substance.

A proposed reaction mechanism for the destruction of ozone by chlorine in the stratosphere is as follows:

Step 1: $Cl\ (g) + O_3\ (g) \rightarrow ClO\ (g) + O_2\ (g)$
Step 2: $ClO\ (g) + O\ (g) \rightarrow Cl\ (g) + O_2\ (g)$

Although you may not be accustomed to a rate law containing the product of a reaction, the rate law for ozone destruction does because the concentration of the product of the reaction has an effect on the reaction rate. These reactions are rare, most rate laws contain only reactants.

270. (D) Increased temperature increases the reaction rate of almost all chemical reactions, whether they are exothermic or endothermic (except for the few reactions that have negative activation energies). Higher temperatures result in more *effective collisions* (collisions with enough energy and with the proper orientation or reactants relative to each other) because increasing the average kinetic energy of the particles in the reaction mixture increases the percentage of particles that will attain an energy that meets or exceed the *activation energy* of the reaction, the energy barrier that must be overcome in order for the reaction to proceed.

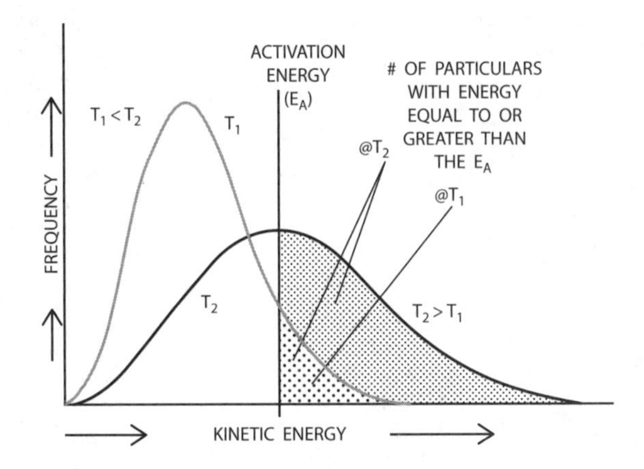

271. (C) You can think of the activation energy of a reaction as the quantity of energy needed for the formation of the transition state of the reaction. It is considered an energy barrier to the reaction's progression. The spark of the lighter (or spark plugs in a combustion engine) produces the heat needed to get the combustion going by supplying the energy needed by the reactants to form the transition state required to form the products (in other words, the activation energy). The combustion of the butane is highly exothermic, so once the reaction starts, the energy it liberates supplies the activation energy for the other butane molecules in the lighter.

You probably don't go to bed at night worrying that your chemistry textbook will turn into carbon dioxide and water while you sleep. The conditions present in your house don't (usually) provide the activation energy needed to start the process. But if you applied the heat of the lighter flames to it (or, if we could get the spark to directly catch one of the pages), you could easily turn your book into carbon dioxide and water (which is not recommended until *after* the AP exam).

272. (D) Increasing the pressure only shifts the equilibrium (and thereby momentarily changes the rate of either the forward or reverse reaction) in a reaction that involves gases, and even then, the number of gaseous moles of reactant and product must differ or the change in pressure has no effect on the system. **Increasing the temperature at which a reaction occurs will almost always increase the rate, whether it is exothermic or endothermic.**

When answering a question asked with a "not," mark each answer choice as true or false (or yes or no). When you're done with the four choices, the one that's false (or no) is the correct answer.

273. (A) This question is tricky. Choice A is *almost* correct—each radioactive species *does* has a particular half-life—but a particular element can have *several* different isotopes, each with its own half-life. All of the other statements are true.

When answering a question asked with a "not," mark each answer choice as true or false (or yes or no). When you're done with the four choices, the one that's false (or no) is the correct answer.

274. (D) The activation energies (E_A) for forward and reverse reactions are typically different. Activation energy is the energy difference between the reactants and the activated complex. If the forward reaction is exothermic, the E_A = energy of activated complex – energy of reactants. For the reverse, endothermic reaction, the activation energy includes the enthalpy change of the reaction (see diagram below). In other words, the E_A of the reverse reaction = (E_A *exothermic* + ΔH). If the E_A of the forward and reverse reactions are the same, the ΔH of the reaction must be zero.

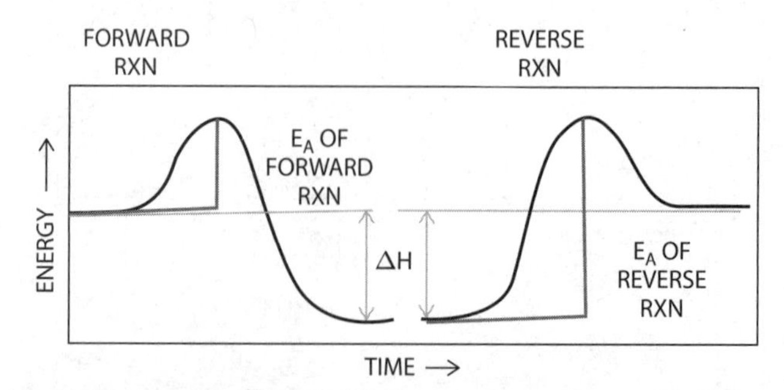

275. (B) A catalyst does not differentiate between the forward and reverse reactions. The activated complex is the same whether it was formed by the forward or reverse reaction, so decreasing the energy of the activated complex will lower the activation energy in both directions. Because a catalyst increases the rates of both forward and reverse reactions equally, **the presence of a catalyst does not change the K_{EQ} for a reaction but it does decrease the time it takes to reach equilibrium.** Importantly, a catalyst does not change the enthalpy (ΔH) or free-energy change (ΔG) of the reaction.

276. (A) The half-life of a first-order reaction is independent of the concentration. It takes between 10 and 15 minutes to decrease the amount of reactant by 50%, which implies the half-life is about 12.5 minutes. It takes another 10 to 15 minutes to decrease the remaining amount of reactant by another 50%, so then you can be confident that the half-life is about 12.5 minutes. Because the concentration clearly decreased throughout the experiment but the half-life does not change, you can be relatively sure this is a first-order reaction (zero-order was not an option).

For a second-order reaction, the half-life varies with concentration. It can often take several half-lives to truly distinguish between first- and second-order reactions by this method, which is why the table has enough data to supply four half-lives (though more may be needed in a real experiment).

277. (C) In order for a chemical reaction to occur, the reactants must not only collide but do so in the **proper orientation relative to one another** and with **enough kinetic energy** to overcome the activation energy of the reaction, which will result in the formation of the activated complex, a transient, high-energy reaction intermediate. When these two criteria are met and the activated complex transitions into product, an **effective collision** has occurred. **Increasing the temperature increases the average kinetic energy of the particles**, but **it does not affect the orientation at which they collide.**

The collision model of chemical reactions is a framework with which to visualize the interactions between reactants as a function of their particulate nature. The diagram below shows the "cone of successful attack." This "cone" indicates the general approach angle an oncoming reactant must have relative to another reactant to produce an effective collision (provided there is enough kinetic energy).

Notice that particle A has a good approach angle and will enter the imaginary cone to displace (or replace) atom X from the molecule. Particle B does not approach the molecule within the space of the cone and so the collision between the two particles will not produce a product. In other words, it will not be effective. Rutherford the dog reluctantly agreed to pose for this drawing to show that patting the pup on the head requires approach within the cone.

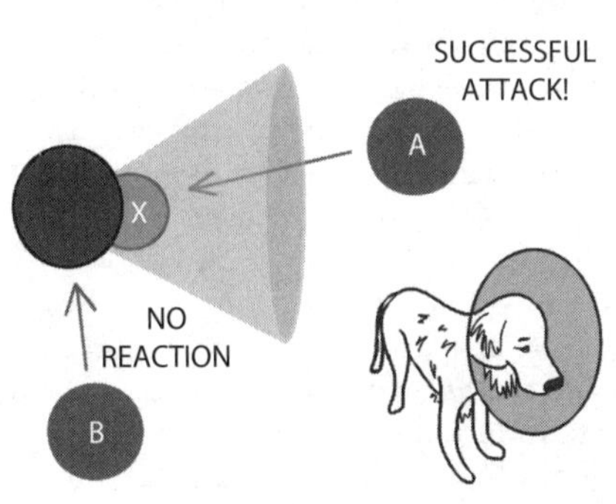

A common misconception regarding the effect of temperature on reaction rate is that increased temperatures increase the frequency of collisions. This is a true; however, **the increased number of collisions with increased temperature does *not* have a significant effect on reaction rate.** The frequency of a two-particle collision (in a gas) is proportional to the square root of the Kelvin temperature (Temperature $\propto KE_{AVE}$ = mass × velocity^2). Increasing the temperature from 22° C to 40° C (an 18° increase) only increases the frequency of collisions by 3% $\left(\sqrt{\left(\frac{313}{295}\right)} = 1.03\right)$. The diagram below answer 270 shows how an increased temperature changes the frequency distribution of the particles. Increasing the average kinetic energy flattens out the curve, in effect spreading out the distribution so **a greater percentage of particles have enough kinetic energy to overcome the activation energy barrier.**

278. (B) The rate law (usually) only contains reactants, so the participation of the product in the reaction, whether it acts to increase or decrease the reaction rate, may not be indicated by the rate law. Some reactions, like the decomposition of ozone (see answer 269), have rate laws with a negative reactant order. **A negative reactant order indicates that the reaction rate decreases with an increased concentration of a particular substance.**

$$2\ O_3\ (g) \rightarrow 3\ O_2\ (g)$$
$$\text{Rate} = k\ [O_3]^2[O_2]^{-1}$$

In this (atypical) case, the product of the reaction is included in the rate law. **The negative reactant order shows that the buildup of the product slows the reaction rate.**

Except for zero-order reactions, **reaction rates change as reactions proceed because the concentration of reactants decreases as the reaction progresses**, which typically lowers the reaction rate. (If you can correctly imagine the mechanism, the collision model strikes again!) **The rate law allows you to quantify how much the change in reactant concentration affects the rate** (by the reaction order) and allows a reasonable calculation of a reaction rate using any combination of reactant concentrations if we know the rate constant for the reaction at that temperature.

Choice A is incorrect because zero order reactions do follow a rate law, it's just that the rate law says that the reaction rate does not depend on the concentration of the reactant(s). Choice C is incorrect because the concentrations of the reactants will always change as long as the reaction is proceeding in (net) one direction and not at equilibrium.

Choice D is tricky because initial reaction rates are easy to measure and may be the only rates you've measured in your chemistry class. The method of measuring initial rates is useful for what a rate law represents, but don't be fooled by understanding a simple algorithmic application of the rate law. You should be aware that there are two types of rate laws: **The differential rate laws** (what we have been working with) **express how reaction rates depend on concentration** while the **integrated rate laws express how concentrations depend on time.** Investigating rate laws with data for initial rates enables a prediction of how concentration will vary as the reaction progresses.

279. (A) Reaction rates are typically measured in one of two ways: the appearance of product over time and/or the disappearance of one or more reactants over time. **When analyzing reaction rate data, it is vital to consider the stoichiometric coefficients to compare relative rates of product appearance or reactant disappearance.** For example, for each mole of NO consumed by this reaction, one mole of H_2 will be consumed as well.

In addition, one mole of H_2O and $\frac{1}{2}$ mole of N_2 will be formed. In the same period of time, N_2 will be formed at $\frac{1}{2}$ the rate as H_2O and $\frac{1}{2}$ the rate at which NO and H_2 are consumed.

The order of a reactant in the rate law indicates the effect of the concentration of that reactant on the initial reaction rate, not the relative rates of consumption of the reactants during the reaction. Remember that the reaction rate applies to the entire reaction, so if, for example, a low concentration of one reactant is "slowing down" a particular reaction, the rate of disappearance of all of the reactants will be affected according to their stoichiometry, not reactant order.

280. (D) It is more difficult to determine the rate law by a reaction mechanism in which an intermediate (NOBr, in this case) is involved in the slow (rate-determining) step.

The first, fast step is at equilibrium, so that the forward *and* reverse reactions are occurring at the same rate.

Write and equate a rate law for the forward and reverse reactions [the prime sign (′) indicates the reverse reaction]:

$$k\,[NO][Br_2] = k'\,[NOBr_2]$$

Solve for $[NOBr_2]$, the intermediate:

$$[NOBr_2] = \frac{k}{k'}\,[NO][Br_2]$$

Write the rate law for the slow (rate-determining) step (include $NOBr_2$ even though it is an intermediate):

$$\text{Rate} = k_2\,[NOBr_2][NO]$$

Substitute $\frac{k}{k'}\,[NO][Br_2]$ for $[NOBr_2]$ (what you solved for above):

$$\text{Rate} = k_2\left(\frac{k}{k'}\right)[NO][Br_2][NO]$$

Simplify:

$$\text{Rate} = k_{RXN}\,[NO]^2[Br_2]$$

$$K_{RXN} = k_2\frac{k}{k'}$$

281. (B) The slowest (rate-determining) step in a multistep reaction determines the rate law of a reaction. Because CO is not involved in the rate-determining step, it is not included in the rate law (eliminating choices C and D). Because 2 NO_2 molecules are necessary for collision (it is bimolecular), the reactant order for NO_2 is 2. (See answers 265, and 280 for detailed explanations of determining rate laws from reaction mechanisms.)

282. (B) This question, like question 274, is asking about the effect of surface area on reaction rate. Two answer choices, A and B, address surface area. An increased reaction rate due to a change in surface area is the result of the greater exposure of surface particles *per unit of volume or mass of the total amount of the substance.* In other words, a block of wood has a larger surface area than a scrap of wood or piece of sawdust, but the increase in volume and/or mass is much greater than the increase in surface area, so the ratio of surface area to volume (or mass) actually decreases. Choice A is incorrect because the larger pieces of wood have a smaller surface area-to-volume ratio, not a larger one.

Assume you had two cubes of wood, one with dimensions $2 \times 2 \times 2$ units and the other with dimensions $4 \times 4 \times 4$ units. **The smaller cube has a surface area-to-volume ratio of 24:8 = 3,** but **the larger cube has a surface area-to-volume ratio of 96:64 = 1.5** (see calculations below). Doubling the size of the cube decreases the surface area-to-volume ratio by $\frac{1}{2}$.

Surface area = Length × Width × # of sides
Small cube = $2 \times 2 \times 6$ = **24 units^2**
Large cube = $4 \times 4 \times 6$ = **96 units^2**

Volume = Length × Width × Height
Small cube = $2 \times 2 \times 2$ = **8 units^3**
Large cube = $4 \times 4 \times 4$ = **64 units^3**

The surface area of the larger piece of wood is four times that of the smaller piece of wood. This makes sense because the cube is basically twice as large, and since area has a unit of length squared, $(2 \times \text{length})^2 = 4$ times the surface area. As you may have expected, the volume of the larger piece of wood is eight times that of the smaller pieces of wood because the unit of volume is length cubed, $(2 \times \text{length})^3 = 8$ times the volume.

If any cube of wood is grinded up into sawdust, the cube and the pile of sawdust would have the same mass, but the pile of sawdust would clearly have greater surface area, therefore more of the wood would be exposed to oxygen during its combustion.

If the solids have the same composition (wood, twigs, sawdust) then there should be no change in composition when they are broken into smaller pieces (why choices C and D are incorrect).

283. (A) The large value of *k* shows that this reaction happens very quickly, but by itself, it is not evidence for or against the proposed reaction mechanism. **Evidence for reaction mechanisms will often come in the form of rates.** For example, an increased concentration of a reactant will usually increase the reaction rate. If a small increase in concentration results in a large increase in the rate, the order of that reactant may be 2 or 3. In the case of ozone and chlorine, the rate at which the reaction occurs increases linearly with the concentration of chlorine, which would be expected from the proposed mechanism. Chlorine is also regenerated by the second reaction, making it a catalyst for the whole reaction.

284. (D) The activated complex that forms during a chemical reaction without a catalyst must be different from the one formed by the same reaction with a catalyst or the activation energy would be the same. The mechanism by which a catalyst lowers the activation energy is the provision of an alternative pathway for the reaction, **a different path of transition**, which **means a different set of intermediates are formed along the way.** See the figure below, in which the running man must run on the trail that goes up and over the mountain while the woman on skis can take a different, lower-energy path. In this metaphor, the skis are the catalyst because they allow the lower-energy path to be taken.

When answering a question asked with a "not," mark each answer choice as true or false (or yes or no). When you're done with the four choices, the one that's false (or no) is the correct answer.

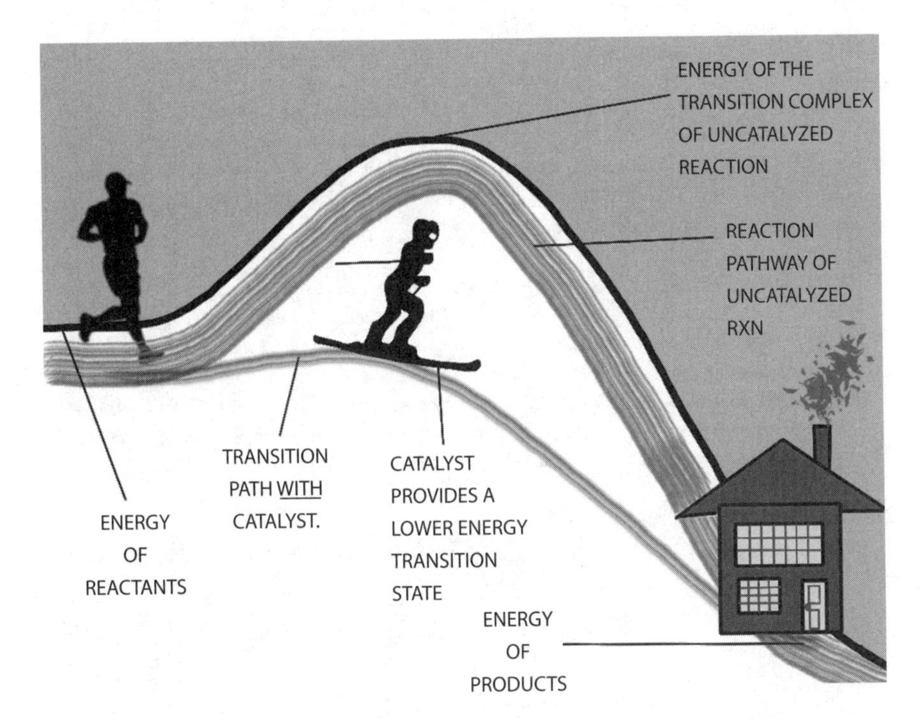

285. (D) The reaction is endothermic because the products have more potential energy than the reactants; however, **the temperature at which the reaction takes place does not affect the difference in potential energy between the reactants and the products**. This is different from the free-energy change of a reaction. The equation for ΔG, $\Delta G = \Delta H - T\Delta S$, includes temperature ($T$). For reactions in which the enthalpy and entropy changes are either both positive or both negative, the temperature determines whether a reaction is thermodynamically favorable (the magnitude and sign of ΔG, the free-energy change of the reaction).

The $\boldsymbol{E_A}$ is always calculated by **subtracting the energy of the reactants from the energy of the activated complex** ∴ 510 – 75 = **435 kJ mol^{-1}**. The **enthalpy change (ΔH)** is calculated by **subtracting the potential energy of the products from the potential energy of the reactants** ∴ approximately 250 kJ mol^{-1} – 75 kJ mol^{-1} = **175 kJ mol^{-1}**.

286. (C) Reaction rate is measured by product formation (or substrate consumption) over time and **is a function of the number of effective collisions that occur in a given period of time.** For some reactions, the actual time course for the molecular events of the reaction is known. But in the absence of specific data, whether each individual reaction occurs faster (choice A) or more successful reactions happen per second but each reaction takes the same amount of time isn't known. In other words, unless additional data regarding the reaction mechanism are provided, there is no way to know if the addition of the catalyst speeds up each individual reaction to occur in less than 50 seconds or if more reactants are turned into product in that time, with each reaction occurring in about 50 seconds.

A catalyst increases reaction rate by lowering the activation energy. A catalyst reduces activation energy by **providing an alternative pathway** for the transition of reactant to product. This transition pathway may be shorter or it may allow more molecules to make it past the (lower) energy barrier in a given period of time.

Imagine you attempt 25 free-throws in 1 minute and you make 10 of them. But when you wear your lucky hat, you make 22 of them. The measured "baskets per minute" increases, though the rate at which you threw the ball stayed the same, since each time you attempted 25 in 1 minute. The lucky hat is the catalyst in this example, but only with regard to rate. The mechanism of its action is unknown. ☺

This question highlights the importance of knowing the limits of your data. The College Board wants you to understand not only the basics of what is known in chemistry, but how it came to be known. In other words, how do we know what we know? The way in which an experiment is designed limits the range of interpretations of the data it produces. "How do they know that?" should be a question you ask yourself continually throughout your chemistry course. The purpose of questions like this is to assess whether you know what the data actually says as opposed to what you think it says.

287. (B) In this case, rate = $k\,[X][Y]^2$

Halving the concentration of X, a first-order reactant, would halve the reaction rate: $\left[\frac{1}{2}\right]^1 = \frac{1}{2}$ R.

Doubling the concentration of Y, a second-order reactant, would quadruple the reaction rate: $[2]^2 = 4$ R.

$$\text{rate} \propto [X][Y]^2 \therefore \text{rate} \propto \left(\frac{1}{2}\right)(4) = 2\text{ R}$$

288. (C) In this case, rate = $k\,[X]^2[Y]$

Halving the concentration of X, a second-order reactant, would decrease the reaction rate to one-fourth because $\left[\frac{1}{2}\right]^2 = \frac{1}{4}$ R.

Doubling the concentration of Y, a first-order reactant, would double the reaction rate: $[2]^1 = 2$ R.

$$\text{rate} \propto [X]^2[Y] \therefore \text{rate} \propto \left(\frac{1}{4}\right)(2) = \frac{1}{2}\text{ R.}$$

289. (A) The rate (proportionality) constant is a general indicator of the reaction rate. The formula for k is given by the Arrhenius equation (see note in italics below), but you can tell a lot from k, and having two or more values for k at different temperature can qualitatively reveal a lot about the reaction.

- **The larger the value of k, the higher the reaction rate** (all else being equal).
- **Increasing temperature increases the value of k.** The greater the difference between k values for a given temperature increase, the greater the dependence of rate on temperature. However, you need two or more values of k to make the comparison.

The College Board expects you to be familiar with the Arrhenius equation. Specifically, you should be able to use it as a qualitative tool to interpret data from experiments on the temperature dependence of a reaction rate in terms of the activation energy needed to reach the transition state. However, calculations involving the Arrhenius equation are considered beyond the scope of the course and the exam.

290. (A) The rate at which a reaction occurs is independent of the free-energy change of the reaction. **Thermodynamic data tell us about the favorability of a reaction** or process, but tell us nothing about its rate. **Kinetic data tell us about the rate of a process** or reaction, but tell us nothing about its favorability.

291. (B) The hydrogen bubbles physically prevent the solution from accessing the zinc surface. Stirring the solution may help to dislodge some of the bubbles but will not be as effective as brushing them off the surface.

292. (A) Statements A and B true, but only A correctly states why radioactive decay displays first-order kinetics. Choice D is not true. **The probability that any particular radioactive atom will decay at *any moment* is $\frac{1}{2}$**—it will decay or it won't—and there's no way to accurately predict anything more about it. The amount of decay that has or hasn't occurred does not change the probability of another particle decaying. However, you can confidently predict that for a sample of that substance, one-half of it will decay in the time period of one half-life (thanks to first-order kinetics).

293. (A) The rate law identifies the reaction order as 1, first-order, so you know the rate is directly proportional to the concentration of A. The unit of the rate constant for a first-order reaction is sec^{-1} or $\frac{1}{sec}$. When the reaction is first-order, the rate is directly proportional to the concentration of one reactant (unless there are reactants with fractional or negative reactant orders), so the interpretation of the rate law is very straightforward:

$$\text{Rate} = 0.001\ [A] \text{ per second}$$

Remember that the concentration of A decreases with time, so the actual number of molecules of A represented by 0.001 decreases with time. Choice B is incorrect because the number of X molecules keeps increasing as the reaction progresses. Although the concentration of X does increase, the percent it increases drops with time because the amount of X keeps increasing (and the rate of conversion of A→X depends on a decreasing concentration of A).

Choice C is incorrect because the decrease in the number of particles is not linear: In each half-life the concentration decreases by 50%. Choice D is incorrect because unless the reaction order is zero (which is not indicated by the rate law of this reaction), the reaction rate changes as the concentration changes.

294. (C) Use the data from the trials to calculate a rate law (as opposed to deduce one from a reaction mechanism). Comparison of trials 2 and 1 (chosen to compare the effect of [NO] on rate while $[O_2]$ remains constant) shows that increasing NO concentration 1.6 times increases the rate 2.6 times. The reactant order of NO must be higher than 1 because if it were 1, a 1.6-fold increase in concentration would increase the rate by 1.6-fold. A reactant order of 2 for NO works better, but do the math to check: $[1.6 \times \text{concentration}]^2 = 2.6 \times \text{rate}$

Trial	[NO] (mol L^{-1})	[O_2] (mol L^{-1})	RATE (M s^{-1})
1	2.4×10^{-2}	3.5×10^{-2}	1.43×10^{-1}
2	1.5×10^{-2}	3.5×10^{-2}	5.60×10^{-2}
3	2.4×10^{-2}	4.5×10^{-2}	1.84×10^{-1}

Apply the same procedure to the find reactant order of O_2. Trials 1 and 3 are the two in which *only* the concentration of O_2 changes. The [O_2] increases 1.3 times (from 0.035 to 0.045 M), which causes the rate to increase by 1.3 times (from 0.143 to 0.184 M s^{-1}). Because the rate increase is directly proportional to the concentration increase, O_2 is a first-order reactant (within the range of concentrations measured).

295. (C) You can substitute data from any one of the trials into the rate law. Since you won't have a calculator for the multiple-choice section of the exam, use the trial that has the easiest numbers to crunch.

For trial 1:

$$\text{Rate} = k[\text{NO}]^2[\text{O}_2]$$
$$k = \frac{\text{rate}}{[\text{NO}]^2[\text{O}_2]}$$
$$k = \frac{0.143}{[0.024]^2[0.035]}$$
$$k = 7{,}093, \text{ or } 7.0 \times 10^3$$

The AP Chemistry exam will not ask a multiple-choice question that relies on the correct answer from a different multiple-choice question, even if they are part of a set of related questions. This question could be a "stand-alone" question in which you are required to write the rate law in order to find the value of the rate constant.

296. (C) The reaction order for NO is 2, so increasing its concentration fivefold would increase the reaction rate by 25 times ($[5x]^2 = 25$-fold increase in rate). You needed to know the reactant order for NO to answer this question, but not the rate law.

297. (D) You need to know the rate law (answer 294; see note in italics below) and plug in the new concentrations and the value of k.

$$\text{Rate} = k[\text{NO}]^2[\text{O}_2]$$
$$\text{Rate} = 7.3 \times 10^3\ [2 \times 10^{-2}]^2[4 \times 10^{-2}]$$
$$\text{Rate} = \mathbf{1.1 \times 10^{-1}\ Ms^{-1}}$$

The AP Chemistry exam will not ask a multiple-choice question that relies on the correct answer from a different multiple-choice question, even if it is part of a set of related questions. This question would be a reasonable "stand-alone" question if you were given the value of k.

298. (C) The half-life of a first-order reaction is independent of concentration. For all other reactant (and reaction) orders, the half-life varies with concentration. This could be used as a diagnostic procedure for a first-order reaction. Experimentally, it may take several half-lives to be able to distinguish between first- and second-order reactions by this method. See figure below.

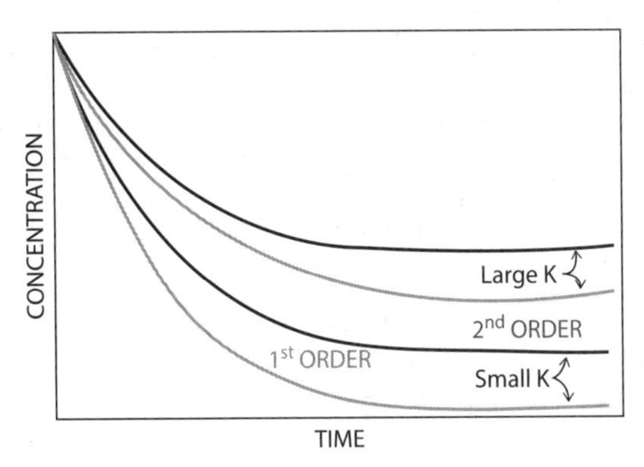

299. (D) The only way a reaction order or rate law can be determined is through experimentation. Only if a reaction mechanism is known *beyond a shadow of a doubt* can the elementary steps provide the molecularity data to construct a rate law. The rate laws written from proposed reaction mechanisms can be used to test the proposed mechanism, but are not considered true rate laws until the trial data confirm it. Reaction order can provide clues to the reaction mechanism as well. The decomposition of dinitrogen pentoxide is actually first-order, but that can't be known with certainty from the reaction as it's written.

A reaction with only one type of reactant that requires the second reactant to form the product is bimolecular ($A + A \rightarrow$ products, rate $= k\,[A]^2$). A reaction in which only one molecule is needed to react (through a collision!) but does not require any of the atoms from the other molecule (in other words, the reaction is a rearrangement of the atoms in the reactant) is unimolecular and first order ($A \rightarrow$ products, rate $= k\,[A]$).

300. (D) The effect that the concentration of a particular reactant has on a reaction cannot be determined until the actual experiment of changing the concentration of that reaction and measuring the rate change that results is performed. In writing the equilibrium expression, the exponents that accompany reactions are taken from the balanced equation, but the equilibrium concentrations must be measured to determine the K_{EQ} for the reaction. Although a rate law can theoretically be determined through a known reaction mechanism, the rate law must be verified by experimentation. **A hypothesis is a critical starting point, but until you have the data you can't possibly know its validity.**

301. (D) The total reaction order is 3, so doubling *both* reactants will increase the reaction rate eightfold: $[2x]^3 = 8x$ increase in rate.

302. (B) The expression does not contain a concentration because **the half-life of first-order reactions is independent of concentration.** The unit of half-life is time, so the larger the value of k, the rate constant, the smaller the time period for the half-life. That means the reaction rate increases with a larger value of k (and a large value of k is a reasonable indicator that the reaction occurs quickly).

303. (B) Reaction mechanisms demonstrate how the rearrangement of atoms actually occurs in a chemical reaction. Most reactions take place by a series of simple, elementary steps. Typically, the elementary reactions are deduced after the rate law for a reaction has been experimentally determined. Afterward, a reaction mechanism that accounts for the rate law is devised. An important guideline in the construction or evaluation of a reaction mechanism is that the elementary steps show how individual atoms and molecules take part in a reaction. **A step that is truly elementary cannot be further broken down into additional elementary steps.**

Choice A is not an ideal reaction mechanism because it's first step is not elementary:

$$NO + O_3 \rightarrow NO_2 + O$$

The two steps that are involved in this reaction are:

$$NO + O_3 \rightarrow NO + O_2 + O$$
$$N + O_2 \rightarrow NO_2$$

That is choice B, and it accounts for the rate law.

Choice C also contains a step that is not elementary:

$$NO + NO \rightarrow 2\ N + O_2$$

The steps involved are:

$$NO + NO \rightarrow 2\ N + 2\ O$$
$$O + O \rightarrow O_2$$

Additionally, this reaction mechanism does not account for the rate law. Step 1 is bimolecular, so the reactant order for NO would be 2 if this were the mechanism.

Choice D is an indirect mechanism that requires two of everything. The first step is bimolecular for O_3, however, so it does not correctly account for the rate law.

Chapter 4: Free-Response Answers

304. The correct graph looks like this:

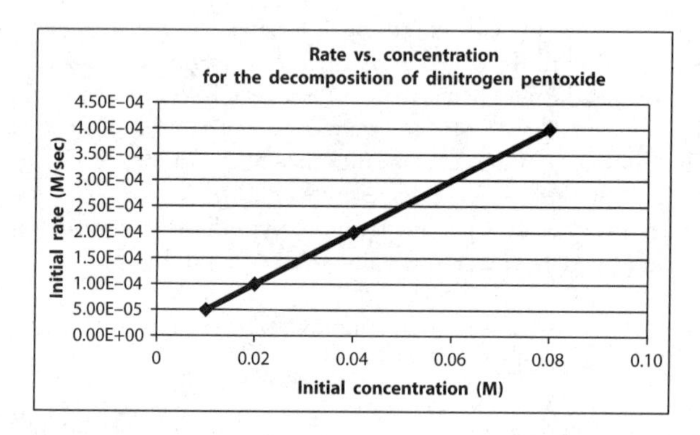

305. Rate = $k\ [N_2O_5]^1$

$$k = 5 \times 10^{-3}\ sec^{-1}$$

The graph of initial rate versus concentration is a straight line, confirming that the reaction is first-order. **The slope represents the value of the rate constant.**

The equation for the line: v = (m)(x) + b

The equation for the line: y = ($\boxed{5 \times 10^{-3}}$) × (-7×10^{-20}) ~~+ b~~

$$\text{Slope} = \frac{5 \times 10^{-3}\ \text{M sec}^{-1}}{\text{M}} \therefore \mathbf{5 \times 10^{-3}\ sec^{-1}}$$

$$\text{Rate} = k\,[N_2O_5] \therefore k = \frac{\text{rate}}{\text{concentration}}$$

$$\text{The slope} = \frac{\Delta y}{\Delta x} \therefore \frac{\text{rate}}{\text{concentration}}$$

306. The reaction is first-order, so the half-life is independent of concentration. There are two ways to attack this problem.

One way is to use the data, the graph, and/or the rate law generated from the data to find the time it takes for the concentration of N_2O_5 to decrease to half the initial concentration. If you choose to work from the data, you can choose any data point: 0.02 M concentration will be our example. One-half the concentration is 0.01 M, which indicates both the amount of N_2O_5 consumed *and* the amount left over. The question becomes, How long does it take for the reaction to consume 0.01 M? To do the math, think about the rate in terms of seconds instead of moles per liter, for example, in one second, 1×10^{-4} moles per liter are consumed (see how the fraction representing the rate is written below). Make sure you select the rate that accompanies the initial concentration you chose, not the half-concentration.

$$0.01\ \text{M} \times \frac{\text{sec}}{1 \times 10^{-4}\ \text{M}} = \mathbf{100\ seconds}$$

Because this is a first-order reaction, the half-life is independent of the concentration, so any data point you choose will give the same half-life.

Another way is to use the formula. You may remember the expression for calculating the half-life of a first-order reaction: $t_{1/2} = 0.693/k$

$$\mathbf{t_{1/2} = \frac{0.693}{k}}$$

$$= \frac{0.693}{(5 \times 10^{-3}\ \text{sec}^{-1})}$$

$$= \mathbf{139\ seconds}$$

The actual value of k for the reaction is $5.2 \times 10^{-3}\ \text{sec}^{-1}$ and the half-life is 130 seconds. The numbers were rounded to make the graphing and calculations easier.

The slope of the line of the graph of the reaction rate versus concentration line is the value of the rate constant, so the relationship between them is inverse. **As the value of the rate constant increases, the slope of the line increases, and the reaction rate increases, decreasing the time it takes to get to half-concentration of reactant.**

307. A > B > C

The larger the activation energy of a reaction, the greater the temperature dependency of the rate.

The Arrhenius equation may be expressed as $\boldsymbol{k = Ae^{-\frac{E_A}{RT}}}$ or $\mathbf{\ln k = -\frac{E_A}{RT} + \ln A}$, where:

- A = the frequency factor. **The frequency factor relates to the frequency of collisions and the probability that the particles will be in a favorable orientation to make the collision effective.** The frequency factor is considered nearly constant for a particular reaction, even at different temperatures (see answer 277 for an explanation of effective collisions and the relationship between temperature and number of collisions).

- **E_A** = the activation energy. All else being equal, **as the magnitude of the activation energy increases, the value of *k* decreases (and reaction rate decreases).**
- **R** = the gas constant (watch the units!)
- **T** = temperature (K)

The AP Chemistry exam will expect you to use the Arrhenius equation to make qualitative predictions about the temperature dependency of a reaction rate and explain this dependence in terms of the activation energy needed to reach the transition state. You will not be asked to perform calculations involving the Arrhenius equation.

308. The area under the curve represents the number of particles. At higher temperatures the frequency distribution of particle speeds is spread out more toward higher speeds (higher kinetic energies). This means **the number of particles with energy at or above the activation energy is greater at higher temperature.** The increased energy of molecular collisions results in a greater percentage of those collisions resulting in a product, in other words, **a higher percentage of collision will be effective**. See answer 270.

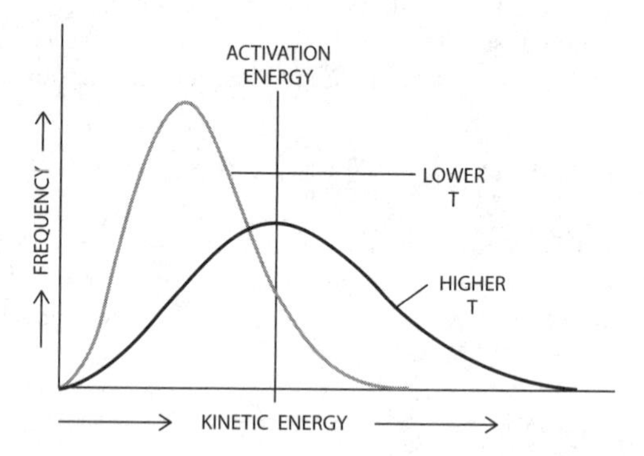

309. (a) Combustion requires two reactants, the fuel (in this case, gasoline) and oxygen. The match provides the activation energy to get the reaction started, but the large release of heat from the reaction provides the activation energy for the next reactants to get past the energy barrier, and so on. This is what makes combustion a self-propagating process.

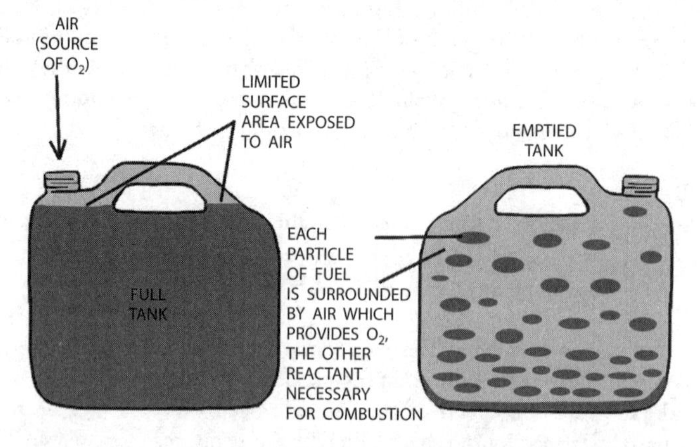

(b) The diagram shows that the full container of fuel has a greater quantity of fuel, but it will burn slowly because only the top layer of fuel has access to the oxygen. With such a large fuel reserve, this process will **last a long time but will proceed at a relatively slow rate**. The container on the right shows that gasoline (octane) has a high vapor pressure inside the emptied container. The volatilized fuel molecules are surrounded by air, which supplies them with all the oxygen they need to react. **The combustion will occur quickly but will also quickly exhaust the fuel store. The emptied container will produce a more violent explosion, but it will be over very quickly.**

The diagram is not necessary to answer the question. It is used to support the explanation. If you draw a diagram to support a written answer on your AP Chemistry exam, make sure you refer to it in your answer.

310. The activation energy for the hydrolysis of sucrose is larger than the reaction of ethyl bromide and hydroxide. Therefore, sucrose hydrolysis rate has greater temperature dependence (see graph under question 307 and the answer).

Generally, the larger the activation energy of a reaction, the greater the temperature dependency of the rate.

Evaluate the difference in the values of k at different temperatures.

Sucrose: 5.9×10^{-4} for a 2° difference
Ethyl bromide and hydroxide: 2.2×10^{-4} for a 5° difference
The temperature dependence on the value of k is greater for sucrose.

The greater the change in the value of k, the greater the dependence of rate on temperature.

Chapter 5: Energy

311. (C) A reaction that is exothermic ($\Delta H < 0$) and increases entropy ($\Delta S > 0$) will be thermodynamically favorable (exergonic) at all temperatures. See answer 312 and the table that follows it.

312. (A) Gibbs free energy is a thermodynamic state function given by the formula $\Delta G = \Delta H - T\Delta S$. This equation elegantly relates the capacity of a process to do work with just three other quantities: the enthalpy change, the entropy change, and the temperature at which the process occurs. Typically, a process is driven by a negative change in enthalpy (the process "loses" heat), a gain in entropy (the process results in more disorder), or both (see table below).

ΔH	ΔS	ΔG
– EXOthermic	+ ↑ entropy	– **exergonic at ALL temperatures**
+ ENDOthermic	– ↓ entropy	+ endergonic at ALL temperatures
+ ENDOthermic	+ ↑ entropy	– ∴ **exergonic at HIGH temperatures** + ∴ endergonic at low temperatures
– EXOthermic	– ↓ entropy	– ∴ **exergonic at LOW temperatures** + ∴ endergonic at high temperatures

The table shows that if the signs of entropy and enthalpy are the same for a particular reaction, it is thermodynamically favorable only above or below a particular temperature. To calculate the specific temperature, you can plug the values of ΔH and ΔS into the equation and use +1 or 0 for the value of ΔG. (Be careful! The unit for ΔS is **J**·mol^{-1}· K^{-1} whereas the unit for ΔH is **kJ**·mol^{-1}.)

313. (A) Free energy (ΔG) is a measure of how much useful (but nonmechanical) work can be done by a system at a constant pressure and temperature. Systems in disequilibrium, i.e., not at equilibrium, have the ability to do work as they move toward equilibrium.

A process is thermodynamically favorable if the ΔG is negative (it can do work) and endergonic (not thermodynamically favorable) if the ΔG is positive (work needs to be done to get the process to happen). In this question, you first needed to identify which reactions could be thermodynamically favorable at 298 K (choice D is endergonic at all temperatures) and then solve for the value of ΔG for each reaction at 298 K (see table below).

EXERGONIC versus ENDERGONIC processes

An **EXERGONIC process** will occur because something in the system is driving it—either a negative enthalpy change, a positive entropy change, or both. No action from the surroundings is needed. An exergonic process can do work!

An **ENDERGONIC process** will not occur unless the surroundings take action (do work) on it.

The **temperature** will determine the spontaneity of a process when the enthalpy change is positive *or* the entropy change is negative (but not both).

The large $-\Delta H$ and $+\Delta S$ of reaction A makes it a reasonable choice even if you didn't have time to solve for the others (and solving for a problem like this means generously rounding!). *Notice that the unit of ΔS has to be converted from J/mol into kJ/mol.*

Reaction	ΔH° (kJ)	ΔS° (kJ·K^{-1})	ΔG° @ 298 K
A	**−566.0**	**−0.1730**	**−514.4**
B	484.0	0.0900	457.2
C	−164.0	0.1490	−208.4
D	23.4	−0.0125	27.1

The ° (naught) symbol means the data were collected under standard conditions: 25° C, 1 atm pressure for gases, and 1 M concentration for solutes.

314. (D) Kinetic Molecular Theory of Gases (KMT) states that **the average kinetic energy (KE) of the particles in a gas is directly proportional to the absolute temperature.** Therefore, since all three gases are at 298 K, they have the same average KE.

315. **(C)** Kinetic Molecular Theory of Gases (KMT) states that the average kinetic energy (KE) of the particles in a gas is directly proportional to the absolute temperature. All three gases are at 298 K, so they all have the same average KE. However, the KE is a function of the mass and velocity of the particles.

$$\mathbf{KE = \frac{1}{2}\,mv^2}$$

Therefore, the mass of the particles is inversely related to the square of the velocity: $m \propto \frac{1}{(v^2)}$

At a given temperature, the gas with the greatest molar mass will have the lowest average particle speeds.

316. **(A)** Kinetic Molecular Theory of Gases (KMT) states that the average kinetic energy (KE) of the particles in a gas is directly proportional to the absolute temperature, so lowering the temperature in the container will decrease the kinetic energy and, therefore, the speed. Although you know the pressure and volume are both related to temperature, **only the pressure will respond to the temperature change because the container volume is fixed** (*rigid* is code for constant volume). **Decreasing the temperature only decreases the pressure in a closed container.** In the atmosphere, a lower temperature causes a higher pressure because the particles get closer together. The atmospheric pressure is the weight of the column of air above a particular area on the surface of the earth, so the more particles that are in the column, the greater the pressure. As for volume, **the gas will always expand (or get compressed) to fit the volume of the container**.

317. **(B)** Solutions are complex, but a few general rules can help you sort out the process.

If a particular solvation process is exothermic, then more and/or stronger interparticle forces of attraction (IPFs) were formed than were broken (remember IPF formation is exothermic). If the solvation process is endothermic, then more and/or stronger IPFs are broken than formed (IPF "breakage" is endothermic).

Process	Energy change
1. The separation of ions in the solid	**IPFs broken** **Endothermic** Energy "required" to separate highly organized and tightly bound charged particles in an ionic solid
2. The separation of particles in the solvent (water)	**IPFs broken** **Endothermic** Energy "required" to break hydrogen bonds (strong dipole-dipole attraction) between water molecules
3. The formation of ion-dipole interactions between the ions and the water molecules	**IPFs formed** between ions and water molecules **Exothermic** Energy "released" when water molecules organize themselves around ions

318. (B) The sum of the three processes must be positive because the ΔH is positive (3.87 kJ mol^{-1}), but the first two processes must be positive and the third must be negative. If their sum is positive, the third process is not exothermic enough to "cover the energy cost" of the enthalpy changes from the first two. See the chart under answer 317.

319. (A) The negative ΔH value indicates the reaction is exothermic in the direction written, so heat would be released by the reaction. The balanced equation is written to show the consumption of one mole of Na and the formation of one mole of NaCl, which would release 411 kJ, so 0.1 mole would release 41.1 kJ (0.1 mol × 411 kJ mol^{-1}) of heat.

320. (B) The negative ΔH value indicates the reaction is exothermic in the direction written, but the question asks us to reverse the reaction to form Cl_2 from NaCl. If the reaction proceeds exothermically as written, it must proceed endothermically in reverse, so the reaction would absorb heat. The reverse of the balanced equation shows that to form 0.5 mole Cl_2 from 1 mole of NaCl would require 411 kJ of energy, so to form 0.05 mole Cl_2 would require $1/10^{th}$ the energy, or 41.1 kJ (.05 is 1/10 of 0.5).

321. (B) The information in the table may have reminded you of **the Born-Haber cycle**, a thermodynamic cycle used to calculate the lattice energy of an ionic compound (which cannot be directly measured by experiment) that applies Hess's law to its enthalpy of formation and other measurable quantities. **The enthalpy of formation of NaCl *(s)* from Na^+ *(s)* and Cl^- *(g)* is equal to the sum of the energies of the individual steps** (see table below).

	Process	Direction of enthalpy change (sign of ΔH)	Explanation
v	Na *(s)* → Na *(g)*	+	The standard state of Na is Na *(s)*. Converting Na into a gas requires energy because metallic bonds must be broken; hence, this process is **endothermic**.
w	Na *(g)* → Na^+ *(g)* + e^-	+	The first ionization energy of Na. All first ionization energies of the elements (and successive I_E, too) are **endothermic**.
x	Cl_2 *(g)* → 2 Cl *(g)*	+	This process breaks the covalent bonds within the chlorine molecule and is therefore **endothermic**.
y	Cl *(g)* + e^- → Cl^- *(g)*	–	The electron affinity of chlorine. Adding an electron to individual chlorine atoms is **exothermic** because it forms a more stable substance. (Cl^- has a full valence shell of electrons whereas Cl^0 has only 7).
z	Na^+ *(g)* + Cl^- *(g)* → NaCl*(s)*	–	The formation of NaCl from its constituent ions is **exothermic**. You can think of this as the "negative lattice energy."

322. **(A)** The fairly large $-\Delta H$ clearly indicates that an enthalpy change drives the reaction. The entropy of the system is likely to increase because a compound is formed from pure substances. A useful oversimplification of entropy changes can be inferred from the third law of thermodynamics, which roughly states that **the definition of zero entropy is a perfect, pure, crystalline solid at 0 K.** Any deviation away from 0 K and/or a pure, perfect, crystalline solid indicates that entropy is increasing. In this case, the number of particle types making up the substance is 2 instead of just 1 for Na *(s)* and 1 for Cl_2 *(g)*.

You may be thinking that because Cl_2 is a gas, the formation of an ionic lattice would result in result in a decrease in entropy. Yes, there is a decrease, but not as much as the increase in entropy caused by the other processes involved. The value can be obtained from the thermodynamic data that is in the appendices of most chemistry textbooks (and available online, of course), which shows the $\Delta S°$ of the reaction is actually positive ($+72.3$ J mol^{-1} K^{-1}), so there is a net gain in entropy.

Entropy

Entropy is a measure of energy dispersal in a process.

Any process that increases the entropy (energy dispersal) in the universe is thermodynamically favorable.

Entropy can also be thought of as **disorder** or the most **statistically probable** microstate(s).

Imagine a deck of cards in which each card has a number from 1 through 52 and all 52 are in numerical order. For simplicity, we will assume this arrangement (numerical order from 1 to 52) is the *only* ordered state of the deck (this is one microstate, and an unlikely one to have been attained by random shuffling).

If the cards are shuffled in any way, they will get "out of order" because there are 52! – 1 ways the cards can be disordered (**8.07×10^{67} ways,** each representing one microstate) and only one way they are considered ordered. If each of those microstates took one second consider, it would take 2.56×10^{60} years, or 1.83×10^{50} times the age of our universe.

If there were only three cards in the deck, there would be 3! – 1 ways (that is, 5, a small number of microstates) the cards could be disordered and one way they could be in order.

What that shows us is that the more (different) components that make up a system, the more potential for disorder and entropy in the system because of the greater number of microstates the system can assume.

323. **(C)** Apply Hess's law:

$$\begin{aligned} 2w &= 2\,\text{Na}\,(g) \rightarrow \cancel{2\,\text{Na}^+\,(g)} + \cancel{2\,e^-} \\ x &= \text{Cl}_2\,(g) \rightarrow \cancel{2\,\text{Cl}\,(g)} \\ 2y &= \cancel{2\,\text{Cl}\,(g)} + \cancel{2\,e^-} \rightarrow \underline{\cancel{2\,\text{Cl}^-\,(g)}} \\ 2z &= \cancel{2\,\text{Na}^+\,(g)} + \underline{\cancel{2\,\text{Cl}^-\,(g)}} \rightarrow 2\,\text{NaCl}\,(s) \end{aligned}$$

$$2w + x + 2y + 2z = 2\,\text{Na}\,(g) + \text{Cl}_2\,(g) \rightarrow 2\,\text{NaCl}\,(s)$$

324. (B)

$mass_{CUP + WATER} - mass_{CUP} = \mathbf{mass_{WATER}}$

$$(156.5 \text{ g}) - (6.40 \text{ g}) = \mathbf{(150.1 \text{ g})}$$

$mass_{CUP + WATER + ICE} - mass_{CUP + WATER} = \mathbf{mass_{ICE}}$

$$(181.56 \text{ g}) - (150.1 \text{ g}) = \mathbf{(31.46 \text{ g})}$$

325. (D) The mass of the ice *measured* would have been greater than the actual amount of ice, so the heat of fusion calculated would be too low. The units of heat of fusion (H_{fus}) = kJ g^{-1} or kJ mol^{-1}. Either way, a larger mass would have made the value in the denominator larger than the actual value and therefore the calculated H_{fus} would be lower than the actual H_{fus}.

326. (B) To calculate the heat of fusion of ice using this method, the specific heat of the ice must be known, since the temperature of the ice must be raised from –20° to 0° C before it melts. In addition, the heat needed to raise the temperature of the ice and to melt it comes from the *loss of heat* of the H_2O *(l)* so the specific heat of H_2O *(l)* must also be known.

$$H_{fus} = [\text{total heat lost by } H_2O\ (l)] - [\text{heat gained by } H_2O\ (s) \text{ to reach } 0° \text{ C}]$$

327. (C) To calculate the ΔH_{RXN} from $\Delta H°_F$ values use the following formula:

$$\mathbf{\Delta H_{RXN} = (\Sigma \Delta H°_F \text{ PRODUCTS}) - (\Sigma \Delta H°_F \text{ REACTANTS})}$$

Watch out for the signs and make sure to multiply the $\Delta H°_F$ of each substance by its stoichiometric coefficient.

$\Delta H_{RXN} = (\Sigma \Delta H°_F \text{ PRODUCTS}) - (\Sigma \Delta H°_F \text{ REACTANTS})$
$\Delta H_{RXN} = (-635.5 + -393.5) - (-1{,}207.1)$
$\Delta H_{RXN} = (-1{,}029) - (-1{,}207.1)$
$\Delta H_{RXN} = 178 \approx \mathbf{180 \text{ kJ mol}^{-1}}$

328. (C) This is a Hess's law problem. Since enthalpy is a state function (a function that is path independent, i.e., it only depends on the final and initial states), it doesn't matter how we get there, only where we started and where we ended up.

If we keep the conversion of graphite and oxygen to carbon dioxide as written but "flip" the conversion of diamond and oxygen to carbon dioxide (and remember to *reverse the sign of* ΔH), the two equations cancel out oxygen and carbon dioxide and their sum leaves us with the conversion of graphite to diamond that we wanted:

$\boxed{C_{(graphite)}} + \boxed{\cancel{O_2 (g)} \rightarrow} \cancel{CO_2 (g)}$ $\Delta H = -393.5$ kJ
$\cancel{CO_2 (g)} \rightarrow C_{(diamond)} + \cancel{O_2 (g)}$ $\Delta H = +395.4$ kJ

$\boxed{C_{(graphite)} \rightarrow C_{(diamond)}}$ $\mathbf{\Delta H = +1.9 \text{ kJ}}$

329. (B) Our world is fairly stable thanks to activation energies. The reason diamond doesn't just "turn into" graphite, even though the reaction is thermodynamically favorable (exergonic), is that the activation energy of the reaction is too high to be met or exceeded under normal conditions.

Choice A is incorrect because although it is true and the lower entropy of diamond (really, the greater entropy of graphite) contributes to the negative $\Delta G°_{298}$ of the reaction, it does not explain *why* diamond doesn't get converted into graphite under standard conditions. The fact that the conversion of diamond to graphite increases entropy prompts the question, why doesn't diamond regularly get converted into graphite?

Choice C is incorrect because, if anything, the C–C bond in graphite should be stronger than the C–C bond in diamond. As the figure under answer 118 shows, each carbon in diamond is covalently bonded to four other carbon atoms and the hybridization of carbon is sp^3. However, the sheets of carbon in graphite are sp^2 hybridized, and the carbon-carbon bonds have a bond order of about 1.3, which makes the bond shorter and stronger. However, the sheets of carbon in graphite are held together by weak dispersion forces (due to delocalized pi electrons) which is makes graphite soft, electrically conductive and a good dry lubricant.

Choice D is incorrect because it's not true (the density of diamond is 3.5 g cm^{-1} and the density of graphite is 2.3 g cm^{-1}) but more importantly, because density has nothing to do with the activation energy.

330. (C) The ΔH_{RXN} is given in kJ mol^{-1} CH_6N_2, but the stoichiometric coefficient of CH_6N_2 in the reaction is 2, which means you don't convert 6 mole H_2O *(g)* to H_2O *(l)*, you only convert 3 moles. That is because two moles of 2 CH_6N_2 *(l)* produce 6 H_2O *(g)*, whereas one mole of CH_6N_2 would only produce three).

$$H_2O\ (g) \rightarrow H_2O\ (l) = -44 \text{ kJ mol}^{-1} \times \mathbf{3\ mol} = \mathbf{-132\ kJ}$$

The negative sign indicates **the condensation of water is exothermic** but that means we will actually get *more* energy out of the combustion of CH_6N_2, so the value of ΔH will be *more negative.*

$$-1{,}303 \text{ kJ mol}^{-1} + (-132 \text{ kJ}) = \mathbf{-1{,}435\ kJ\ mol^{-1}}$$

331. (A) If a reaction lowers the temperature of the container it occurs in, the reaction is *endothermic.* It is absorbing heat from its surroundings (the environment), which is why the both the water in the beaker and the beaker itself decrease in temperature (they are losing kinetic energy while the potential energy of the substances involved increases). The entropy change of the dissolution of a solid into a liquid is usually positive. **The increase in entropy is almost always favorable to solution formation.** There are situations, however, when the addition of a solute results in a more orderly arrangement of solvent particles, which would make the entropy change negative.

332. (A) A thermodynamically favorable reaction has a $-\Delta G$, i.e., the $\Delta G < 0$. If the reaction is thermodynamically favorable *only at low temperatures*, then we know that the reaction is *exothermic* and the entropy change is *positive* (see the table under answer 312). *Be careful! The units of ΔS are J mol^{-1} K^{-1}, not kJ, like ΔG and ΔH.*

$$\Delta G = \Delta H - T\Delta S \text{ and } \Delta G < 0$$

$\therefore \Delta H - T\Delta S < 0$
$(-18{,}000 \text{ J mol}^{-1}) - (300)(\Delta S) < 0$
$300\ \Delta S < 18{,}000$
$\therefore \mathbf{\Delta S < 60 \text{ J mol}^{-1} \text{ K}^{-1}}$

333. (D) Bond energy is a measure of bond strength. It is **the amount of energy required to break 1 mole of that bond.** All **bond energies are positive** numbers because **breaking bonds is always endothermic**. A formula to use to determine the ΔH_{RXN} using bond energies is:

$$\Delta H_{RXN} = (\Sigma \text{ energy of bonds BROKEN}) - (\Sigma \text{ energy of bonds FORMED})$$

*(A simple mnemonic device is "**B-FOR**" (before), bonds **B**roken are accounted for before bonds **FOR**med.)*

Make sure you account for the stoichiometric coefficients in the balanced equation!

$$2\ H_2\ (g) + O_2\ (g) \rightarrow 2\ H_2O\ (l)$$

Bonds broken:

2 moles H–H bonds @ 432 kJ mol^{-1} = 864 kJ
1 mole O=O bonds @ 494 kJ mol^{-1} = 439 kJ
TOTAL energy needed to break bonds = 1,358 kJ

Bonds formed:

4 moles O–H bonds @ 459 kJ mol^{-1} = **1,836 kJ**

The formula already accounts for the exothermicity of bond formation because the minus sign precedes the bond energies of the bonds formed, so *leave the values for bond energy positive.*

$\Delta H_{RXN} = (\Sigma \text{ energy of bonds BROKEN}) - (\Sigma \text{ energy of bonds FORMED})$
$\Delta H_{RXN} = (1{,}358 \text{ kJ}) - (1{,}836) = -478 \text{ kJ per 2 moles } H_2O$
$\therefore$ the formation of one mole H_2O is $-\frac{478}{2}$, or **−239 kJ.**

Alternatively, you could have just performed the calculations for 1 mole H–H bonds and $\frac{1}{2}$ mole O_2 bonds to form 1 mole H_2O.

Break ALL bonds in the reactants and *form all the bonds in the products* unless you are positive of the reaction mechanism. If a bond isn't actually broken during the reaction, its formation will be subtracted in the "formed" term so the "+ broken" and "− formed" values will sum to zero.

334. (C) In this situation, the addition of Ca^{2+} and Cl^- ions *decreases* entropy because the water molecules become more ordered as they form hydration shells around the ions. The entropy change of a system is the sum of the entropy changes of its components; even though the $CaCl_2$ became *less* ordered as it dissolved, the entropy decrease of the water molecules was greater than the entropy increase of $CaCl_2$.

335. (D) Wood is mostly made of lignin and cellulose. Lignin is a complex chemical compound ($C_9H_{10}O_2$, $C_{10}H_{12}O_3$, $C_{11}H_{14}O_4$) and cellulose is a large, linear polymer of glucose ($C_6H_{12}O_6$). Their combustion with O_2 produces CO_2 and H_2O as follows (for 1 glucose):

$$C_6H_{12}O_6\,(s) + 6\,O_2\,(g) \rightarrow 6\,CO_2\,(g) + 6\,H_2O\,(g)$$

The combustion of wood (or petroleum products) is highly exothermic (why we use it for heat, light, and work) and produces six more moles of gas than it consumed, which results in increased entropy.

336. (C) Precipitation is the reverse of dissolving and is considered to decrease the entropy of a solution (unless the dissolution has a negative entropy change).

337. (D) The reaction is at equilibrium ∴ $\Delta G = 0$. Use the Gibbs free-energy equation:

$$\Delta G = \Delta H - T\Delta S$$
$$\mathbf{\Delta G = 0 \therefore \Delta H = T\Delta S}$$

338. (D) If the process is adiabatic and reversible, there is no heat flow ($\Delta H = 0$) and the system is in equilibrium ($\Delta G = 0$).

$$\Delta G = \Delta H - T\Delta S$$
$$\Delta G = 0 \therefore \Delta S = \frac{\Delta H}{T}$$
$$\Delta H = 0 \therefore \mathbf{\Delta S = 0}$$

All adiabatic, reversible processes are isentropic ($\Delta S = 0$, there is no change in entropy).

However, you can reason through the problem without the $\Delta S = \frac{\Delta H}{T}$ equation. The entropy of the gas in a cylinder is increased when its temperature is increased; however, the entropy in the cylinder is decreased when its volume is reduced. When we compress the cylinder of gas without heat exchange, its volume is reduced while its temperature increases, so its entropy remains unchanged.

This has an important application: work can be done by the system *only* if no heat is transferred.

ΔH versus q

Total energy (E) = $q + w$

q = heat

w = work

If volume is constant, no mechanical work can be done ∴ $\mathbf{\Delta E = q}$

If pressure is constant E is not necessarily equal to q.

Enter enthalpy:

$$\mathbf{\Delta H = \Delta E + P\Delta V}$$

∴ at constant P (pressure) and V (volume), when no mechanical work is done, $\Delta H = q$

ΔE = heat transferred in a process where $\Delta V = 0$

ΔH = heat transferred in a process where $\Delta P = 0$

339. (D) The ΔH_{RXN} can be calculated using the following formula:

$$\mathbf{\Delta H_{RXN} = [(\Sigma\ \Delta H^\circ_F\ products) - (\Sigma \Delta H^\circ_F\ reactants)]}$$

Remember:

- The ΔH°_F of elements in their standard states is zero.
- Multiply the ΔH°_F of the compound by the stoichiometric coefficient in the balanced equation.

Let $x = \Delta H^\circ_F\ CO_2$

$$\mathbf{\Delta H_{RXN} = [(\Sigma\ \Delta H^\circ_F\ products) - (\Sigma \Delta H^\circ_F\ reactants)]}$$

$$-1{,}367 = [(2x) + 3(-286)] - (-278 + 0)$$
$$-1{,}367 = (2x - 858) + 278$$
$$2x = -787$$
$$\mathbf{x = -393.5\ kJ\ mol^{-1}}$$

340. (B)

$$\Delta H_{RXN} = [(\Sigma\ \Delta H^\circ_F\ \text{products}) - (\Sigma \Delta H^\circ_F \text{reactants})]$$
$$\Delta H_{rRXN} = 9.7 - 2(34)$$
$$\mathbf{\Delta H_{RXN} = -58.3\ kJ}$$

341. (D) This is a Hess's law problem. **Because enthalpy is a state function** (it only depends on final and initial states), **you can combine the two reactions of known enthalpy changes in a way in which the same reactants yield the same products and the sum of their enthalpies will be the ΔH_{RXN} of the reaction of interest.**

The reaction of interest is 2 CO *(g)* → C *(s)* + CO_2 *(g)*.

2 CO *(g)* → C *(s)* + CO_2 *(g)*

CO *(g)* + ~~H_2 *(g)*~~ → C *(s)* + ~~H_2O *(g)*~~	$\Delta H^\circ_{298} = -131$ kJ
~~H_2O *(g)*~~ + CO *(g)* → CO_2 *(g)* + ~~H_2 *(g)*~~	$\Delta H^\circ_{298} = -41$ kJ
	$\Delta H^\circ_{298} = -172$ kJ

Both reactions needed to be reversed (therefore, reverse sign of ΔH°_{298}) in order to cancel the intermediates and end up with the correct forward reaction.

Hess's law works because enthalpy is a state function; therefore, enthalpy changes are path independent. It doesn't matter how reactants turn into products; the enthalpy change will always be the same (as long as the reactants and products are the same).

342. (D) The K_P of a gaseous system at equilibrium will NOT change with pressure changes. When the pressure on a gaseous system in equilibrium increases, **the equilibrium shifts** to favor the side of the reaction with the least number of moles of gas. For a pressure increase to shift the equilibrium, there must be a different number of moles of gas in the products and the reactants. However, when the new equilibrium is established, the value of K_P will be the same. Even though the partial pressures of the individual gases will change, their ratio (as defined by the equilibrium expression) will have the same value of K_P. **For a particular reaction, the numerical values of K_P, K_{EQ}, and K_{SP} only change with temperature.**

Reaction X:

1 mole gas reactants < 2 moles gas products ∴ ↑ P shifts equilibrium to favor reactants

Reaction Y:

2 moles gas reactants = 2 moles gas products ∴ ↑ P has no effect on equilibrium

Reaction Z:

2 moles gas reactants < 1 in the products ∴ ↑ P shifts equilibrium to favor reactants

343. (A) The K_P increases with increasing temperature for endothermic reactions (X and Y) **and decreases with increasing temperature for exothermic reactions** (Z). If you imagine heat as a product of an exothermic reaction, adding more heat by increasing temperature pushes the equilibrium to the left, favoring the reactants. The major difference between adding more heat and adding more of a chemical substance is that increasing the temperature is accomplished by continually adding more heat, so the reaction continues to get "pushed" to the left as long as the heat is being added (or removed), whereas adding more of a chemical product transiently pushes the equilibrium to the left, but then the previous equilibrium reestablishes itself.

You needed to know that the ΔH of reaction Z was exothermic to answer this question correctly. The AP Chemistry exam will not ask a multiple-choice question that relies on a correct answer from a different multiple-choice question, even if it is part of a set of related questions. This question could be a "stand-alone" question if the ΔH of reaction Z was provided in the data table.

344. (D) Gases are the most entropic state of matter and solids are the least entropic. Converting a gas to solid decreases entropy ∴ $\Delta S = S_{FINAL} - S_{INITIAL} < 0$.

Choices A and B are both incorrect because the entropy change is negative: the products are "more ordered" or less entropic than the reactants. Choice A correctly states that there are more species of product than reactant, and that *could* mean the products have greater entropy, but one of the products is a solid, which is less entropic than a gas. Choice B incorrectly states that there are more moles of products. There are two moles of products and two moles of reactants.

Choice C is alluring because we calibrate our definition of zero entropy with a pure, crystalline solid at 0 K. However, elements don't necessarily have lower entropy than compounds. The table below shows that elemental, molecular oxygen has a higher entropy than carbon monoxide. Solid carbon has the lowest entropy, but the entropy of diamond is not that much lower than the entropy of graphite. Carbon dioxide wins the entropy competition, but not by much. The high entropy of carbon monoxide, carbon dioxide, and oxygen are all due to the same factor: They are all gases.

Substance	$S°$ ($J \cdot K^{-1} \cdot mol^{-1}$)
C *(diamond)*	2
C *(graphite)*	6
CO *(g)*	198
CO_2 *(g)*	214
O_2 *(g)*	205

345. (D) The thermodynamic data in the table only tell us about energy and entropy changes in chemical reactions, not about reaction rates. In the absence of actual rates, values such as the rate constant and activation energy provide information about the rates of a particular reaction and can be used to compare and predict reaction rates.

346. (B) The graph below shows the effect of temperature on the distribution of particle speeds in a gas. **The *x*-axis may be called velocity, speed, or kinetic energy and the *y*-axis may be called frequency, probability, or particle number.** The Kinetic Molecular Theory of Gases (KMT) states that the temperature of a gas is directly proportional to the average kinetic energy of the particles in the gas. **Because it is an average, it can only tell us something about a population of particles, not an individual particle within the population.** Each particle can be moving within a range of speeds at any moment.

The higher the temperature, the larger the range of speeds the population has. Because the particles are colliding, the speed of an individual particle can change. One of the axioms of KMT is that collisions are elastic. Therefore, kinetic energy is conserved. If one particle loses speed to a collision, the other must pick up that speed.

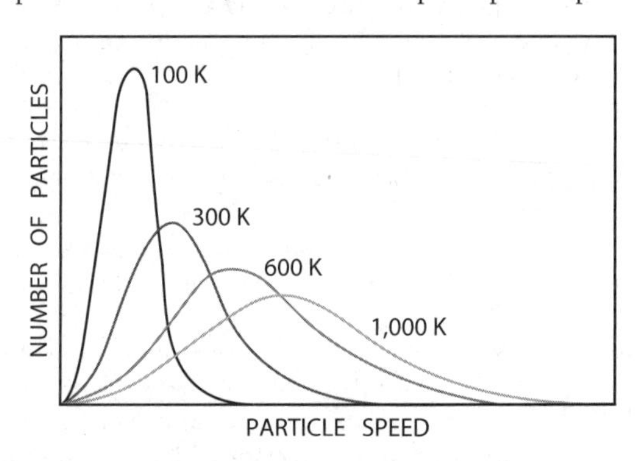

The importance of the graph (above) is that it shows us that it is not just the **average speed** of the particles that changes with changes in temperature, but their **distribution** changes as well. Notice the right end of the 1,000 K curve, which shows the distribution of speeds at the highest temperature. The peak is shorter and flatter with the largest percentage of particles at the highest velocities. The lowest temperature curve (100 K), all the way to the left, shows that most of the particles move at a lower speed and within a narrower range of speeds, with few of them moving with high velocity.

This graph offers insight into the effect of temperature on the rates of chemical reactions. **The higher the temperature, the greater percentage of particles will have or exceed the activation energy required for the reaction to proceed past its energy barrier.**

347. (B) The *y*-axis represents the frequency, or percent of particles, at each speed. See answer 346.

348. (B) Choice A is incorrect because it invalidates conservation of energy. Choice C incorrectly interprets the data; the specific heat (or heat capacity) of water is higher than any of the metals, which is why the water resists temperature changes. The energy transfer is a function of the masses, specific heats and temperatures of the substances, not just the temperature (why choice D is incorrect).

The basic assumption of calorimetry is that energy is conserved (the first law of thermodynamics): the heat lost or gained by the process being evaluated is equal to heat gained or lost by the water. The final temperature is the temperature at which thermal equilibrium has been reached. (The zeroth law of thermodynamics states that heat moves from a substance of higher temperature to a substance of lower temperature until thermal equilibrium.)

349. (A) The metal that produced the highest temperature change in the water has the highest specific heat because the same mass of that metal at the same temperature transferred the most heat to the water.

Metal	ΔT_{WATER} (°C)	**Heat gained by water (q, in J)** Using $q = mc\Delta T$, $m = 100$ g water, $c = 4.18$ J/g °C	ΔT_{METAL} (°C)	**Specific heat of metal (c, in J g^{-1}°C^{-1})** Using $c = \frac{q}{m}\Delta T$, $m = 50$ g metal
Al	7.28	3,047	67.7	0.90
Fe	3.83	1,602	71.2	0.45
Cu	3.34	1,298	71.7	0.39
Au	1.15	480	73.9	0.13

350. (D) The Dulong-Petit law (proposed in 1819) states that the heat capacity per weight comes close to a constant value when expressed as a molar quantity. In other words, the heat capacity of most solid substances is approximately 25 J mol^{-1} K^{-1}. This value is approximately $3R$ where R is the gas constant (8.314 J mol^{-1} K^{-1}). This relationship fails for light atoms close to room temperature and for all atoms at extremely low temperatures.

Metal	**Molar mass (g mol^{-1})**	**Number of moles in 50 g**	**Heat lost to water by 50 g**	**Specific heat (c) in J mol^{-1} K^{-1}** $c = \frac{q}{(\text{\# moles})(\Delta T_{METAL})}$
Al	26.980	1.850	3,047	24.3
Fe	55.847	0.895	1,602	25.1
Cu	63.546	0.787	1,298	24.5
Au	196.970	0.254	480	25.4

351. (C) The vaporization of 0.250 kg of water required 565 kJ of heat. The fusion (melting) of 1 mole of water requires an average of 1.5 moles of hydrogen bonds to be broken. The evaporation of 1 mole of water requires an average of 2.5 moles of hydrogen bonds to be broken.

Lowest entropy **Lowest potential energy** **SOLID (ICE)**	→	**LIQUID WATER**	→	**Highest entropy** **Greatest potential energy** **WATER VAPOR**
4 hydrogen bonds per water molecule		2–3 hydrogen bonds per water molecule		No hydrogen bonds

352. (B) The density of liquid water is greater than that of solid water, ice (ice floats!). The process of elimination could be used to answer this question even if you didn't think of that right away. Because the temperature does not change during a phase change, the average kinetic energy of the particles in the substance undergoing the phase change is constant. Hydrogen bonds are broken during melting (and evaporation, see answer 351). According to conservation of energy, the potential energy must increase during melting and vaporization.

353. (C) Less than 100 kJ of energy was absorbed to melt 0.250 kg of water $\therefore \frac{100}{0.250} =$ 400 kJ kg^{-1}. Choice C, 300 kJ kg^{-1}, is the closest number that is less than 400. The actual heat of fusion of water is 334 kJ kg^{-1}.

354. (B) Approximately 600 kJ of energy was absorbed to vaporize 0.250 kg of water $\therefore$ 600/0.250 = 2,400 kJ kg^{-1}. The actual heat of vaporization of water is 2,260 kJ kg^{-1}.

355. (A) Less than 100 kJ of energy was absorbed to increase the temperature of 0.250 kg of water vapor by 150° C $\therefore c = \frac{q}{m\Delta T} = \frac{100}{(0.250 \times 150)} \approx 2.7$ kJ kg^{-1}°C^{-1}. The actual value must be lower because the heat absorbed was less than 100 kJ. The actual specific heat of water vapor is 2.1 kJ kg^{-1} °C^{-1}.

356. (D) An average of 1.5 moles of hydrogen bonds per mole of water are broken during fusion (4 moles H bonds per mole ice – 2.5 moles H bonds per mole water = 1.5 moles H bonds broken during fusion), whereas an average of 2.5 moles of hydrogen bonds are broken (per mole of water) during vaporization (see answer 351). Water molecules *do* move closer together during fusion and do move farther apart during vaporization, but this fact alone does not explain the difference in energy requirements between fusion and vaporization.

357. (B) With the information in the graph accompanying questions 351–358 and the phase change flowchart under answer 351, the enthalpy of hydrogen bond formation can be calculated. (It will have the same magnitude but opposite sign of the enthalpy of *breaking* hydrogen bonds.) The heating curve of water is not showing the enthalpy change involved in forming water from H_2 and O_2 gas; it is simply changing state, so we cannot calculate the enthalpy of formation from the data. Because there is no time component, there is no way to know the rate at which heat is being added. Finally, there is no volume data with which to calculate density $\left(\text{density} = \frac{\text{mass}}{\text{volume}}\right)$.

358. (A) The absolute temperature of a substance is proportional to the average kinetic energy (KE) of the particles in the substance. (This is such a useful relationship you may want to consider replacing the word *temperature* with *average kinetic energy* in your mind while you're doing chemistry.) If the temperature is increasing, the average KE of the particles is increasing. **During phase changes, the temperature remains constant; therefore, there is no change in KE.** The heat added is used to increase the **potential energy (*PE*)** of the particles, which are changing positions relative to each other during phase changes. Melting, evaporation, and sublimation are endothermic processes that result in higher entropy states of matter.

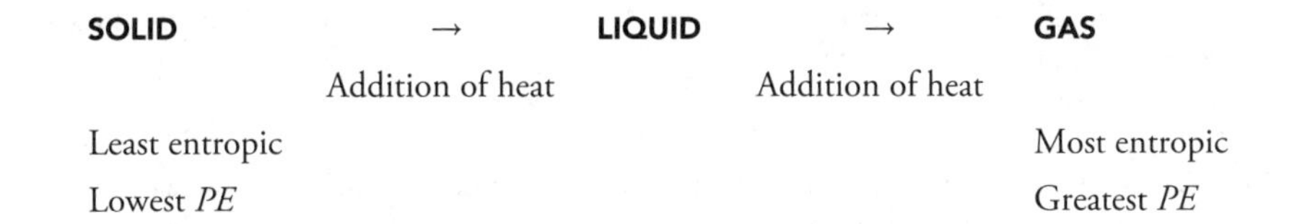

Questions asked in the negative are not common but do occur on the AP Chemistry exam, and they can be tricky. When answering a question asked with a "not," mark each answer choice as true or false (or yes or no). When you're done with the four choices, the one that's false (or no) is the correct answer.

359. (A) Questions 359–361 refer to a graph of **the Lennard-Jones potential, a mathematical model of the interaction between two neutral atoms or molecules.** The depth of the well represents the distance between the particles where the potential energy is at a minimum. The negative of this value is the **bond energy** (the amount of energy required to break the bond). See figure below answer 360.

360. (C) The distance represented is the average bond length. As shown by the leftmost double arrow, at distances shorter than the bond length there is a net repulsion between the atoms, and at longer distances, a net attraction. At very large distances the attraction is negligible. See figure below.

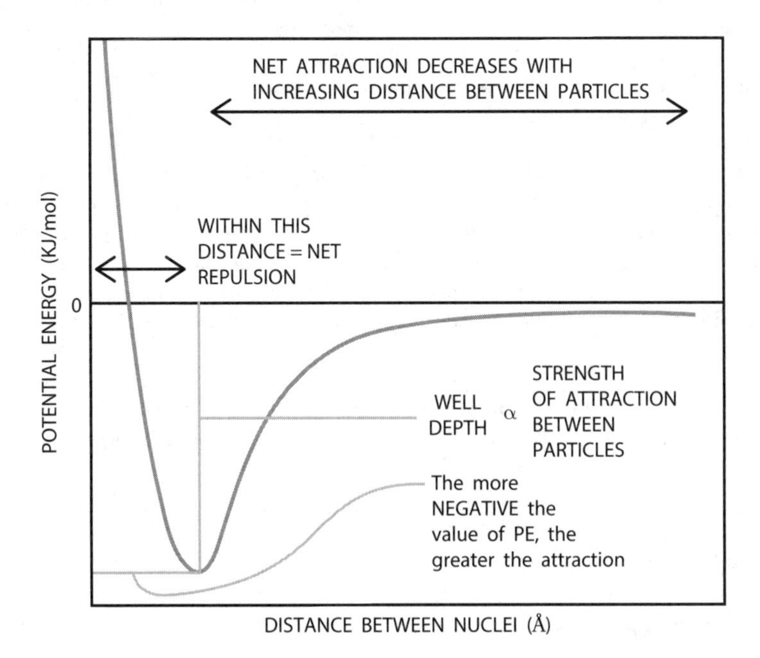

361. (B) The distance between the particles increases from left to right (see diagram below answer 360). The figure shows that there is net repulsion when the two particles are very close and then at larger distances, the net force is attractive. The area on the curve called the well, the big dip, indicates the distance of the "ideal bond length," the lowest-energy arrangement of the particles.

362. (D) The C—C and N—N bonds are the shortest in the table, so the distance between the well and the left vertical axis should be the smallest (graphs D and B, respectively). The C—C bond has the largest bond energy in the table, and in the graphs, the deepest well represents the largest negative bond energy (graph D).

363. (A) VSEPR (valence shell electron-pair repulsion) is a model for predicting molecular shapes from Lewis structures. **Valence bond theory** and **molecular orbital theory** (not covered in depth in the AP Chemistry syllabus) both use quantum mechanics to explain chemical bonding, including the prediction of molecular shapes. Lewis (in 1916) proposed that chemical bonds were the shared valence electrons. Lewis structures are representations of the molecules formed by the sharing of electrons. They are still useful in chemistry. Heisenberg's uncertainty principle, a foundational principle of quantum mechanics, states that there is a fundamental limitation to how precisely we can know both the momentum and position of a particle at a given time. However, it does not limit our measurements of bond lengths and energies. **Bond lengths can be measured by x-ray diffraction in solids or by microwave spectroscopy in gases. Bond energy is defined as the average of all the gas-phase bond dissociation energies for bonds of a specific type in the same kind of molecule (usually at 298 K).** It involves some measurement and some calculation. For example, measuring the hydrogen–oxygen bond in water requires breaking up water molecules into oxygen and hydrogen atoms (or radicals, but not O_2 and H_2, which means that the process would form bonds, too). Because there are two O–H bonds in water, that value would need to be divided by 2.

364. (D) See the diagram with answer 360. The two atoms would be approaching from a large distance, so start on the right side of the graph. As they approach each other, the electron-electron repulsion would eventually be overcome by the attraction of the electrons of one atom for the nucleus of another. Choice A is incorrect. It is an attempt to confuse the strong nuclear force with the repulsive behavior of negative-negative charges. Choice B is "kind of" correct. It is an incomplete and imprecise version of choice D. Choice C is just incorrect.

365. (D) In all three reactions, an H—H and an X—X bond must be broken and an H—X bond formed. You can consider this in terms of bond energies by rearranging the formula for ΔH_{RXN}:

$$\Delta H_{RXN} = (\Sigma \text{ energy of bonds BROKEN}) - (\Sigma \text{ energy of bonds FORMED})$$

$$\text{Energy of H—X bond formed} = (\text{energy of bonds H—H + X—X}) - \Delta H_{RXN}$$

Be careful of the stoichiometry! There are only $\frac{1}{2}$ moles of each reactant so you need to divide the energies of bonds broken in half.

Bond	H-H bond energy (kJ mol^{-1})	X–X bond energy (kJ mol^{-1})	Energy of bonds H–H + X–X (kJ mol^{-1})	Energy of bonds broken ÷ 2 (kJ)	ΔH_{RXN} (kJ mol^{-1})	Energy of bonds broken – ΔH = H–X bond energy (kJ)
H–F	436	155	591	**295.5**	−269	**565**
H–Cl	436	242	678	**339.0**	−92	**431**
H–Br	436	193	629	**314.5**	−51	**366**

You don't necessarily need to do all the math. The essence of the problem is that there are two values that are "competing" with one another: the energy to break the X–X bond and the energy to form the H–X bond.

Bond	Bond Energy (kJ mol^{-1})
H–F	567
H–Cl	431
H–Br	366

In the formation of HF, the energy needed to break the F–F bond is quite low, only 155 kJ mol^{-1}, but the energy that is released upon H–F bond formation is great—567 kJ mol^{-1}—resulting in the largest enthalpy change of the reactions listed. The Cl–Cl bond was the most energetically expensive to break at 242 kJ mol^{-1}; however, the formation of the H–Cl bond did not liberate as much energy as the formation of the H–F bond, resulting in a lower enthalpy change. The actual bond energies are listed above. Remember that different tables may have slightly different values and rounding errors in calculations may result in different answers.

Generally, an answer is considered correct (or at least reasonable) if it is within 5% of the accepted value.

366. (B) The universe tends to favor low energy (high stability) and highly disordered (entropic, statistically probable) states. The low energy "preference" of the universe is implied in the formula to calculate the enthalpy change of a reaction: $\Delta H_{RXN} = (\Sigma$ energy of bonds BROKEN) – (Σ energy of bonds FORMED).

The higher the bond energy of the bonds formed, the greater the negative enthalpy change. This makes sense: **the more stable the products (and the less stable the reactants) the greater the driving force of the reaction.** In the case of the decomposition of nitrogen trichloride, both entropy and enthalpy changes drive the reaction. The entropy change is positive because there are $\frac{5}{2}$ moles of gaseous products produced for every 1 mole of gaseous reactant consumed. **The enthalpy change is driven by the high bond energy and great bond stability of the nitrogen molecule.** The triple bond in nitrogen has a bond energy of 941 kJ mol^{-1}! Nitrogen trichloride, like ammonia, has a trigonal pyramidal shape. It is a yellow oil that explodes easily with great violence. However, it is rarely encountered as a pure substance because it is sensitive to light, heat, and mechanical shock.

Nitrogen Trichloride

Nitrogen trichloride has a very interesting past. The French chemist Pierre Dulong (of the famous Petit-Dulong law from answer 350) first prepared nitrogen trichloride in 1812 and ended up losing two fingers and an eye in explosions resulting from his work. Humphrey Davy, another great chemist, was temporarily blinded by an explosion involving nitrogen trichloride. This led him to hire Michael Faraday, giving him an unprecedented opportunity to work in a lab (he was a self-educated book-binder). Faraday went on to become one of the greatest (if not *the* greatest) experimental scientists of all time!

367. (D) The mixing of two gases at a constant temperature must have an enthalpy change of zero ($\Delta H = 0$). If no reaction occurs and the gases experience no forces of attraction or repulsion, then the entropy of the system must increase (**$\Delta S > 0$**). (See answer 322 for an in-depth explanation of entropy.)

$$\Delta G = \Delta H - T\Delta S \text{ and } \Delta H = 0 \therefore \Delta G = -T\Delta S$$

$$\therefore \boldsymbol{\Delta S = -\Delta \frac{G}{T}}$$

Because $\Delta S > 0$, **ΔG must be negative.**
In other words, **entropy-driven work can be done by this process!**

Chapter 5: Free-Response Answers

368. 211.5 kJ ∴ Endothermic

The method you use to calculate an enthalpy change will be determined by the data you are given. When provided with the standard enthalpies of formation you will need to use the following formula:

$\boldsymbol{\Delta H^\circ_{RXN} = \Sigma\ (\Delta H^\circ_F \text{ products}) - \Sigma\ (\Delta H^\circ_F \text{ reactants})}$
$\Delta H^\circ_{RXN} = [(-46.1 \times 2) + (-393.5) + (-241.8)] - [-939.0]$
$\Delta H^\circ_{RXN} = [-727.5] - [-939]$
$\Delta H^\circ_{RXN} =$ **211.5 kJ ∴ Endothermic**

369. 216 kJ ∴ Endothermic

The method you use to calculate an enthalpy change will be determined by the data you are given. When provided with bond energies you will need to use the following formula:

$\Delta H_{RXN} =$ **(Σ energy of bonds BROKEN) – (Σ energy of bonds FORMED)**

Ammonium carbonate is an ionic compound formed by 2 ammonium ions and 1 carbonate ion. In the reaction, the carbonate ion loses 1 oxygen atom (leaving a carbon dioxide molecule), and 2 ammonium ions each lose 1 proton (leaving two ammonia molecules). The oxygen from the carbonate ion and the protons from ammonium combine to form a water molecule. See figure below.

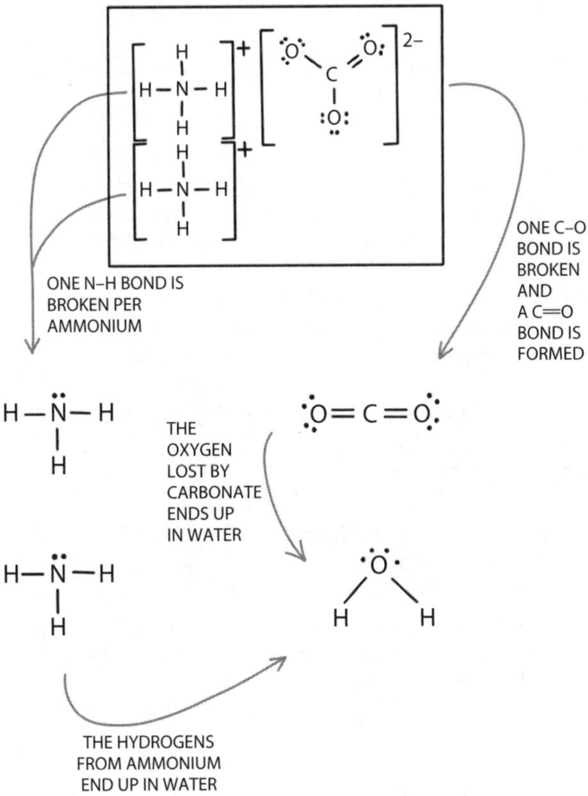

In a question like this, all the bonds do not need to be broken and re-formed. Because the magnitude of the energy change is the same whether the bond is broken or formed, the two equal and opposite values should cancel each other out. For example, breaking all 8 N–H bonds (requires 3,128 kJ) and forming 6 (yields 2,346 kJ) of them produces an energy change of 782 kJ, but just breaking 2 also yields an energy change of 782 kJ.

If you opt for this approach:
Break ALL bonds:

N–H (×8) = 391 × 8 = 3,128 kJ
C–O (×2) = 360 × 2 = 720 kJ
C=O (×1) = 802 × 1 = 802 kJ
TOTAL = 4,650 kJ absorbed
Bond breakage is always endothermic.

Form ALL bonds:
N–H (×6) = 391 × 6 = 2,346 kJ
C=O (×2) = 802 × 2 = 1,604 kJ
O–H (×2) = 463 × 2 = 926 kJ
TOTAL = 4,876 kJ released

Bond formation is always exothermic.

ΔH°_{RXN} **= (Σ energy of bonds BROKEN) – (Σ energy of bonds FORMED)**
ΔH°_{RXN} **= (4,650 kJ) – (4,876 kJ)**
ΔH°_{RXN} **= – 226 kJ**

If you feel comfortable doing so, you can save yourself a little time and a few calculations by breaking only the bonds that are present in the reactants but not in the products and forming only those bonds that are in the products but not the reactants:

Break 2 N–H bonds (one from each ammonium ion) = 391 × 2 = 782 kJ
Break 2 C–O bonds (in the carbonate ion, keep the C=O bond) = 360 × 2 = 720 kJ
TOTAL = 1,502 kJ absorbed

Form 1 C=O bond (there's 2 C=O in CO_2) = 802 × 1 = 802 kJ
Form 2 O–H bonds (water) = 463 × 2 = 926 kJ
TOTAL = 1,728 kJ released

ΔH°_{RXN} **= (Σ energy of bonds BROKEN) – (Σ energy of bonds FORMED)**
ΔH°_{RXN} **= (1,502 kJ) – (1,728 kJ)**
ΔH°_{RXN} **= – 226 kJ**

Remember the subtraction process gives the bonds formed the correct sign for the exothermic nature of bond formation, so **keep the values of all bond energies positive when applying this formula.**

370. 211.5 kJ and 216 kJ, a difference of 2.1%.

Generally, two answers are considered to be in agreement if they are within 5% of each other.

You probably learned at least three methods for calculating the enthalpy change of a reaction in your AP Chemistry class:

1. Using bond energies (energy of bonds broken – formed)
2. Using standard enthalpies of formation (ΔH°_F products – ΔH°_F reactants)
3. Hess's law

They should all provide the same answer within about 5% of each other. Calculation differences may be due to rounding errors or the differences in values provided in the tables.

The actual enthalpy change of a specific reactant (or reactants) converted to a specific product (or products) will be the same no matter what process is used to do it or whatever calculation is performed.

371. $\mathbf{P = 3.42}$ **atm**

Convert grams of reactants to moles.

$$10.0 \text{ g } (NH_4)CO_3 \times \frac{1 \text{ mol } (NH_4)CO_3}{96.1 \text{ grams } (NH_4)CO_3} = 0.104 \text{ mol } (NH_4)CO_3$$

Use stoichiometry to convert moles of reactant into moles of gaseous products formed.

From the balanced equation we know that for every mole of $(NH_4)CO_3$ consumed, four moles of gas are produced.

$$0.104 \text{ mol } (NH_4)CO_3 \times \frac{4 \text{ moles gas}}{1 \text{ mol } (NH_4)CO_3} = 0.417 \text{ mol gas}$$

Apply the ideal gas law to determine the pressure.

$PV = nRT$
$P(5) = (0.417)(0.0821)(500)$
$P = 3.42$ atm

372. $\mathbf{P_{NH_3} = 1.71}$ **atm,** $\mathbf{P_{CO_2} = 0.853}$ **atm,** $\mathbf{P_{H_2O} = 0.853}$ **atm**

Total number of moles gas (from answer 371) = 0.417 mol
Total pressure (from answer 371) = **3.42 atm**

$$\therefore \mathbf{\frac{8.20 \text{ atm}}{mol^{-1}}}.$$

$$\frac{1}{2} \text{ mol gas are } NH_3 \therefore \frac{1}{2} \times 0.417 \text{ moles gas} = \mathbf{0.209 \text{ mol } NH_3}$$

$$0.209 \text{ mol } NH_3 \times \frac{8.20 \text{ atm}}{\text{mol}} = \mathbf{1.71 \text{ atm } NH_3}$$

CO_2 and H_2O are present in equimolar quantities; therefore, they have the same partial pressure.

$$\frac{1}{4} \text{ moles are } CO_2 \therefore \frac{1}{4} \times 0.417 \text{ moles gas} = \mathbf{0.104 \text{ mol } CO_2}$$

$$\frac{1}{4} \text{ moles are } H_2O \therefore \frac{1}{4} \times 0.417 \text{ moles gas} = \mathbf{0.104 \text{ mol } H_2O}$$

$$0.104 \text{ mol } CO_2 \times \frac{8.20 \text{ atm}}{\text{mol}} = \mathbf{0.853 \text{ atm } CO_2}$$

$$0.104 \text{ mol } H_2O \times \frac{8.20 \text{ atm}}{\text{mol}} = \mathbf{0.853 \text{ atm } H_2O}$$

373. 5,139 kJ

The method you use to calculate an enthalpy change will be determined by the data you are given. When provided with bond energies you will need to use the following formula:

$$\mathbf{\Delta H_{RXN} = (\Sigma \text{ energy of bonds BROKEN}) - (\Sigma \text{ energy of bonds FORMED})}$$

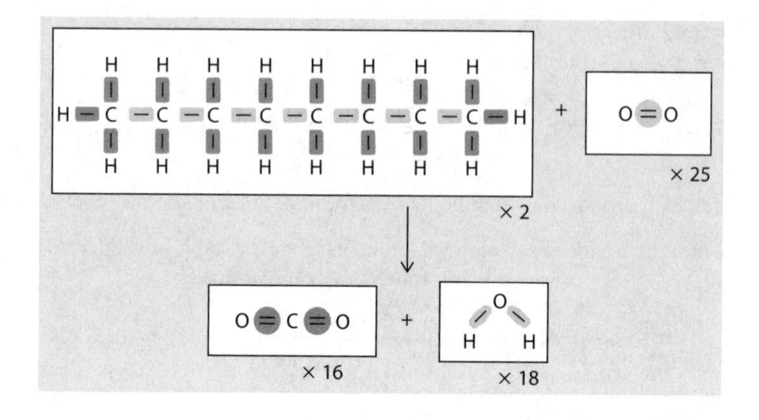

$$2\ C_8H_{18}\ (l) + 25\ O_2\ (g) \rightarrow 16\ CO_2\ (g) + 18\ H_2O\ (g) + \text{heat}$$

It is easier to use fractional coefficients ∴ $C_8H_{18} + \frac{25}{2} O_2 \rightarrow 8\ CO_2 + 9\ H_2O$

Bonds broken: 16,051 kJ

7.0 moles C−C bonds = 7(347) = 2,429 kJ
18.0 moles C−H bonds = 18(413) = 7,434 kJ
12.5 moles O=O bonds = 12(495) = 6,188 kJ
Total = 16,051 kJ
Bond breakage is always endothermic.

Bonds formed: 21,190 kJ

16 moles C=O bonds = 16(799) = 12,784 kJ
18 moles H−O bonds = 18(467) = 8,406 kJ
Total = 21,190 kJ
Bond formation is always exothermic.

$$\Delta H_{RXN} = (\Sigma \text{ energy of bonds BROKEN}) - (\Sigma \text{ energy of bonds FORMED})$$
$$\Delta H^\circ_{RXN} = (16{,}051) - (21{,}190) = \mathbf{5{,}139\ kJ}$$

Remember the subtraction process gives the bonds formed the correct sign for the exothermic nature of bond formation, so **keep the values of all bond energies positive when applying this formula.**

374. 5,574 kJ

It is easier to use fractional coefficients ∴ $C_8H_{18} + \frac{25}{2} O_2 \rightarrow 8\ CO_2 + 9\ H_2O$

$$\begin{aligned}\Delta H^\circ_{RXN} &= \Sigma\ (\Delta H^\circ_F \text{ products}) - \Sigma\ (\Delta H^\circ_F \text{ reactants}) \\ &= [(-393.5 \times 8) + (-241.8 \times 9)] - (-250) \\ &= (-5{,}324) - (-250) \\ &= \mathbf{-5{,}574\ kJ}\end{aligned}$$

The difference between this value (5,574 kJ) and the answer calculated in answer 373 (5,139 kJ) is 8.5%. That is not within the acceptable 5% limit, so technically, our answers are not the same. The accepted value of the combustion of octane varies between 5,330 and

5,530 kJ mol^{-1}, a difference of 3.8%. The 3.8% difference in accepted values complies with the 5% rule of thumb, but also indicates that *even the accepted value is not known with great precision*. In reality, deviations between values require some explanation. For example, **the energy of a particular bond can vary depending on the compound it's in**. For example, not all C–O bonds have the same bond energy. Tables of bond energies are typically averages of all the bond energies for that particular bond used in the construction of the table. Bond lengths are also averages as bonds are typically thought of as being "springy". However, many of the reasons for the deviations in these types of problems are beyond the scope of the what you need to know for the AP Chemistry exam. However, the following table may help illustrate the magnitude of percent error involved in these calculations.

$$\text{Percent error} = [|\text{accepted} - \text{calculated}| \div \text{accepted}] \times 100$$

Accepted value of combustion (kJ mol^{-1})	Percent error value of answer 373 (5,139 kJ mol^{-1})	Percent error value of answer 374 (5,574 kJ mol^{-1})
5,330	3.6%	4.6%
5,530	7.1%	0.8%

375. Energy is stored in chemical bonds. The energy of the products is less than the energy of the reactants, so the conversion of reactants to products results in the conversion of energy into heat.

The number of gas particles increases from 25 moles of oxygen gas to 34 moles of carbon dioxide and water gases, which forcefully expand due to the high temperature at which they are formed. The expansion of gases results in an increased volume due to the movable nature of the piston. The increased pressure of the gases pushes the piston up by exerting a force on it.

$$\text{Pressure} = \frac{\text{Force}}{\text{Area}} = \frac{\text{Force}}{\text{Length}} \times \text{Length}$$

$$\text{Volume} = \text{Length} \times \text{Length} \times \text{Length}$$

$$P\Delta V = \left(\frac{\text{Force}}{\cancel{\text{Length}} \times \cancel{\text{Length}}}\right)\left(\frac{\text{Length} \times \cancel{\text{Length}} \times \cancel{\text{Length}}}{1}\right)$$

$$= \text{Force} \times \text{Length}$$

$$= \textbf{Force} \times \textbf{Distance}$$

$$= \textbf{Work}$$

Because of the negative enthalpy change and the positive entropy change, the free-energy change is very negative.

376. Enthalpy change: The value of ΔH **is very negative**. The potential energy of the products is less than the potential energy of the reactants, so the conversion of reactants to products results in the conversion of potential energy into heat.

Entropy change: The value of ΔS **is very positive**. The number of gas particles increases from 25 moles of oxygen gas to 34 moles of carbon dioxide and water gases. The high temperature further increases entropy.

Free-energy change:

$\Delta G = \Delta H - T\Delta S$

ΔH is very negative
ΔS is very positive
T is very large
$\therefore$ **ΔG is very large and very negative**

377. $H_{fus} = 334 \text{ J g}^{-1}$
Calculate the mass of water.

$$\textbf{Mass}_{WATER} = (\text{Mass}_{CUP + WATER}) - (\text{Mass}_{CUP})$$
$$\mathbf{150.1\ g} = 156.50 \text{ g} - 6.40 \text{ g}$$

Calculate the change in temperature of the water.

$$\Delta T = T_{FINAL} - T_{INITIAL}$$
$$\mathbf{-16.75°\ C} = 13.25° \text{ C} - 30° \text{ C}$$

Calculate the heat lost by the water (and gained by the ice).

$$q = mc\Delta T$$
$$\mathbf{10{,}508\ J} = (150.10)(4.18)(16.75)$$

The assumption is that **all energy lost by the water is used by the ice to melt** (and only to melt).

Find the mass of ice.

$$\textbf{Mass}_{ICE} = (\text{Mass}_{CUP + WATER + ICE}) - (\text{Mass}_{CUP + WATER})$$
$$\mathbf{31.46\ g} = 187.96 \text{ g} - 156.50 \text{ g}$$

Calculate the heat of fusion (H_{fus}) of ice using the calculated value of q, the energy gained by the water.

$$q = m(H_{fus})$$
$$H_{fus} = \frac{q}{m}$$
$$H_{fus} = 10{,}508 \text{ J}/31.46 \text{ g}$$
$$\mathbf{H_{fus} = 334\ J/g}$$

378. The technique of calorimetry is based on the **conservation of energy. The calorimeter is assumed to be thermally isolated from the surroundings.** Another assumption is that **all energy lost by the water is used by the ice to melt the ice to 0° C but not raise the temperature of the melted ice.** As the ice begins to melt, the melted water absorbs heat from the original water until the two pools of water reach a new thermal equilibrium. If you had waited until the final lowest temperature was reached (in other words, not immediately after all the ice had melted), you would have had to account for the heat required to raise the temperature of the melted water.

The assumptions are reasonable but the quality of data obtained from any technique is only as good as the experimental design and the skill of the person performing the experiment. If the procedure is done with good technique, losses will be minimal. There will be losses, particularly from the lid (which is why the lid should be insulated and have small, tight openings to form a seal around the thermometer and stirring rod). Air, which will be present between the surface of the liquid and the ceiling of the lid, has a low specific heat. However, it can actually act as an insulator once it changes temperature (assuming there's no exchange with the air outside the calorimeter).

Notice that at 0° C (273 K), $\Delta G = 0$. At 0° C, the melting/freezing of water is at equilibrium. Any process at equilibrium has a ΔG of zero.

$$\begin{aligned}\Delta G &= \Delta H - T\Delta S\\ &= (6.012) - (273)(0.022)\\ &= 6.01 - (6.01)\\ &= \mathbf{0\ kJ\ mol^{-1}}\end{aligned}$$

And the higher the temperature, the more negative the free energy change and therefore the more thermodynamically favorable the process! Here's the calculation for the process **at room temperature (25° C)**.

$$\begin{aligned}\Delta G &= \Delta H - T\Delta S\\ &= (6.012) - (298)(0.022)\\ &= 6.01 - (6.56)\\ &= \mathbf{-0.55\ kJ\ mol^{-1}}\end{aligned}$$

379. The Gibbs free-energy equation contains the **three factors that contribute to the spontaneity of a process: enthalpy change, entropy change, and the temperature at which the process occurs.**

(a) $(334\ \text{J g}^{-1}) \times (18\ \text{g mol}^{-1}) \times (1\ \text{kJ}/1{,}000\ \text{J}) = \mathbf{6.012\ kJ\ mol^{-1}}$
(b) To calculate the ΔG use $\Delta G = \Delta H - T\Delta S$.

$$\begin{aligned}\Delta H &= 6.012\ \text{kJ mol}^{-1}\\ T &= 274\ \text{K}\\ \Delta S &= S_{FINAL} - S_{INITIAL}\\ &= 70 - 48 = 22\ \text{J mol}^{-1}\ \text{K}^{-1}\\ &\therefore 0.022\ \text{kJ mol}^{-1}\ \text{K}^{-1}\\ \mathbf{\Delta G} &= \mathbf{\Delta H - T\Delta S}\\ &= (6.01) - (274)(0.022)\\ &= 6.01 - 6.03\\ &= \mathbf{-0.02\ kJ\ mol^{-1}}\end{aligned}$$

Endothermic processes can be thermodynamically favorable if the entropy change is positive and the process occurs at a high enough temperature. The temperature required depends on the value of ΔH.

The Postulates of Kinetic Molecular Theory

- A gas consists of a collection of small particles traveling in constant, straight-line motion that obey Newton's laws of motion.
- The volume of the particles of the gas is negligible relative to its container.
- Collisions between gas particles are perfectly elastic (in other words, kinetic energy is conserved in collisions).
- There are no forces of attraction or repulsion between the particles.
- The average kinetic energy of the gas is proportional to the absolute temperature (or the kinetic energy of a particle is $\frac{3}{2}k\text{T}$, where T is the absolute temperature and k is the Boltzmann constant).

380. The postulates of Kinetic Molecular Theory are listed in the box above. **Pressure is defined as force/area .** The pressure of a gas in a closed container is measured by the pressure exerted by the gas particles on the walls of the container. **At a constant volume, increasing the temperature increases the average kinetic energy of the particles,** which, according to the Maxwell-Boltzmann distribution of particle speeds (see answer 346), means that a high frequency of particles are moving at high speeds and therefore colliding with each other, and the walls of the container, with greater force. **Because the volume is constant,** the area of the container wall is the same, so there is **a greater force distributed over the same area** and therefore a greater gas pressure.

381. The postulates of KMT are listed in the box above answer 380. Increasing the temperature increases the average kinetic energy of the particles, which, according to the Maxwell-Boltzmann distribution of particle speeds (see answer 346), means that a higher frequency of particles are moving at high speeds and colliding with each other and the walls of the container with greater force in the same area of container wall; therefore, a greater pressure is exerted on the walls. **To maintain a constant pressure,** however, **the container must be flexible enough so that it can be moved or stretched by the increased force exerted by the gas.**

As the volume of the gas increases, the distance between the gas particles increases. This is the result of the higher speed collisions between the particles. Because **kinetic energy is conserved in collisions,** the **higher speed of the collision results in a higher speed rebound, pushing the particles farther apart** with each collision.

382. (a) The kinetic energy of helium at 600 K is twice that of xenon at 300 K (calculations below).

The kinetic energy (*KE*) of a particle in either gas is given by the formula $KE = \frac{3}{2}kT$ where T is the absolute temperature and k is the Boltzmann constant. For simplicity, I'll ignore the Boltzmann constant; therefore, units are not included.

Container at 600 K	Container at 300 K
$KE \propto \frac{3}{2}\mathrm{T}$	$KE \propto \frac{3}{2}\mathrm{T}$
$KE \propto \frac{3}{2}(600)$	$KE \propto \frac{3}{2}(300)$
$\boldsymbol{KE \propto 900}$	$\boldsymbol{KE \propto 450}$

After the gases reach equilibrium, the energy of the gas will be the average of the two. In compartment A, helium was at the higher temperature; therefore, there was a larger range of particle speeds (see figure below, left) due to the low molar mass of helium. The high temperature and low molar mass of helium also means that there is a large range of particle speeds (see figure below, right). A higher percentage of helium particles will be moving at high speeds.

The higher molar mass and lower temperature of xenon means that most of the particles in compartment B are moving at a lower speed and there is a narrow range of particle speeds. Upon mixing, both will be at the same temperature, somewhere in between 600 and 300 K (the specific heat of the gases is needed to calculate the actual temperature), so the particle distribution will look like the bottom left figure. **At any temperature, the gas of higher mass will have a narrower particle distribution.** For example, at a particular temperature (see the following figure, left) practically all the xenon atoms are moving between speeds of just above 0 and 500 m/s whereas the speeds of the helium atoms at the same temperature range between just above 0 and 2,500 ms^{-1} and more than half of them are moving at a speed at or above 1,250 ms^{-1}.

(b) If xenon were at a higher temperature, the particle speed distribution would be greater than for xenon at the lower temperature. The lower temperature helium would have a tighter distribution than higher temperature helium, but its lower mass gives it a wider range of speeds than a heavier gas at the same temperature. **After mixing, the temperature of the gases would be the same, so the particle speed distribution would be predicted by the graph at the left (below), where particle speed distribution of two (or more) gases at a particular temperature is determined only as a function of their masses.**

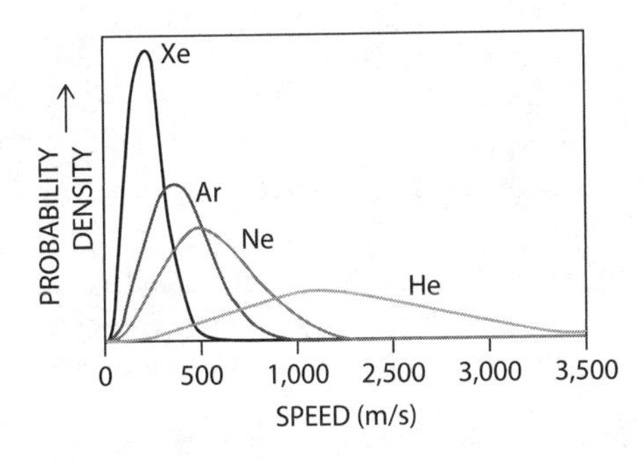

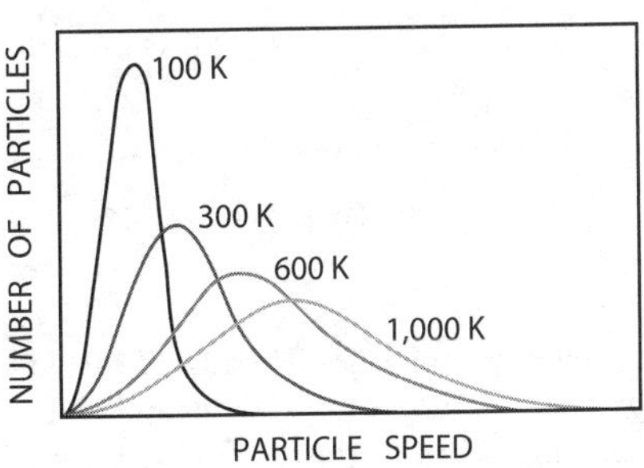

The bottom line: At equilibrium, the same temperature means the same average kinetic energy (the actual temperature reached cannot be determined by the information provided in this question) but **the smaller particles will always have a wider range of speed distribution and a larger fraction moving at higher speeds.**

383. Gas expansion is endothermic. Gases absorb heat in order to expand. The heat is absorbed from the immediate environment of the gas, in this case, the cartridge it was compressed in. Metals typically have low specific heats, so the cartridge quickly loses heat to the rapidly expanding gas.

Frosting occurs because the water vapor in the atmosphere condenses and freezes on the surface of the cartridge, indicating the temperature of the cartridge drops (momentarily) to below freezing! The amount of heat "removed" from the cartridge depends on the amount of gas stored in it and the pressure difference between the cartridge and the atmosphere.

Be careful when using compressed gas to blow dust from computer keyboards or other electronic devices as it may cause them to momentarily "freeze up."

384. Supersaturated solutions are extremely unstable. They are systems of high potential energy. Crystallization stabilizes and lowers the potential energy, releasing heat.

385.

(a) The dissolution of ammonium nitrate in water is endothermic. It absorbs heat from the water to dissolve, lowering the temperature of the water. (See answer 317 for a detailed explanation of solution enthalpy.)

(b) Use $q = mc\Delta T$ to calculate **the amount of energy needed to be removed to lower the temperature of water by 20° C.** Watch your units! The mass of water is given in grams, the specific heat is given in kJ °C^{-1} kg^{-1}, and the ΔH_{SOL} is given in kJ mol^{-1}. I converted grams of water to kilograms.

$$\begin{aligned} q &= mc\Delta T \\ &= (0.250)(4.18)(20) \\ &= \mathbf{20.9\ kJ} \end{aligned}$$

Calculate the number of moles of NH_4NO_3 needed to dissolve to absorb that much energy.

$$20.9 \text{ kJ} \times \frac{1 \text{ mole}}{25.7 \text{ kJ}} = \mathbf{0.81\ moles\ NH_4NO_3}$$

Convert moles of NH_4NO_3 to mass.

$$0.81 \text{ moles } NH_4NO_3 \times \frac{80 \text{ g}}{1 \text{ mole}} = \mathbf{64.8\ grams\ NH_4NO_3}$$

386. Heat is a notoriously difficult concept to define, but its behavior can be understood through its properties. **Heat is a form of energy that arises from the random motion of particles.** It moves from a body of higher temperature to a body of lower temperature until the temperature of the two bodies is the same.

Heat transfer between two bodies requires contact between the bodies. For example, the transfer of heat from a hot cup of coffee to the cup or to the air requires them to be in contact with the coffee. The coffee molecules, with their higher kinetic energy, transfer some of their kinetic energy to the lower kinetic energy molecules of the air or cup through collisions with them. The temperature of the coffee gets lower as the average kinetic energy decreases and the temperature of the cup and the air around it increases as they gain kinetic energy from the coffee.

In the Dewar flask, the near-vacuum space between the compartments prevents heat from being transferred from the environment to the inner flask or from the inner flask to the environment. It is the lack of particles in the space that separates and insulates the inner flask from the environment.

Theoretically, the Dewar flask can keep the inner flask temperature constant indefinitely, but in reality, having a removable lid provides a leak for heat transfer. A laboratory-grade Dewar flask has a vacuum lid as well as a vacuum flask, but the lid still has to be removable. That means the two separate vacuum compartments are joined but because they are not continuous, a tiny bit of heat will be transferred where the lid and the flask make contact.

387. $\mathbf{N_2 > NF_3 > NH_3 > NI_3}$

Many factors affect the stability of a molecule. Generally, **a compound with a negative enthalpy of formation is stable relative to the elements from which it was formed.** However, the triple bond in molecular nitrogen has a bond energy of 941 kJ mol^{-1}, making nitrogen more stable than its compounds. The negative enthalpies of formation of NH_3 and NF_3 are good predictors of their general stability. The $\Delta H°_F$ of NF_3 is significantly more negative than that of NH_3, so it is reasonable to predict it is more stable (under standard conditions).

The relative electronegativity values of the elements in a compound can be used to predict stability when other data are available and the compounds are related. Very generally, the greater the difference in electronegativity, the greater the stability of the compound. This is not a "rule"! The differences are shown below and indicate the lower stabilities of NCl_3, NBr_3 and NI_3.

Elements	Electronegativity difference
N & H	0.84
N & F	0.94
N & Cl	0.12
N & Br	0.08
N & I	0.38

The positive $\Delta H°_F$ values of NCl_3, NBr_3, and NI_3 also indicate these compounds are not very stable. Finally, the bond energies of these compounds are quite low. In fact, NCl_3, NBr_3, and NI_3 are considered very unstable, highly energetic compounds. Google them! They are mighty and impressive compounds! (See sidebar accompanying answer 366 for more information on NCl_3 and some unfortunate accidents resulting from it).

Chapter 6: Equilibrium

388. (B) A reaction or process is at **equilibrium** when **the rates of the forward and reverse reactions are equal, not zero** (answer choice A). Choice C is incorrect because equal concentrations of products and reactants are *not* criteria for equilibrium, but unchanging concentrations of reactants and products is an indicator that equilibrium has been reached. **Negative enthalpy changes and positive entropy changes are the driving forces that move systems toward equilibrium.** A chemical or physical process will occur for as long as one or both of these driving forces are present. Once a system is at equilibrium, however, its entropy is at a maximum and its energy is at a *minimum* (why choice D is incorrect).

389. (B) The free-energy change, ΔG, is the energy available to do work. The maximum free-energy change is 257 kJ because the reaction "loses" that free energy, so that amount of work can be done on something else using that energy. It's the maximum amount of work because often some of the energy usable to do work is converted to forms of energy that are not usable for work, such as heat.

Almost all reaction rates increase with temperature regardless of the enthalpy change (positive or negative) of the reaction. **The activation energy is a major determinant of the temperature dependency of a reaction.** Only reactions with a negative activation energy will have a decreased rate of reaction at higher temperatures, and those reactions are very rare.

Both the enthalpy and free-energy changes are negative, which indicate the process is thermodynamic favorable. However, the entropy change is also negative which means *entropy decreases* as the reaction proceeds, which is thermodynamically unfavorable. In order for a reaction to be *thermodynamically favorable* under any conditions, it must have a negative enthalpy change and a positive entropy change.

390. **(A)** At the temperature specified (500° C), $K_{EQ} = 1.45 \times 10^{-5}$, so the reactants, and the process that produces reactants (the reverse reaction), are favored. A good strategy to approach this type of problem is to calculate the reaction quotient (*Q*) to determine how far away from equilibrium the process is when three moles of each reactant and product are added.

$$Q = \frac{[NH_3]^2}{[N_2][H_2]^3}$$

$$= \frac{[3]^2}{[3][3]^3}$$

$$= \frac{9}{81} = \frac{1}{9}, \text{ or } \mathbf{0.111}$$

Q (= 0.111) $\gg K_{EQ}$ (= 0.0000145), so when the gases are combined in equimolar amounts, there is net movement in the reverse direction. Because $Q \neq K$, answer choice D cannot be correct.

Choices B and C have the same result, so if one is true, the other must be as well. (In the forward reaction, the formation of 2 moles of gas from 4 moles of gas will result in a decrease in pressure.) That means neither can be the correct answer to the problem.

391. **(B)** **Qualitative information about the enthalpy change of a reaction can be obtained by comparing the value of the equilibrium constant, *K*, at different temperatures.** A single value of *K* is a valuable clue as to whether or not the reaction is thermodynamically favorable ($-\Delta G$), but it can't tell you if a reaction is endothermic or exothermic. However, **if the magnitude of *K* *increases* with increasing temperature, the reaction is *endothermic*. If the magnitude of *K* *decreases* with increasing temperature, the reaction is *exothermic*.**

This can be reasoned using **Le Chatelier's principle**, which states that **for a system at equilibrium to which a stress is applied, the system will respond to relieve the applied stress**, at least as much as it can. The stress can be a change in the temperature, pressure, volume, and/or concentration. Le Chatelier's principle is an excellent reasoning tool for systems at equilibrium, of which there are many! Here's a simple way of viewing the temperature change in the reaction though Le Chatelier's lens:

$$\mathbf{N_2\ (g) + 3\ H_2\ (g) \rightarrow 2\ NH_3\ (g) + heat}$$

The temperature is increased by adding heat, so the stress is added "on the right" (of course that is not what is actually happening since the reaction is contained in one vessel) and stress causes "pushing" to the left. Temperature changes are unlike concentration changes, which are usually temporary—more of a substance is added, for example, disturbing the equilibrium, which will then return to its former state. **The increase in temperature is constant because to maintain the temperature, more heat must be continually added (or removed), so the reaction is "pushed and held" to the left.** Adding more NH_3, in contrast, would push the reaction to the left until the reverse reaction slowed down enough and the forward reaction sped up enough to reach the same equilibrium state as when the NH_3 was added.

The reverse reaction looks like this:

$$\mathbf{2\ NH_3\ (g) + heat \rightarrow N_2\ (g) + 3\ H_2\ (g)}$$

Here we see that increasing the temperature pushes (and holds) the reaction to the right, but the result is the same, an increased amount of N_2 and H_2. But as written, the products are favored, so the value of K would increase with increasing temperature.

392. (A) This buffer region of the titration curve is the part of the curve before the half-equivalence point. The species of highest concentration in the solution are HA and A^-. **At the half-equivalence point, [HA] = [A–] (and pH = pK_A).** Because the point of interest in the question precedes the half-equivalence point, and therefore pH < pK_A, then [HA] > [A^-].

The value of the pK_A of a compound is determined by its molecular structure. The quantitative behavior of an acid or base in solution can only be understood if its pK_A value is known.

393. (D) Equilibrium has been reached when the rate of a forward process is equal to the rate of the reverse process, making the system appear as **unchanging on a macroscopic level**. The concentrations of products and reactants does not change, but they are not necessarily equal. If they are, it's just a coincidence. There's only one temperature at which a K_P applies, so if a particular reaction has an equal concentration of products and reactants at equilibrium, it won't at a different temperature.

The K_P is a ratio of products and reactants. In this case, $K_P = \frac{[NO_2]^2}{[N_2O_4]}$. The small value of K_P indicates that the equilibrium of the system under the present conditions favors the reactants (and the reverse reaction), but the stoichiometry makes the exact relationship unclear until we **do the math**. The units can be any unit of pressure, probably atmospheres. Because there were no values of pressure in the question and other combinations of pressure can be "crunched" in the same way to give the value of 0.212, I've left out the units here to show just the relationship between the products and reactants.

$$\text{Let } x = [N_2O_4]$$
$$2x = [NO_2]$$
$$\mathbf{K_P = \frac{[NO_2]^2}{[N_2O_4]}}$$
$$0.212 = \frac{(2x)^2}{x} = \frac{4x^2}{x} = 4x$$
$$\mathbf{x = [N_2O_4] = 0.530}$$
$$\mathbf{2x = [NO_2] = 0.106}$$

394. (D) You don't need to do the math on this problem, but if you can, it's good to practice (see below). You can figure this out qualitatively. The pH of a 1 M concentration of a strong base is 14. Therefore, the pH of a 0.1 M concentration of a strong base is 13. Caffeine isn't *that* weak of a base—its K_B is in the range of 10^{-4} (it's a stronger base than ammonia)—but it's at a concentration of 0.1 M, a significant concentration. The pH would be lower than the pH of a 0.1 M strong base, but not by that much because this base is on the stronger side of weak.

Use the Brønsted-Lowry definition of a base.

$$C_8H_{10}N_4O_2 + H_2O \leftrightharpoons C_8H_{10}N_4O_2H^+ + OH^-$$
$$\text{Let } x = [OH^-] = [C_8H_{10}N_4O_2H^+]$$

Because the magnitude of the K_B is 10^{-4} (or lower), the value of x can be ignored in the denominator (also, importantly, because a 0.1 M concentration of caffeine is large compared with the $K_{B \text{ of caffeine}}$).

$$K_B = \frac{[C_8H_{10}N_4O_2H^+][OH^-]}{[C_8H_{10}N_4O_2]}$$
$$= \frac{(x)(x)}{(0.1 - x)}$$
$$= \frac{x^2}{0.1}$$
$$4 \times 10^{-3} = x^2$$

$x = 0.018$ ∴ **$[OH^-]$ is 0.018** ∴ pOH = 1.74 ∴ **pH = 12.3**

395. (D) Le Chatelier's principle states that if a chemical system at equilibrium experiences a change in concentration, volume, temperature, or pressure (total or partial), the equilibrium will shift to counteract the change and a new equilibrium will be established. To increase the amount of MgO produced, a disturbance to the equilibrium must be applied that will counteract it by producing *more* MgO. In other words, "pulling" the equilibrium toward MgO can also be thought of as "pushing" the equilibrium away from Mg and O_2.

Removing a reactant or product will typically cause the system to replace the removed substance; however, removing (or adding) Mg *(s)* or MgO *(s)* will not cause any change in the equilibrium because they are solids. Solids don't have a "concentration" and they are not represented in the equilibrium expression (or they are expressed as the value 1).

Increasing the pressure of a gaseous system at equilibrium will cause the equilibrium to shift to favor the side of the reaction with the least moles of gas. It follows then that decreasing the pressure of a gaseous system at equilibrium shifts the equilibrium to favor the side of the reaction with the most moles of gas.

Adding more O_2 *(g)* will increase MgO formation because the system will consume it in its effort to counteract the increase in O_2 pressure.

Choice A, using Mg filings instead of strips, will increase the reaction rate but will *not* affect the equilibrium. Because the reaction is exothermic, increasing the temperature will favor the reactants (This is why choice B is incorrect). The standard heat of formation of MgO represents the reaction, which is the formation of MgO from its elements (although the reaction shows the production of 2 moles whereas the standard heat of formation value applies to the formation of 1 mole). Choice D would result in an increased pressure inside the reaction vessel.

396. (D) The reaction is endothermic because heat (energy) is on the reactant side of the equation, so increasing the temperature would shift the equilibrium to the right. Lowering the temperature is the same as removing heat, in a way, removing a reactant that would cause the equilibrium to respond by replacing that reactant.

Choices A and B are the same. Increasing the pressure is often accomplished by decreasing the volume of the container. Either way, the result is the same: the equilibrium would be "pushed" to favor the products, of which there are 2 moles (there are 3 moles of gaseous reactants). Choice C would also "push" the equilibrium toward more product formation.

397. **(B)** The equation tells us that for every *one* F_2 that reacts, *two* F are produced. If we put the numbers of particles in the box into the equilibrium expression we get $\frac{[F]^2}{[F_2]} = \frac{4^2}{9} = 1.8$.

398. **(B)** An equilibrium constant of 2.5×10^3 is quite large, so the products are greatly favored in this reaction. Y^- is the proton acceptor in the forward reaction and X^- plays that role in the reverse reaction. Since the forward reaction is favored much more than the reverse, we can infer that Y^- is a stronger base than X^-. We can also infer that HX is a stronger acid than HY, but that is not an answer choice.

399. **(A)** **The percent dissociation of weak acids decreases with increasing hydrogen ion concentration.** Notice in the graph below that the percent ionization of acetic acid decreases at higher concentrations, and they aren't even very high concentrations! See answer 404 for an example of how high concentrations of a strong acid can affect ionization.

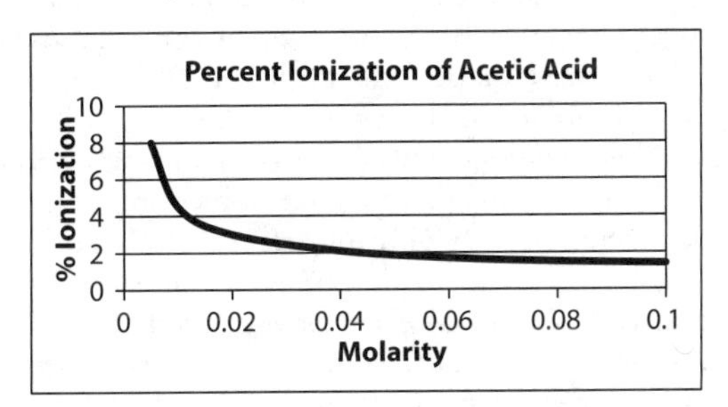

400. **(B)** The **strength of oxyacids** increases with

- increasing electronegativity of the central atom (the atom the O–H attaches to).
- other atoms of high electronegativity in the compound.
- an increasing number of oxygen atoms.

401. **(C)** Although choice C is not the most direct way to consider acid strength, strong acids have weak conjugate bases, which is why they dissociate and stay dissociated under most conditions. Weak acids don't completely dissociate unless they are "pushed" to by the addition of a base. Even in a titration of a weak acid with a strong base, the weak acid will still pick up protons from water, making the pH at equivalence greater than 7. The strongest conjugate base will be the weakest acid.

See answer 400 for a list of the three main factors affecting the strength of oxyacids. If choice A were true, the K_A of HOCl would be less than the K_A of HOBr. In these oxyacids, the proton is not attached to the halogen, it is attached to the oxygen atom. The size of the halogen *does* matter in the binary acid HF, HCl, HBr, and HI. In these acids, the strength correlates with the radius of the halogen: HI > HBr > HCl > HF (see table below).

Binary acid	K_A	Bond energy (kJ mol^{-1})	Electronegativity difference
HF	6.6×10^{-4}	567	1.78
HCl	1.3×10^{6}	431	0.96
HBr	1.0×10^{9}	366	0.76
HI	3.2×10^{9}	299	0.46

Choice B confuses the relationship of the O—H bond strength and the dissociation of the proton. A stronger bond between the oxygen and hydrogen would make the acid weaker (see table above to compare the bond strength of the binary acids with their K_A). Be careful not to equate the breaking of a bond with the dissociation of the proton from an acid. They are not the same thing! But the strength of the bond does have a negative correlation with acid strength in the binary acids. **Generally, the greater the difference in electronegativity between the atoms in a bond, the stronger the bond.**

Acid strength is independent of concentration; a particular acid has a K_A and a pK_A value, both of which are indicators of acid strength. The concentration of the acid will determine the pH of the solution, but a dilute solution of a strong acid can have a higher pH (be less acidic) than a high concentration of a weak acid. The K_A of hypoiodous acid (HIO) is 2×10^{-11}.

402. (D) The two curves shown on one graph illustrate an interesting point: the curves are the same except for the length of the steep portion of the curve. This is because the relative concentration difference between the acid and the base is the same. In both titrations, the base is 10 times more concentrated than the acid.

The pK_A of an acid is, experimentally, the pH at the half-equivalence point. It is measured experimentally under standard conditions, which are *not* the conditions shown. Regardless of that fact, there's typically **only one pK_A value for a particular acid**, so that should be a clue. Even if you only thought of pK_A as the negative log of the K_A, there is still only one value for each acid (at a particular temperature). The important point of this question is that in many of the calculations you do in chemistry, **there are certain times we consider the value of an unknown negligible**, and that is an accepted practice to obtain an approximate value, but it is true only under certain conditions. At certain concentrations, values we might consider negligible out of habit are not actually negligible. (See answer 405.)

Notice that the concentration difference between the two solutions is that the more concentrated solutions are 100 times more concentrated; the log of 100 is 2. Or, the less concentrated solution is so by $\frac{1}{100}$, the log of which is −2. And the pH difference in the buffer region is 2. That's not a coincidence. The pH of the buffer region depends on concentration. If the concentration differed by $10\left(\text{or } \frac{1}{10}\right)$, the pH difference in the buffer region would be 1.

403. (C) The weakly acidic solution contains a large number of unionized molecules at a given pH. In a solution of strong acid, the concentration of acid IS the concentration of hydrogen ions in the solution (at most concentrations you'll see in AP Chemistry, but see question and answer 404 for exceptions), but for a weak acid there are a lot of hydrogen ions that remain on the unionized acid molecules that will only be "pulled off" after the hydrogen ions in solution have been neutralized by the added base. In a sense, there is a reservoir of undissociated acid.

For example, a solution with a pH of 2.5 has a hydrogen ion (or hydronium ion) concentration of 0.0032 M. A 0.0032 M HCl solution would provide the hydrogen ions to produce a solution of that pH, but a 0.585 M concentration of acetic acid ($K_A = 1.76 \times 10^{-5}$) is needed to dissociate that many protons. That means the 0.5818 M undissociated acetic acid molecules are still in solution. If you had 1 L of each solution and had a 1 M concentration of a strong base, you'd need 3.2 mL to neutralize the HCl solution and 585 mL to neutralize the acetic acid solution:

$M_{HCl}V_{HCl} = M_{NaOH}V_{NaOH}$
$(0.0032\ M)(1\ L) = (1\ M)(V_{NaOH})$
$V_{NaOH} = 0.0032\ L = \mathbf{3.2\ mL}$

$M_{Acetic}V_{Acetic} = M_{NaOH}V_{NaOH}$
$(0.585)(1\ L) = (1\ M)(V_{NaOH})$
$V_{NaOH} = 0.585\ L = \mathbf{585\ mL}$

404. (A) There is only one peak on the 5 M condition at 1,049 cm^{-1}, indicating that the nitrate ion is the only species the technique is detecting. The peak gets smaller with increasing concentrations and other peaks appear at different cm^{-1}, indicating other species are present at concentrations detectable by the technique.

We can think of this in terms of Le Chatelier's principle:

$$HA \leftrightharpoons H^+ + A^-$$

Remember that **equilibrium is a condition that exists under a limited set of conditions.** For example, adding OH^- to this situation would remove H^+ from the solution (because it would combine with OH^- to make water) and that would promote further ionization of HA. This is what happens in a titration of a weak acid. If H^+ is added, the equilibrium will shift to consume it. The A^- (conjugate base) will "pick up" the H^+ ion, forming HA.

Now consider the extent of dissociation not as the strength of the acid, but as the strength of the conjugate base. Under most conditions, the conjugate base of a strong acid is very weak, but if we provide a high enough concentration of H^+ the base will "pick them up." **There is always a set of conditions that will take a reaction to completion, take its reverse reaction to completion, or hold the reaction in equilibrium**. The challenge to the chemist is to work out what those conditions are.

The AP Chemistry exam expects you to be able to reason about data that are totally unfamiliar to you. Never heard of Raman scattering? Most AP Chemistry students haven't, and that's why the College Board may ask about something like it. They don't want to assess only the specific facts you learned; they want to assess your ability to reason about chemistry with limited information.

405. (D) The Raman scattering shows that the lowest concentration (5 M) has the greatest ionization. (It appears the nitrate ion is the only species present.) The temperature at which this data was collected is not known, but **the table of K_A values shows that the K_A decreases with increasing temperature,** so there is less ionization at higher temperatures. This suggests the ionization of nitric acid is exothermic.

At 25° C, a 1 M HNO_3 solution 96.5% of HNO_3 dissociates:

$$K_A = 26.8$$
$$HNO_3 \rightleftharpoons H^+ + NO_3^-$$
$$\text{Let } x = [H^+] = [NO_3^-] \therefore [HNO_3] = 1 - x$$
$$K_A = \frac{[H^+][NO_3^-]}{[HNO_3]}$$
$$26 = \frac{x^2}{(1 - x)}$$
$$\boldsymbol{x = 0.965\ M}$$

406. (A) It is important to remember that even though pH is the negative log of the hydrogen ion concentration, the amount produced by the autoionization of water is equal to that of hydroxide ions, so as long as they are present in equal concentrations, the solution is neutral. The K_W of water at 25° C is 1×10^{-14} and is the basis of the pH scale, but the K_W of water changes with temperature, as most equilibrium constants will. **The autoionization of water is endothermic**, so higher temperatures push the equilibrium to the right, the ionized state. The K_W increases with increasing temperature:

Temperature (°C)	K_W	$[H^+]$ and $[OH^-]$
0	1.1×10^{-15}	3.3×10^{-8}
10	2.9×10^{-15}	5.4×10^{-8}
25	$\mathbf{1.0 \times 10^{-14}}$	$\mathbf{1.0 \times 10^{-7}}$
50	5.5×10^{-14}	2.3×10^{-7}
100	5.1×10^{-13}	7.1×10^{-7}

As long as the $[H^+] = [OH^-]$ the solution is neutral, regardless of the pH. It is the excess of $[H^+]$ that makes a solution acidic.

If you've ever wondered why questions (and data) often specifically mention 25° C as the temperature at which they occur in the problem, this is likely the reason. **Temperature is a major factor in most chemical processes.**

Questions asked in the negative are not common but do occur on the AP Chemistry exam, and they can be tricky. When answering a question asked with a "not," mark each answer choice as true or false (or yes or no). When you're done with the four choices, the one that's false (or no) is the correct answer.

407. (C) See the table in answer 406. **The K_W of water increases with increasing temperature, meaning there is greater ionization.** Generally, from low to moderate concentrations of ions, conductivity increases in a relatively linear way. Conductivity at high concentrations depends on the number of ions present. The pH of water changes with temperature as long as the K_W changes. If the ionization of water were exothermic, the pH would increase at higher temperatures. Choice D would need to specifically state that pH *decreases* with increasing temperature to support the statement that the autoionization of water is endothermic.

408. (A) Pure, liquid water, at any temperature, has an equal concentration of H^+ and OH^- ions. See answer 406. When answering a question asked with a "not," mark each answer choice as true or false (or yes or no). When you're done with the four choices, the one that's false (or no) is the correct answer.

409. (C) This is a simple "plug and chug" problem. All you need to do is substitute the concentrations into the equilibrium expression.

$$K_{EQ} = \frac{\text{products}}{\text{reactants}} = \frac{[Y]^3[Z]}{[X]^2} = \frac{125 \times 2}{16} \approx 16.$$

410. (D) The only determinant of the vapor pressure of a particular substance at equilibrium with its liquid is the temperature. The amount of surface area will affect the rates of evaporation (and condensation), and the volume of the container will affect how many moles of vapor are present, but not the *vapor pressure at equilibrium.*

411. (B) The vapor pressure of a liquid is determined at equilibrium at a particular temperature, so we can consider this as we would an equilibrium question. Just as the K_{EQ} of an endothermic reaction *increases* with increasing temperature (and the K_{EQ} of exothermic reactions *decreases* with increasing temperature), **higher temperatures favor vaporization over condensation until a new equilibrium is reached**, and it will include a larger fraction of vapor than the equilibrium condition at a lower temperature. Choice A is not at all true; the system in choice B is not at equilibrium. For a system at equilibrium (and at constant temperature), the rates of forward and reverse processes is the same, so D is not true. Finally, the ΔH of a forward and reverse process (or reaction) are equal in magnitude, but opposite in sign.

412. (C) At atmospheric pressures, N_2, O_2, and Ar condense at –195° C, –183° C, and –186° C, respectively. CO_2 does not condense into a liquid at atmospheric pressures, but becomes a solid at about –80° C.

413. (A) For proton-transfer reactions at equilibrium, the stronger acid will transfer the proton to the stronger base, yielding the weaker base and the weaker acid.

CH_3COOH	+	HCO_3^-	$\rightleftharpoons$	CH_3COO^-	+	H_2CO_3
stronger acid $K_A = 1.76 \times 10^{-5}$		**stronger base** (the conjugate base of the weaker acid)		**weaker** conjugate **base**		**weaker** **conjugate acid**

414. (A) At halfway to equivalence (and way before!), the strong acid is completely dissociated, but in the weak acid, there is still undissociated acid (see table below). If the weak and strong acid solutions were at the same pH, the solution of weak acid would be much more concentrated to produce the same concentration of hydrogen ions, so the volume of base needed to neutralize it would be higher than that required to neutralize the strong acid (choice B).

Acid-base neutralizations occur rapidly, regardless of the strength of the acid or base (choice C).

The buffer region of the curve precedes the equivalence point. Because the buffer region represents the point in the titration where the solution resists pH changes, **the higher the buffering capacity of the solution, the less steep the curve** will be.

Main Species Halfway to Equivalence

Strong acid	Weak acid
H_3O^+	H_3O^+
Anions from the acid	Anions from the acid
	Undissociated acid
Cations from the base	Cations from the base

415. (D) The pK_A is an indication of the extent of ionization of an acid (or base) at a particular pH. The pK_A is the pH at which half the molecules are ionized (ionized acids are deprotonated and ionized bases are protonated), so at pHs lower than the pK_A (a higher [H^+]), a greater percent of the molecules will be deprotonated. At pHs higher than the pK_A (a lower [H^+]), a greater percent are protonated. **The larger the value of pK_A, the smaller the extent of dissociation at any given pH.** But be careful: notice that the amino group (pK_A between 8.80 and 10.6) did not dissociate at all, it ionized by picking up a proton!

A useful rule of thumb is as follows:

At a pH < pK_A the proton is "on" more than 50% of the molecules.
At a pH > pK_A the proton is "off" more than 50% of the molecules.

At pH 7.4, amino acids have **protonated amino groups** (pH < pK_A ∴ the proton is "on" ∴ $NH_2 \rightarrow NH_3^+$) and **deprotonated carboxylic acid groups** (pH > pK_A ∴ the proton is "off" ∴ $COOH \rightarrow COO^-$).

416. (D) An increased number of hydrogen atoms does *not* make an oxyacid a stronger acid. If an acid has more than one dissociable proton, it is a *polyprotic acid.* The K_A of the first H^+ is the highest, and the K_A decreases for each successive H^+ that dissociates.

The **strength of oxyacids** increases with

- increasing electronegativity of the central atom (the atom the O–H attaches to)
- other atoms of high electronegativity in the compound
- an increasing number of oxygen atoms.
- When answering a question asked with a "not," mark each answer choice as true or false (or yes or no). When you're done with the four choices, the one that's false (or no) is the correct answer.

417. (C) K_{SP} is the solubility product constant, the equilibrium constant for a solid in equilibrium with its ions. It is measured in a saturated solution at equilibrium. The **molar solubility** is the number of moles that can be dissolved to produce 1 L of a saturated solution of that substance. What simplifies K_{SP} calculations is that there is always a solid reactant in these problems, but *solids are not included in equilibrium expressions, so they will never have a denominator.*

$$BaSO_4\,(s) \rightleftharpoons Ba^{2+}\,(aq) + SO_4^{2-}\,(aq)$$
Let $x = [Ba^{2+}] = [SO_4^{2-}]$ (They are produced in a 1:1 ratio.)
$$K_{SP} = [Ba^{2+}][SO_4^{2-}]$$
$$K_{SP} = x^2 = 1.1 \times 10^{-10}$$
$$\boldsymbol{x \approx 1 \times 10^{-5}}$$

Question 418 is a slightly more complicated variation of this problem.

418. (D) $CaF_2\,(s) \leftrightharpoons Ca^{2+}\,(aq) + 2\,F^-\,(aq)$

Let $x = [Ca^{2+}]$
F^- and Ca^{2+} are produced in a 2:1 ratio ∴ let $2x = [F^-]$.
Because K_{SP} is an equilibrium constant, the [F^-] must be raised to the exponent that is its stoichiometric coefficient, 2.

$$K_{SP} = [Ca^{2+}][2\ F^-]^2$$
$$K_{SP} = 4x^3$$
$$4.0 \times 10^{-11} = 4x^3 \therefore x^3 = 1.0 \times 10^{-11} \therefore x = \mathbf{(1.0 \times 10^{-11})^{1/3}\ M}$$

Remember that the $[Ca^{2+}] = x$, but the $[F^-] = 2x$.
Question 417 is a less complicated version of this problem.

419. (B) At low concentrations, NH_3 is a weak base and produces hydroxide ions, OH^-, that combine with Ni^{2+} to form $Ni(OH)_2$, which is normally insoluble. You don't have to know too many solubility rules, but you should be able to infer the insolubility of $Ni(OH)_2$ from the question. (You do need to know that **all sodium, potassium, ammonium, and nitrate salts are water soluble.**) The excess NH_3 allows it to form a stable, soluble, complex ion with Ni^{2+} instead of precipitating out with the OH^-.

The question does not require specific knowledge of nickel or hydroxide compounds. Only one answer choice makes sense. The strength of a base (or acid) is intrinsic to the compound; it does not depend on concentration (choice A). Adding more base to the solution will only raise the pH, so if choice C were true more precipitate would be formed. (You are not expected to know that much detail about nickel solubility.) Finally, there is no redox (oxidation-reduction) reaction happening (choice D incorrectly states that Ni^{2+} is reduced). Either way, the reduction of Ni^{2+} would form solid Ni, which would precipitate, not become more soluble.

420. (C) Before HF is added, Ba^{2+} and F^- are in equilibrium. The addition of HF, a weak acid, will increase the H^+ concentration of the solution (and lower the pH). The increased $[H^+]$ will shift the equilibrium of the BaF_2 to favor its dissociation (and therefore its solubility) by providing the F^- already in the solution with more H^+ ions to bind to form HF. F^- is a good conjugate base, so it will take up a large fraction of the H^+ ions.

The H^+ and F^- ions in the dilute solution being added are already at equilibrium, so adding them to the BaF_2 solution will not increase the F^- concentration to the final solution as much as it will increase the H^+ concentration. That's because the F^- that is being added will bond to the H^+ ions to form HF, further lowering the F^- concentration and further pulling the solubility equilibrium to the right, favoring dissociation and increased solubility. By the time the two solutions have completely mixed and a new equilibrium is achieved, the final F^- concentration is lower, allowing more Ba^{2+} ions to be dissolved.

421. (A) The solubility of $Fe(OH)_2$ decreases with increased pH because the higher concentrations of OH^- ions pushes the reaction toward the formation of solid $Fe(OH)_2$. Decreasing the pH *increases* its solubility because the extra H^+ ions in solution will combine with the hydroxides to produce H_2O, pushing the reaction to the left. **The value of K_{SP} does not depend on the pH directly**, but since one of the aqueous products of the reaction is OH^-, the K_{SP} already takes the pH into account.

In this case, OH^- is a **common ion**, an ion "added" (you can imagine that a base was added to increase the pH) that is already in the equilibrium mixture. The **common ion effect** is a shift in equilibrium caused by the addition of an ion that reacts with a component of the equilibrium mixture or the precipitate.

The AP Chemistry exam will not ask you to perform computations of solubility as a function of pH. Questions 420 and 421 show a mechanism of how acids and bases can affect solubility, which is something you should understand.

422. (D) The values for K_{A1}, K_{A2}, and K_{A3} decrease successively, so we expect the anion of the 3rd deprotonation to be present in the lowest concentration. We can also consider that only a small fraction of $H_3C_6H_5O_7$ will lose a proton, leaving $H_2C_6H_5O_7^-$ behind. Let's say 1% to make it simple, although it's actually much lower than that. Out of the small number of $H_2C_6H_5O_7^-$ ions, only a small fraction of those will lose the second proton, leaving $HC_6H_5O_7^{2-}$ behind. Again, let's say 1% for simplicity. That means that 1% of 1% (0.01%) of the original $H_3C_6H_5O_7$ loses two protons. Finally, if 1% of $HC_6H_5O_7^{2-}$ loses the third proton, only 1% of 0.01% of the original $H_3C_6H_5O_7$ would have lost all three protons (0.0001%).

Mathematical calculations of the concentrations of each species of the dissociation of a polyprotic acid will not be assessed on the AP chemistry exam.

423. (B) The low pH at the start indicates the analyte was an acid. The two steep portions of the curve indicate there were two dissociable protons; therefore, the acid was diprotic.

The AP Chemistry exam will not ask you to calculate the concentration of each species present in the titration curve for polyprotic acids, but they will for monoprotic acids.

424. (D) The pH scale is logarithmic, so a 1-point pH change means a 10-fold change in the hydrogen ion concentration. In this case, 10 times more hydrogen ions (the pH decreased, so the hydrogen ion concentration decreased) also produces 10 times more A^- ions. Since both resulted from the dissociation of HA, you could also think of this as there being $\frac{1}{10}$ the amount of HA.

425. (C) The compounds with the highest K_{SP} values are the most soluble; therefore, $Ca(OH)_2$ is the most soluble and Sb_2S_3 is the least soluble of the compounds listed (not HgS, as choice B states). Choice A is incorrect because it reverses the relationship between K_{SP} and solubility. It states that the solubility of AgCl is greater than the solubility of $CaCO_3$; however, the K_{SP} of AgCl (1.6×10^{-10}) is less than the K_{SP} of $CaCO_3$ (8.7×10^{-9}).
Although the dissolution of Sb_2S_3 in water is clearly not favorable, the exo- or endothermicity of the process is not predictable with this information (choice D). Remember that K_{SP} is an equilibrium constant and equilibrium constants allow the calculation of the free energy of a solution as it moves toward equilibrium ($\Delta G = \Delta G° + RT \ln K_{EQ}$, or in this case, $\ln K_{SP}$).

426. (D) For titration reactions, the strength of the acid or base in the titrant solution is not as important as the concentration. The titrant with highest concentration will require the lowest volume. All acid-base neutralizations occur quickly and the titration drives the reaction to completion, so the concentration of acid or base determines the concentration of H^+ or OH^- in the solution, regardless of the percent ionization of the solution prior to titration. (See answer 403 for a detailed explanation.) You probably remember applying the formula $M_AV_A = M_BV_B$ to strong and weak acids and bases alike.

427. (D) A buffer consists of a weak acid and its conjugate base or a weak base and its conjugate acid. **The general rule for choosing the weak acid or base for a buffer solution is to choose the species with a pK_A (or pK_B) value that is the pH of the buffer (+/− 1 point).** The best acid to prepare this buffer would have a pK_A of 4. Mathematically, the pK_A

is the negative log of the K_A, so $HCH_3H_5O_3$ would be an ideal acid. The $K_A = 1.4 \times 10^{-4}$. The negative log of this number is approximately 4 (the negative log of 10^{-4} is 4). The calculated value is 3.9, well within the +/− 1 point range (see table below). HNO_2 also has a comparable K_A (and therefore pK_A, too), but the one choice that includes it has it mismatched with another acid (choice B). Choice A has it as a conjugate base with a different acid of a completely different K_A. Choice C is incorrect because ClO^- is not the conjugate base of CH_3COOH and the pK_A of its parent acid (7.5) is completely out of the range of effective buffering for our desired pH.

Acid	K_A	pK_A
HNO_2	4.00×10^{-4}	3.4
HClO	3.00×10^{-8}	7.5
CH_3COOH	1.76×10^{-5}	4.8
$HCH_3H_5O_3$	1.40×10^{-4}	3.9

The AP Chemistry exam will not ask you to calculate changes in pH resulting from the addition of an acid or a base to a buffer.

428. (C) Point X is where the concentrations of A and B are equal. The reaction reaches equilibrium at about 1.0 second and that is when the rate of the forward reaction (A → B) is equal to the rate of the reverse reaction (B → A). Since choices A and B can't both be correct, neither can be true.

The rate of the forward reaction continues to decrease until about 1 second. The rate of a forward reaction can be calculated by the disappearance of A or the appearance of B. On a graph of concentration versus time, this is done by determining the slope of the tangent to a point of the concentration-versus-time curve. As the concentrations stabilize, the opposing rates become equal. Because the reaction started with only A, the rate of B → A will never exceed the rate of A → B. The reason is that the highest rate B → A will achieve is at equilibrium, when the rate of the reverse reaction is equal to the rate of the forward reaction.

429. (D) Le Chatelier's principle says that when a chemical system at equilibrium experiences a change in concentration, the equilibrium shifts to counteract the change imposed on it. By adding B, the equilibrium would shift to favor the reverse reaction, B → A, in order to consume the excess B. In other words, the higher concentration of B increases the rate of the reverse reaction, but as the reverse reaction consumes B, it slows down.

The concentration of A would increase, increasing the rate of the forward reaction to produce B, and thereby reducing the concentration of A. From the graph you can see that the forward reaction is favored (at equilibrium, the concentration of B is higher than A), so when more A is formed the rate of the forward reaction would increase to produce B until, once again, the forward and reverse reaction rates are the same.

Equilibrium is, in a sense, a series of never-ending corrections to concentrations, volumes, or pressures.

430. (B) The positive values of both ΔG and $\Delta G°$ indicate that the process is not thermodynamically favorable in the direction written or under either of the conditions given. Work can only be done by thermodynamically favorable processes.

The free-energy change of reactions depends on the nature of reactants and products, but is also dependent on the temperature, concentrations, pH, and pressure. The standard-state free-energy change, the $\Delta G°$, is measured under specific conditions (25° C, 1 M and 1 atm) but can be applied to other conditions by the formula $\Delta G = \Delta G° + RT \ln K_{EQ}$.

431. (B) The amount of NH_3 will increase as N_2 is added because a system at equilibrium will respond to the increase in N_2 by consuming it (and forming NH_3). For every molecule of N_2 that is added, three H_2 will be consumed, so the amount of H_2 will decrease. Adding more gas to the tank will cause the pressure to increase, which will cause the system to respond by reducing the pressure (by favoring the reverse reaction because the reactants have a lower number of moles of gas). However, the total pressure will still increase (or at least stay the same) as more gas is added.

432. (A) Only temperature changes can change the value of K_P. Pressure changes in a gaseous system will *shift* the equilibrium until the new equilibrium is reached, but the ratio of the partial pressures of the products and reactants will stay the same. The actual partial pressures of each gas (at equilibrium) may be different, but their ratio as calculated by the equilibrium expression will result in the same value of K_P.

433. (A) A thermodynamically unstable compound is one in which the $\Delta G°_F$ is positive. The reaction shows the standard free energy of formation ($\Delta G°_F$) of benzene, even though the question does not specifically state it (not to be confused with the standard enthalpy of formation). Both the standard free energy of formation and the standard enthalpy of formation refer to the formation of one mole of a compound or non-standard-state element under standard conditions from standard-state elements. **The $\Delta G°_F$ of a substance is a measure of its stability relative to its elements under standard conditions.** Remember that **all reactions have a set of conditions in which they will proceed to completion, but those conditions may not be easily attainable.**

The value of $\Delta G°$ (the "°" indicates that standard-state conditions apply) is +124 kJ. A positive value indicates the reverse reaction is thermodynamically favorable, that is, the decomposition of benzene under standard conditions is the thermodynamically favorable process. **A thermodynamically unstable compound is one in which the $\Delta G°_F$ is positive, but that does not mean that the reaction will actually happen under the conditions present**. The kinetics of the reaction may be very slow (the activation energy may be very high) so that in a laboratory at 25° C, benzene can be kept indefinitely without any decomposition at all!

434. (C) The equation $\boldsymbol{\Delta G° = -RT \ln K_{EQ}}$ shows the relationship between $\Delta G°$ and K_{EQ}. It also states that the standard-state free-energy change of the process can be calculated if the K_{EQ} is known and that the point of equilibrium for a reaction is a function of the standard-state free-energy change. The ΔG (or $\Delta G°$) is equal to zero when a system is at equilibrium, but the K_{EQ} can't be zero unless the reaction as written produces no product at all.

Choice D does not relate $\Delta G°$ and K_{EQ}. $\Delta G°$ does represent the free energy change of a reaction, under standard conditions, but that reaction may proceed to completion under those conditions. The free energy change is a thermodynamic function, and although it is related to equilibrium, by itself it does not tell us anything about whether or not a reaction proceeds to equilibrium or goes to completion. That is the job of the K_{EQ}.

Chapter 6: Free-Response Answers

435.

$\boxed{CH_3CH(OH)COOH}\ (aq) + H_2O\ (l) \rightleftharpoons CH_3CH(OH)COO^-\ (aq) + H_3O^+ (aq)$

$CH_3CH(OH)COOH$ is the acid and $CH_3CH(OH)COO^-$ is the conjugate base.
Conjugate acid-base pairs differ by *one proton*.
For the forward reaction, **the acid and base will be on the left** side of the arrow. **Conjugate acids and bases will be on the right.**

436. (a) Calculate the $[H^+]$ from the pH.

$$pH = 2.23 \therefore -\log [H^+] = 2.23 \therefore 10^{-2.23} = [H^+]$$
$$[H^+] = 0.00589\ M = \mathbf{5.89 \times 10^{-3}\ M}$$

Use the general reaction for a monoprotic acid ($HA \rightleftharpoons H^+ + A^-$) to **calculate the concentrations of the other species**. The original concentration of the solution is 0.250 M

$$[H^+] = \mathbf{[A^-] = 5.89 \times 10^{-3}\ M}$$
$$\mathbf{[HA]} = 0.250 - 5.89 \times 10^{-3} = \mathbf{0.244\ M}$$

(b) Use the $[H^+]$ to calculate the K_A.

$$K_A = \frac{[H^+][A^-]}{[HA]}$$

$$K_A = \frac{[5.89 \times 10^{-3}]^2}{[0.244]} = \mathbf{1.42 \times 10^{-4}}$$

437. Calculate the number of moles of lactic acid needed to make 50.0 mL (0.050 L) of a 0.250 M solution.

$$0.05\ L \times \frac{0.25\ mole}{L} = \mathbf{0.125\ mole\ lactic\ acid}$$

Convert moles of lactic acid the mass.

$$\text{Molar mass of lactic acid} = 90.08\ g/mole$$

$$0.125\ mole \times \frac{90.08\ g}{mole} = \mathbf{1.13\ grams\ lactic\ acid}$$

438.

1. **False.**

$$M_AV_A = M_BV_B$$
$$(0.250)(50.0) = (0.250)(50.0)$$
$$V_B = 50.0\ mL$$

The molarity required to neutralize is only related to the number of H^+ and OH^- ions available on the acid and base, not the relative strengths of the acid and base.

2. **False.**
This is like a titration of a weak acid with a strong base (or a strong base with a weak acid). The number of moles of H^+ and OH^- are the same, but the conjugate base of the weak acid is strong enough to get protonated in solution, lowering the H^+ concentration and thereby raising the pH above 7.

3. **True.**
   Two solutions with the same pH have the same concentration of $[H^+]$ ions, but the lactic acid doesn't completely dissociate, so a greater concentration of lactic acid is needed to produce the same concentration of hydrogen ions. HNO_3 is a strong acid and it will completely dissociate, so the "initial" concentration of $HNO_3 = [H^+]$.
4. **True.**
   There are six (or seven) strong acids: HCl, HBr, HI, HNO_3, H_2SO_4, $HClO_4$, and some consider $HClO_3$ strong. The rest of the acids are weak.

439. There are two moles of hydroxide ions per mole $Sr(OH)_2$.

$$\mathbf{M_AV_A = M_BV_B(2)}$$
$$M_A(0.50) = (0.20)(15.6)(2)$$
$$\mathbf{M_A = 0.125\ M}$$

440. Calculate the concentration of Sr(OH)2, a strong base, so there is complete dissociation:

$$Sr(OH)_2 \rightarrow Sr^{2+} + 2\ OH^-$$
$$0.200\ M\ Sr(OH)_2 = \mathbf{0.400\ M\ OH^-}$$

Calculate the pOH of the solution using the OH^- concentration.

$$\mathbf{pOH} = -\log\,[OH] = -\log 0.40 = \mathbf{0.398}$$

Calculate the pH from pOH.

$$pH + pOH = 14 \therefore 14 - pOH = pH$$
$$\mathbf{pH} = 14 - 0.398 = \mathbf{13.6}$$

Alternatively, you could use the K_W of water to calculate the $[H^+]$ from the $[OH^-]$.

$$K_W = 1.0 \times 10^{-14} = [H^+][OH^-]$$
$$[H^+] = \frac{1.0 \times 10^{-14}}{[OH^-]} = \frac{1.0 \times 10^{-14}}{0.400}$$
$$\mathbf{[H^+] = 2.5 \times 10^-14} \therefore \mathbf{pH} = -\log 2.5 \times 10^{-14} = \mathbf{13.6}$$

441. Calculate the $[H^+]$ using the K_A you calculated for lactic acid in answer 436(b).

$$\mathbf{K_A = 1.42 \times 10^{-4}}$$
$$HA \rightleftharpoons H^+ + A^-$$
$$\text{Let } x = [H^+] = [A^-]$$
$$\text{Let } [HA] = 0.125 - x \text{ (assume } x \text{ is negligible)}$$
$$K_A = \frac{[H^+][A^-]}{[HA]}$$
$$1.38 \times 10^{-4} = \frac{x^2}{0.125}$$
$$x = \mathbf{[H^+]} = 0.00415 = \mathbf{4.15 \times 10^{-3}\ M}$$

Calculate the pH from the $[H^+]$.

$$\mathbf{pH} = -\log [H^+] = -\log 0.00415 = \mathbf{2.38}$$

Is x Really Negligible?

You can check that the x really is negligible by plugging it back into the original formula and solving for it (from answer 441):

$$K_A = \frac{(x^2)}{(0.125 - x)}$$

$$= \frac{(4.15 \times 10^{-3})}{(0.125 - 4.15 \times 10^{-3})}$$

$$= \frac{1.72 \times 10^{-5}}{0.12085}$$

$$= \mathbf{1.42 \times 10^{-4}}$$

The value you obtain should be within 5% of the value that assumed x was negligible. In this case the two values were ~2.8% different, so the assumption that x was negligible was acceptable.

442. The pH at equivalence will be greater than 7. **At equivalence, the number of moles of hydroxide ions *added* is equal to the number of moles of lactic acid present (regardless of how much was originally ionized in solution).** (Lactic acid is a monoprotic acid, so there's one mole of hydrogen ions per mole of lactic acid.) The conjugate base of lactic acid is fairly strong since lactic acid is fairly weak, so the anion of the acid, lactate, can pick up hydrogen ions in the solution from water, decreasing the hydrogen ion concentration and increasing the pH.

443. Indicator C. The pK_A of the indicator should be in the vicinity of 1 pH point of the equivalence point. The pK_A of an acid (or base) is the pH at which half the molecules present are ionized (for an acid that means deprotonated; for a base it means protonated). It is the ionization of the indicator that causes a color change.

A good rule of thumb is: **The pK_A INDICATOR = +/− 1 of pH at equivalence.**

444. K_C = 164

Set up an "ICE Box":

	H_2	I_2	2 HI
Initial	2.00	2.00	0.00
Change	$-x$	$-x$	$+2x$
Equilibrium	$2.00 - x$ ∴ **0.270 moles**	$2.00 - x$ ∴ **0.270 moles**	$2x$ = **3.46 moles**

Solve for *x*.

$$2x = \text{number of moles of HI at equilibrium} = 3.46$$
$$\therefore \mathbf{x = 1.73\ moles}$$

Since all the gases are in a 1 L container, the number of moles per liter is the same as the number of moles in the container, and can be considered the "concentration" of the gases in mol L^{-1}.

$$K_C = \frac{[HI]^2}{[H_2][I_2]}$$
$$K_C = \frac{[3.46]^2}{[0.27][0.27]}$$
$$= \frac{11.97}{0.0729}$$
$$\mathbf{K_C = 164}$$

Because there are three significant digits in all of the data used in the calculations, our answer should also contain three significant digits.

445. Four moles of gas are added to the container.

$$P_{TOTAL}V = nRT$$
$$P_{TOTAL}(1) = (4)(0.0821)(500)$$
$$\mathbf{P_{TOTAL} = 164\ atm}$$

H_2 and I_2 each make up 50% of the gases ∴ 82 atm each.

At equilibrium, 4 moles of gas are present (3.46 moles HI + 0.270 moles H_2 + 0.270 moles I_2).

Four moles of gas at 164 atm ∴ each mole of gas exerts 41 atm of pressure.

H_2 and I_2

$$0.270\ \text{mole} \times \frac{41\ \text{atm}}{\text{mole}} = \mathbf{11\ atm\ each}$$

HI

$$3.46\ \text{moles} \times \frac{41\ \text{atm}}{\text{mole}} = \mathbf{141.9\ atm}$$

446. $K_P = 166$

$$K_P = \frac{[HI]^2}{[H_2][I_2]}$$
$$K_P = \frac{[141.9]^2}{[11][11]}$$
$$\mathbf{K_P = 166}$$

The K_C and K_P values are very similar (less than 2% different). **The calculation of K_C uses the number of moles per L of the gases. The calculation of the K_P uses the partial pressures of the gases.** You can convert between the two using the formula $\mathbf{K_P = K_C(RT)^{\Delta n}}$ where R is the gas constant, T is the temperature and Δn is the change in the number of moles of gas in the reaction (number moles of gaseous product − number moles of gaseous reactant). For this problem, $\Delta n = 0$. **When $\Delta n = 0$, $K_P = K_C$,** so the 2% difference is not significant.

Because there are three significant digits in all of the data used in the calculations, our answer should also contain three significant digits.

447. The partial pressures would be half of those of the 1 L container.

$$P_{TOTAL}V = nRT$$
$$P_{TOTAL}(2) = (4)(0.0821)(500)$$
$$\mathbf{P_{TOTAL} = 82\ atm}$$

Four moles of gas at 82 atm ∴ **each mole of gas exerts 20.5 atm** of pressure
H_2 and I_2

$$0.270\ \text{mole} \times \frac{20.5\ \text{atm}}{\text{mole}} = \mathbf{5.54\ atm\ each}$$

HI

$$3.46\ \text{moles} \times \frac{20.5\ \text{atm}}{\text{mole}} = \mathbf{70.9\ atm}$$

448. Exothermic.
If the reaction were **exothermic:**

$$A + B \leftrightharpoons C + \textbf{heat}$$

If the reaction were **endothermic:**

$$A + B + \textbf{heat} \leftrightharpoons C$$

Le Chatelier's principle states that **for a system at equilibrium to which a stress is applied, the system will respond to relieve the applied stress.** The stress in this case is an increased temperature. Temperature is increased by adding heat, so if the reaction were exothermic, the stress would be applied "on the right" and the system would respond by "pushing" to the left, favoring the reactants and lowering the K_C of the reaction. If the reaction were endothermic, the stress (heat) would be applied "on the left" (bottom reaction) and the system would respond by "pushing" right, favoring the products and increasing the K_C.

Temperature changes are unlike concentration changes, which are usually temporary—more of a substance is added, for example, disturbing the equilibrium, which will eventually return to its former state. **The increase in temperature is constant because to maintain the temperature, more heat must be continually added (or removed), so the reaction is "pushed and held" to the left (or right).**

449. Decrease.
Calculate Q_C and compare to K_C.

$$Q_C = \frac{[HI]^2}{[H_2][I_2]}$$

$$Q_C = \frac{(1)}{(1)(1)} = 1$$

$$\therefore Q < K_C$$

Therefore, the reaction will proceed in the forward direction and the amount of H_2 will decrease.

450. **(a)** $K_P = 0.70$

Set up an "ICE box" to calculate equilibrium partial pressures.

	PCl_5	PCl_3	Cl_2
Initial	1.40	0.00	0.00
Change	−0.70	+0.70	+0.70
Equilibrium	**0.70**	**0.70**	**0.70**

Use equilibrium partial pressures to calculate the K_P.

$$K_P = \frac{[PCl_3][Cl_2]}{[PCl_5]}$$

$$K_P = \frac{(0.70)(0.70)}{(0.70)}$$

$$\mathbf{K_P = 0.70}$$

(b) Total pressure = 0.70 + 0.70 + 0.70 = 2.1 atm

451. The partial pressures of each gas will be the same at equilibrium because for each PCl_5 molecule that decomposes, one PCl_3 and one Cl_2 form. Your drawing should be similar to the one below:

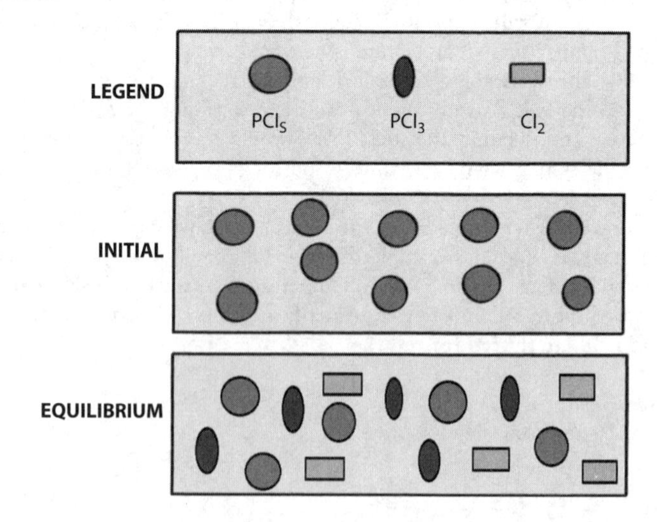

452. The pressure changes in the system as it reaches equilibrium:

Partial pressure PCl_5	Partial pressure PCl_3	Partial pressure Cl_2	Total pressure
1.40	0.00	0.00	**0.00**
1.30	0.10	0.10	**1.50**
1.20	0.20	0.20	**1.60**
1.10	0.30	0.30	**1.70**

1.00	0.40	0.40	**1.80**
0.90	0.50	0.50	**1.90**
0.80	0.60	0.60	**2.00**
0.70	0.70	0.70	**2.10**

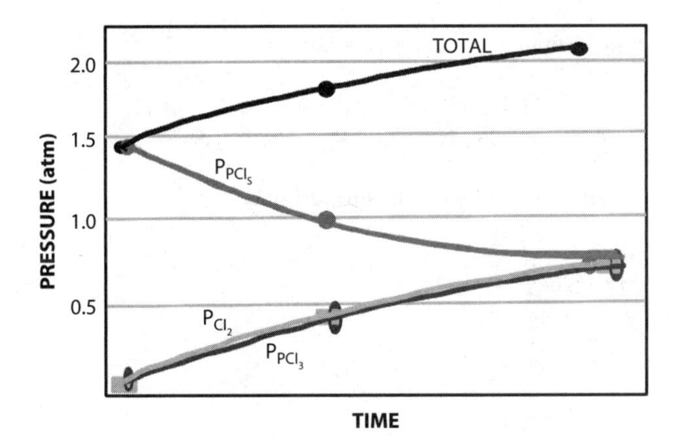

453 and **454.** Your graphs should look similar to the graphs below, shown together to compare the rates of the forward and reverse reactions when additional Cl_2 is added to the system.

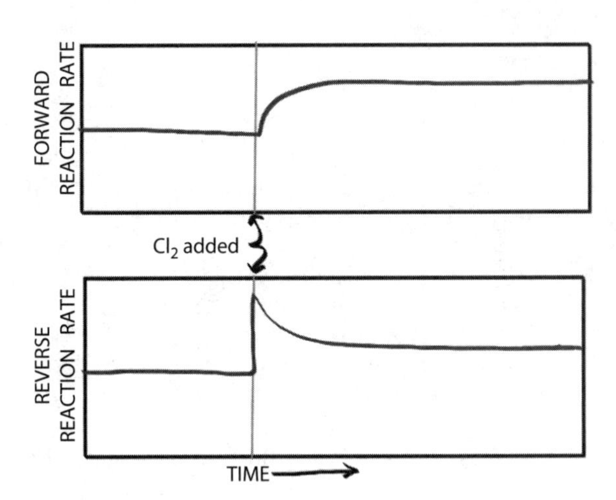

455. Macroscopic indications of equilibrium include:

- no color change over time (may be viewed by the naked eye or by spectrophotometer readings)
- no change in gas volume over time
- no change in gas pressure over time
- stable temperature over time

456. $\boldsymbol{K_P = 0.15}$

Set up an "ICE box" to calculate the number of moles of each species at equilibrium.

	N_2O_4	$2\ NO_2$
Initial	2.0	0
Change	$-x$	$+2x$
Equilibrium	$\mathbf{2.0 - x = 1.9}$	$\mathbf{0 + 2x = 0.24}$

$$\textbf{Moles of NO}_2 = 2x = 0.243 \therefore x = \mathbf{0.12}$$
$$\textbf{Moles of N}_2\textbf{O}_4 = 2.0 - 0.12 = \mathbf{1.9}$$

Find partial pressures of both gases at equilibrium.

N_2O_4

$$PV = nRT$$
$$P(5) = (1.9)(0.0821)(298)$$
$$\mathbf{P = 9.3\ atm}$$

NO_2

$$PV = nRT$$
$$P(5) = (0.24)(0.0821)(298)$$
$$\mathbf{P = 1.2\ atm}$$

Calculate $\boldsymbol{K_P}$**.**

$$K_P = \frac{[NO_2]^2}{[N_2O_4]}$$

$$K_P = \frac{[1.2]^2}{[9.3]}$$

$$\boldsymbol{K_P = 0.15}$$

Note that the 2.0 moles of N_2O_4 and the number of moles of NO_2 measured at equilibrium (0.24) limits the number of significant digits to 2.

457. (a) Calculate $\boldsymbol{\Delta G°}$ from the standard free-energy changes in the table.

$$\Delta G° = [(\Sigma\Delta G°_F \text{ products}) - (\Sigma\Delta G°_F \text{ reactants})]$$
$$= [(51.3 \times 2) - (97.9)]$$
$$= 102.6 - 97.9$$
$$\boldsymbol{\Delta G° = 4.70\ kJ}$$

(b) No. The $\boldsymbol{\Delta G° = 4.70}$ **kJ**, a positive value. The reverse reaction is thermodynamically favorable.

458. $\boldsymbol{K_P = 0.15}$

Use $\Delta G° = -RT \ln K$

You must use the gas constant with the correct units: 8.314 J $K^{-1}mol^{-1}$.

The units of ΔG are in kJ, so you must either convert 8.314 to kJ or convert 4.7 to J. We will do the latter: 4.70 kJ = 4.70×10^3 J.

$$4{,}700 = -(8.314)(298)\ln K_P$$
$$\ln K_P = -1.897$$
$$\mathbf{K_P = 0.150}$$

459. The value of K for any reverse reaction is the inverse of the K value of the forward reaction.

$$K_{REVERSE} = \frac{1}{K_{FORWARD}}$$

$$K = \frac{1}{0.15} = \mathbf{6.7}$$

460. The region indicated is the "halfway-to-equivalence" point, where half of the weak acid molecules are ionized. That means out of 3 molecules of weak acid, at least 1 should remain unionized. Recall that the pK_A = pH at the half-equivalence point under certain conditions, but there is no indication that these conditions are present, so as long as 1 or 2 of the acid molecules are ionized and 2 or 1 is not, the answer is, at least with regards to the acid, reasonable.

The point circled on the graph is near the buffer region. The strong base would be completely dissociated. The hydrogen ion that was dissociated from the acid and the hydroxide ion that dissociated from the base combine to form water. Notice in the following diagram that all 3 particles of base are dissociated, but only 1 of the 3 acid particles has dissociated. If you showed 2 dissociated, that's fine, but not all 3.

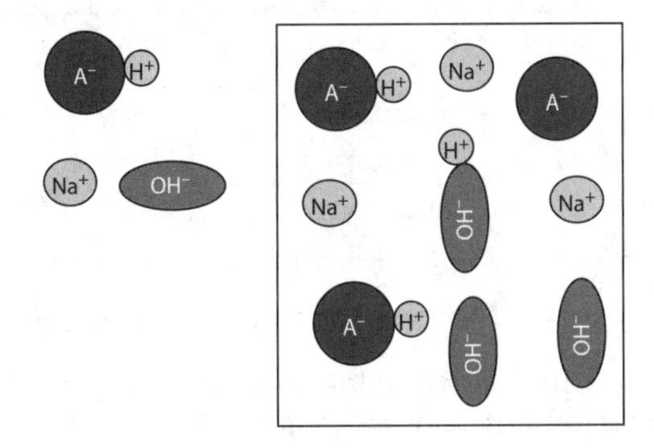

461. A buffer is composed of a weak acid and the salt of its conjugate base. To obtain buffering capacity of 5, **a weak acid with a pK_A within 1 point of the desired pH** should be chosen along with the salt of its conjugate base, or just its conjugate base. In this case, acetic acid, CH_3COOH, has the pK_A closest to the desired pH (see answer 463 for why this works). An example of an acceptable acetate salt would be sodium acetate (CH_3COONa), but **any salt whose cation is a weak conjugate base is acceptable**. (Cations that can pick up hydrogen ions or hydroxides from solution, like Fe^{2+} or Fe^{3+}, would *not* be acceptable, as iron ions can pick up hydroxides from solution and make it acidic!)

462. A buffer contains both ionized and unionized forms of the acid at the pH at which it will be used (which is why the pK_A of the acid should be within 1 pH point of the desired pH of the buffer; see answer 463 and box that follows it). The diagram below shows 2 of the 4 acetic acid molecules have ionized. (That is the origin of the H^+ ions.) The other 4 acetate ions were added as sodium acetate, a conjugate base of acetic acid. (That is the origin of the 4 Na^+ ions.)

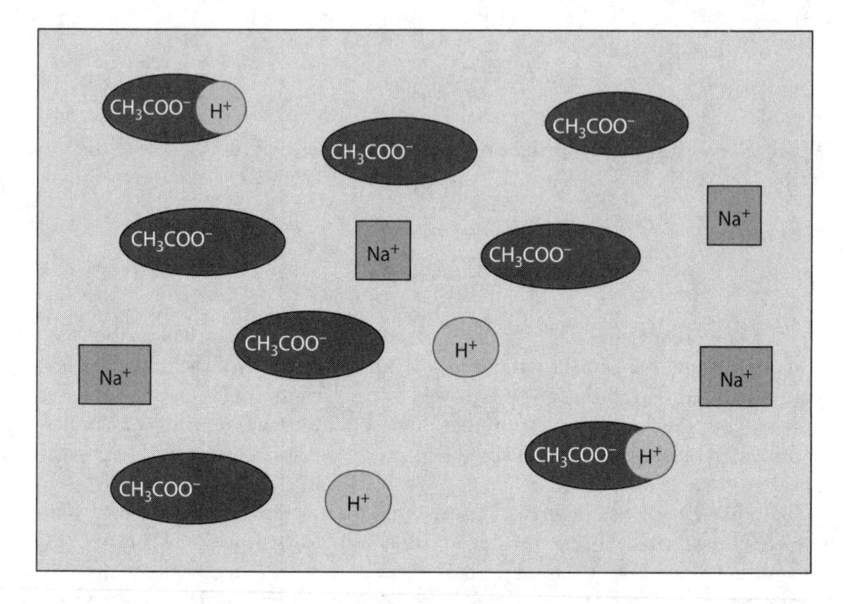

463. When the pH of the solution is equal (or very close) to the pK_A of the ionizing group on the acid, the buffer capacity is optimum. The ratio of weak acid to conjugate base should be about 9:1. Again, you can use the Henderson-Hasselbalch equation to understand why:

$$\text{pH} = \text{p}K_A + \log \frac{[\text{conjugate base}]}{[\text{acid}]}$$
$$= \text{p}K_A + \log {}^1\!/_9 = \text{p}K_A + (-0.95) \approx \mathbf{p}\boldsymbol{K_A} \mathbf{+ 1}$$
$$= \text{p}K_A + \log {}^9\!/_1 = \text{p}K_A - 0.95 \approx \mathbf{p}\boldsymbol{K_A} \mathbf{- 1}$$

The Henderson-Hasselbalch Equation

The Henderson-Hasselbalch equation provides a convenient way to reason about buffers.

$$\text{pH} = \text{p}K_A + \log \frac{[\text{conjugate base}]}{[\text{acid}]}$$

When the [conjugate base] = [acid], the ratio = 1.

The log of 1 = 0 ∴ pH = pK_A.

Computing the change in pH from the addition of an acid or base to a buffer will not be assessed on the AP Chemistry exam. Neither will the derivation of the Henderson-Hasselbalch equation.

464. **(a)** Both solutions have the same pH (2.5) and therefore the same $[H^+]$:

$$\text{pH} = -\log [H^+] \therefore [H^+] = 10^{-\text{pH}} = 10^{-2.5} = \mathbf{0.0032\ M}$$
$$\mathbf{HCl = strong\ acid} \therefore [\text{HCl}] = [H^+] = \mathbf{0.0032\ M}$$

Acetic acid, K_A:

$$CH_3COOH \leftrightharpoons CH_3COO^- + H^+$$
$$[H^+] = [CH_3COO^-] = 0.0032\ M$$
$$\text{Let } x = [CH_3COOH]_{INITIAL}$$

$$\text{Let } x - 0.0032 = [CH_3COOH]_{FINAL}$$

$$K_A = \frac{[CH_3COO^-][H^+]}{[CH_3COOH]}$$

$$1.8 \times 10^{-5} = \frac{[0.0032][0.0032]}{[x - 0.0032]}$$

$$x = 0.75\ M$$

$$\mathbf{[CH_3COOH]_{FINAL}} = x - 0.0032 = \mathbf{0.75\ M}$$

(b) Percent ionization:

$$\mathbf{HCl = 100\%}\ \text{(strong acid)}$$
$$\mathbf{CH_3COOH} = \frac{[\text{ionized acid}]}{[\text{unionized acid}]} = \frac{0.0032}{0.75} = \mathbf{4.3\%}$$

465.

$$\mathbf{M_{HCl}V_{HCl} = M_{NaOH}V_{NaOH}}$$
$$(0.0032)(0.50) = (0.010)V_{NaOH}$$
$$\mathbf{V_{NaOH} = 0.16\ L}$$

$$\mathbf{M_{Acetic}V_{Acetic} = M_{NaOH}V_{NaOH}}$$
$$(0.75)(0.50) = (0.010)V_{NaOH}$$
$$\mathbf{V_{NaOH} = 37.5\ L}$$

The volume of 0.010 M NaOH required to titrate acetic acid is 234 times the volume required to titrate an equal volume and pH of hydrochloric acid!

466. Le Chatelier's principle states that for a system at equilibrium to which a stress is applied, the system will respond to relieve the applied stress. Adding HCl increases the $[H^+]$ concentration, so to "relieve the stress" the acetate ions will "pick up" the extra hydrogen ions, forming more (unionized) acetic acid. This is called "pushing the equilibrium to the left." See the reaction below.

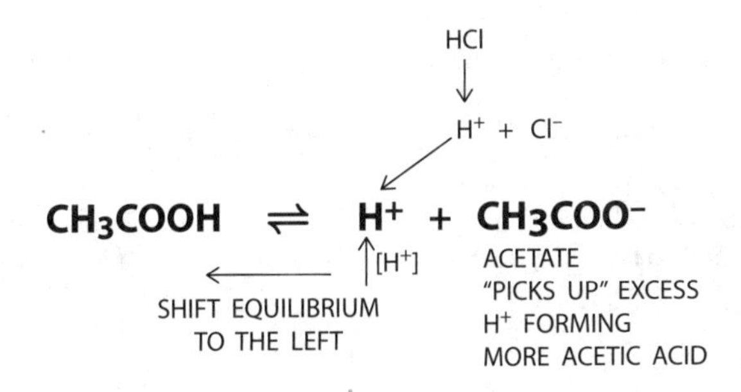

467. Le Chatelier's principle states that for a system at equilibrium to which a stress is applied, the system will respond to relieve the applied stress. Adding NaOH increases the $[OH^-]$ concentration. The hydroxide ions will react with hydrogen ions, forming water, which reduces the $[H^+]$. To "relieve the stress," more acetic acid will dissociate (ionize) to replace the lost H^+ ions. This is known as "pushing" or "pulling" the equilibrium to the right.

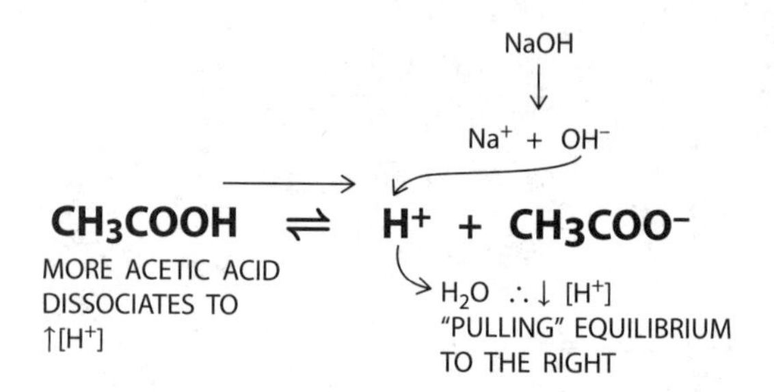

468. NH_4Cl is the better choice. Both salts are completely soluble, but the F^- ion is a good conjugate base, so it will pick up protons in solution. That may be helpful in some buffers, but in this one, the compound is supposed to act as a conjugate acid. The ammonium ion is the "active" substance here. The Cl^- or F^- is a spectator ion. Cl^- is a very, very weak base and will not pick up hydrogen ions from solution under most laboratory conditions.

469. 6.5×10^8 kg

Calculate the volume of NH_3 produced.

$$1.4 \times 10^{12}\ \text{L H}_2 \times \frac{2\ \text{L NH}_3}{3\ \text{L H}_2} = \mathbf{9.3 \times 10^{11}\ L\ NH_3}$$

Convert liters of NH_3 to moles of NH_3 at 1 atm and 298 K.

$$PV = nRT$$
$$(1)(9.3 \times 10^{11}\ \text{L}) = n(0.0821)(298)$$
$$n = \mathbf{3.8 \times 10^{10}\ moles\ NH_3}$$

Convert moles NH_3 to grams.

$$3.8 \times 10^{10}\ \text{moles NH}_3 \times \frac{17\ \text{g}}{\text{mole}} = \mathbf{6.5 \times 10^{11}\ g\ NH_3}$$

Convert grams NH_3 to kilograms.

$$6.5 \times 10^{11}\ \text{g} \times \frac{1\ \text{kg}}{1000\ \text{g}} = \mathbf{6.5 \times 10^{8}\ kg}$$

470. The reaction is exothermic ($\Delta H° = -92.2$ kJ), so increasing the temperature would shift the equilibrium to favor the reactants. To favor the products, a **low temperature** must be used.

A low temperature would result in a low reaction rate because the activation energy is high (230–420 kJ mol^{-1}). The large activation energy indicates a very temperature-sensitive reaction. A **catalyst** is needed to increase the reaction rate.

There are fewer moles of gas in the products than the reactants, so a **high pressure** would push the equilibrium to favor the products. **Removing the product** would also help shift the equilibrium to the right. **Continually adding reactants would keep the pressures of H_2 and N_2 high**, which would further increase the pressure and shift the equilibrium to the right.

471. All the ionizable groups of anserine will have their protons "on" at pH 2, a pH that is lower than all the pK_A values of those groups.

All the ionizable groups of anserine will have their protons "off" at pH 12, a pH that is higher than all the pK_A values of those groups.

See answer 415 for an explanation of the pH−pK_A relationship.

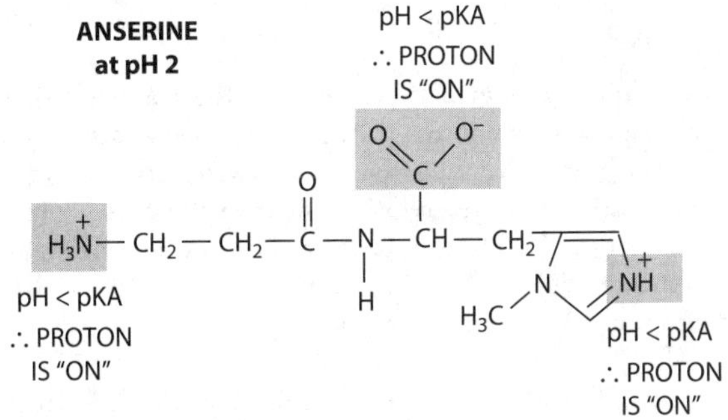

ANSERINE at pH 2

pH < pKA ∴ PROTON IS "ON"

$H_3\overset{+}{N}—CH_2—CH_2—C(=O)—N(H)—CH(COO^-)—CH_2$–(imidazole ring with $N–CH_3$ and $\overset{+}{N}H$)

pH < pKA ∴ PROTON IS "ON"

pH < pKA ∴ PROTON IS "ON"

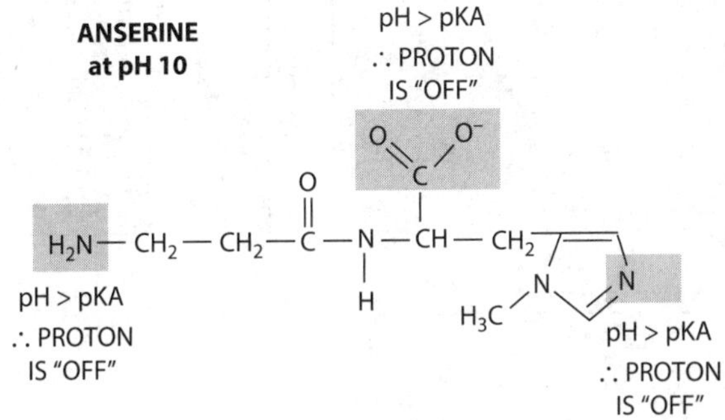

ANSERINE at pH 10

pH > pKA ∴ PROTON IS "OFF"

$H_2N—CH_2—CH_2—C(=O)—N(H)—CH(COO^-)—CH_2$–(imidazole ring with $N–CH_3$ and N)

pH > pKA ∴ PROTON IS "OFF"

pH > pKA ∴ PROTON IS "OFF"

472. The imidazole group (pK_A = 7.04). The buffer capacity is optimal when the pH of the solution equals (or is very close to) the pK_A of the ionizing group. See answer 463 for an explanation and calculation.

473. Your sketches should look like the following:

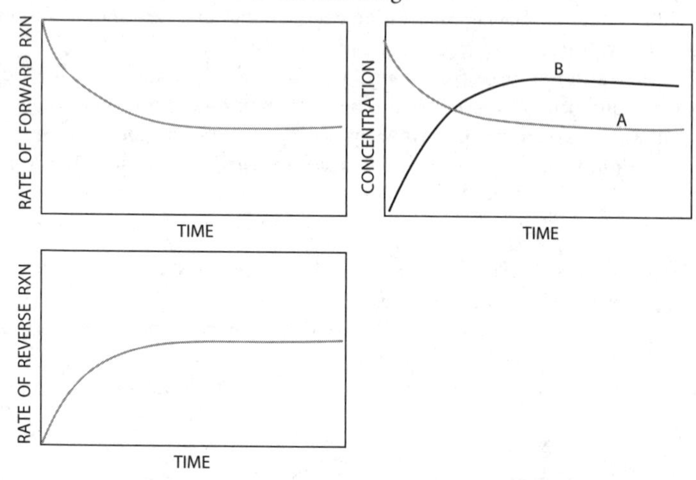

The forward reaction begins at a high rate and slows as reactant A is consumed. The reverse reaction begins after enough products, B, have accumulated and increases as more B is produced. Eventually, the rate of the forward reaction slows and the rate of the reverse reaction increases until their **relative rates are equal**, the point of equilibrium.

Importantly, the relative concentrations of product and reactant may or may not be equal (they're probably *not!*) but the equilibrium state of a reaction may be identified by an unchanging concentration of product or reactant.

474. Your drawing should look like the following. Be sure to include labels where needed.

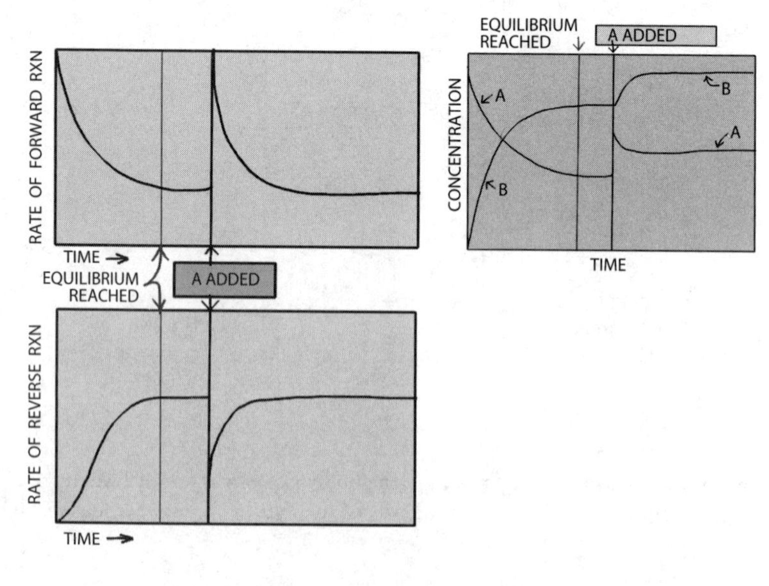

The forward reaction begins at a high rate and slows as reactant A is consumed. The reverse reaction begins after enough products, B, have accumulated and increases as more B is produced.

The addition of A increases the rate of the forward reaction because there is now a greater concentration of A. Le Chatelier's principle states that for a system at equilibrium to which a stress is applied, the system will respond to relieve the stress and restore equilibrium. In this case, the addition of A prompts the removal of A via an increased rate of the forward reaction.

475. Your sketch should look like the following. Be sure to include labels.

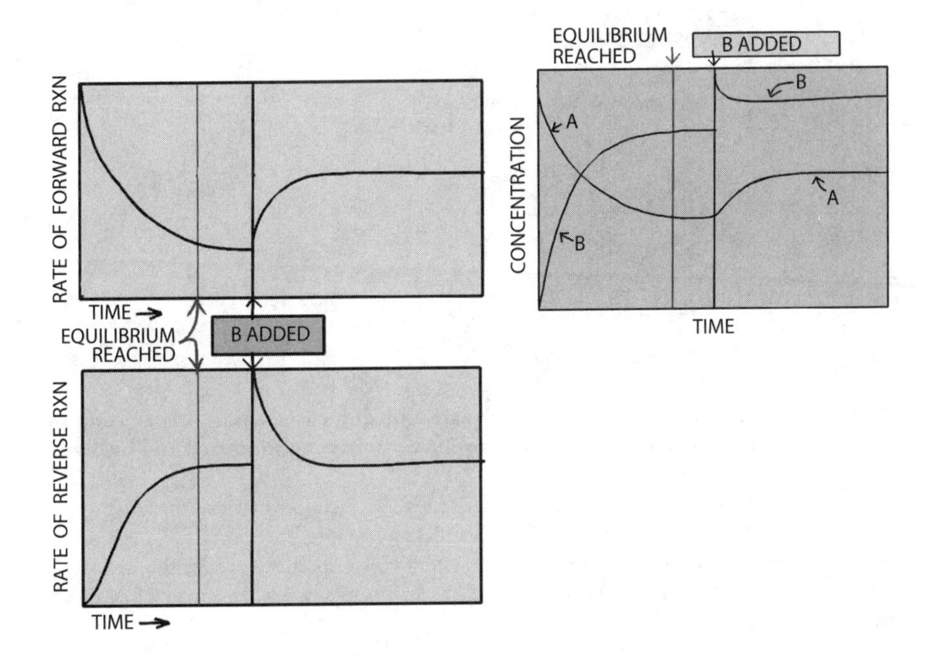

The forward reaction begins at a high rate and slows as reactant A is consumed. The reverse reaction begins after enough products, B, have accumulated and increases as more B is produced.

The addition of B increases the rate of the reverse reaction because there is now a greater concentration of B. Le Chatelier's principle states that for a system at equilibrium to which a stress is applied, the system will respond to relieve the stress and restore equilibrium. In this case, the addition of B prompts the removal of B via an increased rate of the reverse reaction.

476. Your sketch should look like the following. Be sure to include labels where needed.

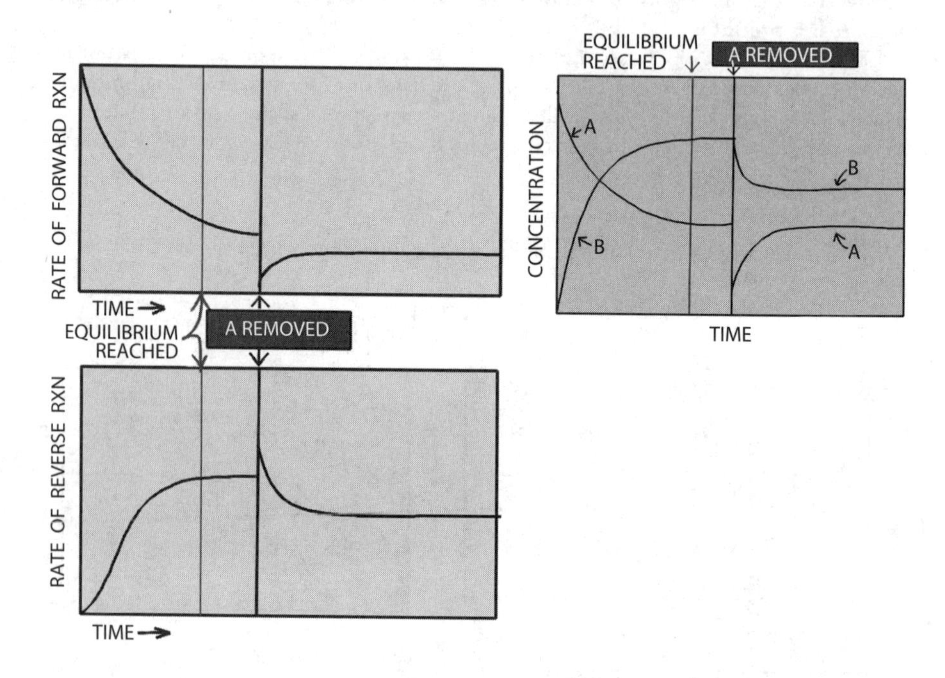

The forward reaction begins at a high rate and slows as reactant A is consumed. The reverse reaction begins after enough products, B, have accumulated and increases as more B is produced.

The removal of A increases the rate of the reverse reaction because there is now a *relatively* greater concentration of B. Le Chatelier's principle states that for a system at equilibrium to which a stress is applied, the system will respond to relieve the stress and restore equilibrium. In this case, the removal of A prompts the replacement of A via an increased rate of the reverse reaction.

477. Irreversible reactions proceed to completion because the energy requirements of the reverse reaction are prohibitive. Examples include precipitation reactions, strong acid-base reactions, strong redox reactions, and reactions that produce a gas (especially if the gas is allowed to escape).

Technically, all other reactions are reversible; however, **the conditions under which the reaction occurs may be optimized to encourage the reaction to proceed to completion**. For example, increasing the temperature under which an endothermic reaction occurs increases the value of the equilibrium constant. For temperature-sensitive endothermic reactions, an increased temperature may allow the reaction to proceed to completion.

478. Indications that a reaction has reached equilibrium:

- **Concentration** of reactants and products does not change over time.
- Forward and reverse **reaction rates** are equal.

- **Color** is constant.
- **Pressure** is constant in a reaction that produces or consumes gases.
- **Temperature** remains constant.

479. **(a)** The second equation is a rearrangement of the first. Using the first equation:

An **endergonic** reaction has a **positive $\Delta G°$**, so **the exponent of K will be negative** $\therefore$ $K < 1$. This means that in an endergonic reaction, the reactants are favored.

An **exergonic** reaction has a **negative $\Delta G°$**, so the **exponent of K is positive** $\therefore$ $K > 1$. That means in an exergonic reaction, the products are favored.

The larger the magnitude of K, the more greatly the reactants or products are favored in endergonic versus exergonic reactions, respectively.

The term RT is a measure of thermal energy. Thermal energy at room temperature is approximately **2.4 kJ/mol**. This number can be used to relate enthalpy and entropy changes to the magnitude of K. If the $\Delta G°$ of a reaction is approximately -2.42, the exponent $\left(\frac{-\Delta G°}{RT}\right)$ is equal to 1 and the value of K is 2.72. If the magnitude of $\Delta G°$ is large, the value of K deviates from 1 substantially. For example, a reaction with a $\Delta G°$ of -250 kJ/mol has a K value of 1.7×10^{45} (it proceeds to completion).

(b) The term ***exothermic*** **relates to enthalpy changes** whereas the term ***exergonic*** **relates to free-energy changes.** Exothermic reactions give off heat, but they are not necessarily thermodynamically favorable. Exergonic reactions are often exothermic, but they may be endothermic (but occur at a high temperature). Exergonic reactions are thermodynamically favorable (they can do work).

Chapter 7: Mixed Free-Response Questions

480. Since both solutions have the same volume and concentration:

$$\frac{0.20 \text{ mole solute}}{\text{L}} \times (0.150 \text{ L}) = \mathbf{0.030 \text{ mole solute}} \ (CoCl_2 \text{ or } H_2CO_3)$$

(a) $CoCl_2 = 129.9 \text{ g mol}^{-1}$

$$0.030 \text{ mole } CoCl_2 \times \frac{129.9 \text{ g}}{\text{mole}} = \mathbf{3.9 \text{ grams } CoCl_2}$$

(b) $H_2CO_3 = 62 \text{ g mol}^{-1}$

$$0.03 \text{ mole } H_2CO_3 \times \frac{62 \text{ g}}{\text{mole}} = \mathbf{1.86 \text{ grams } H_2CO_3}$$

481. **(a)** Preparing a standard solution:

Carefully and accurately mass 1.86 grams of H_2CO_3.
Add dry solute to a 150 mL volumetric flask.
Add enough distilled water to solubilize the H_2CO_3 and mix thoroughly.
Add distilled water slowly until the 150 mL line on the flask is reached.

(b) A standard solution is a solution of a precisely known concentration. Standard solutions are **used in titrations to determine the concentration of other substances.** The accuracy of the measurement depends on the reliability of the standard solution. The volume of the standard solution used to reach equivalence in the titration will depend on the concentration. If the concentration is not known precisely, the calculated concentration of the analyte will be imprecise.

482. H_2CO_3 is a weak acid, so the pH of a 0.200 M solution at 25° C can be measured and the value compared to the pH calculated from the K_A values of H_2CO_3 and HCO_3^-.

The AP exam will not ask you to calculate the pH of di- or triprotic acids.

Or, electrical conductivity can be measured and compared to the calculated conductivity of a solution using the percent ionization of H_2CO_3 and HCO_3^- at 0.200 M concentrations at 25° C.

483. Your drawings should have the following elements.

(a) Note the ions of Co^{2+} and Cl^- are drawn somewhat to scale (the Cl^- ions are larger, but the actual size difference won't matter). The water molecules are arranged so that the partially positive hydrogen atoms are facing the anions in solution and the partially negative oxygen is oriented toward the cations.

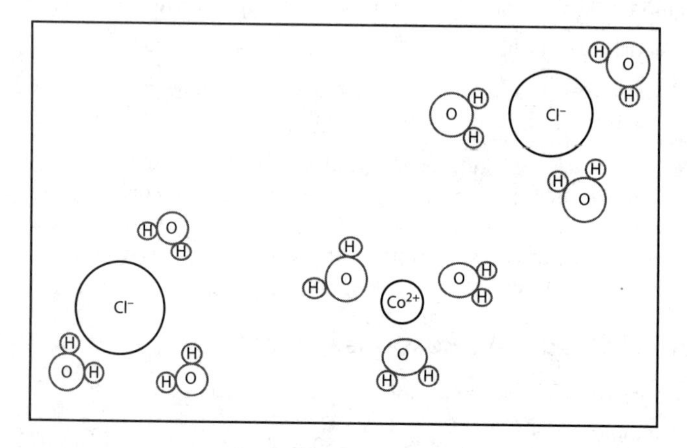

(b) Notice that one H_2CO_3 remains dissociated and the other combined with the chloride ions. The $CoCl_2$ is very soluble, whereas the H^+ and Cl^- ions do not combine.

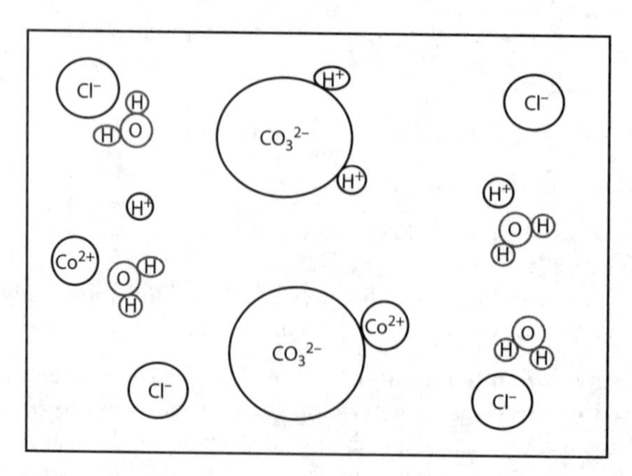

484. **(a)** As the sample dries, it loses mass. The sample will continue to lose mass until all the water has evaporated. By 40 minutes the sample's mass has stabilized around 77.9 g. Slight variations in the mass ($\frac{1}{100}$ of a gram) after 40 minutes may be caused by the temperature of the sample. With a warm sample, the higher temperature of the sample as compared to the surrounding air may produce small air currents that could slightly alter the apparent mass of the sample (often, it will appear lighter). If the sample was allowed to cool before massing, it may attract water from the atmosphere (increasing its mass).

(b) The color of the sample should be between purple and blue. Light purple or red is a reasonable prediction. This intermediate color occurs because the sample is not completely hydrated nor completely dehydrated. Color is produced because some wavelengths of light are absorbed by the crystal and others are reflected.

In a crystal of intermediate hydration, some parts will reflect deep purple wavelengths and others will reflect sky blue wavelengths. The combined effect of these two colors is ultimately determined by the ratio of the reflected wavelengths (i.e., the amount of hydrated versus dry crystal) and the ways the waves interfere with each other after they've been reflected and hit your retina. At 15 minutes, the sample should weigh between 102 and 107 g, which means about half of the water has been lost (see part c), so the color of the crystals should be midway between purple and blue.

(c) Calculate the average mass of the dry sample.

$$\frac{77.92 + 77.96 + 77.94}{3} = \mathbf{77.94\ grams\ CoCl_2}$$

Calculate the number of moles of $CoCl_2$ in that mass.

$$77.94\ g \times \frac{1\ mole}{129.9\ g} = \mathbf{0.60\ moles\ CoCl_2}$$

Calculate the mass of the water lost.

Mass of hydrated salt − mass of dry salt = mass of water lost

$$142.74 - 77.94 = \mathbf{64.8\ grams\ H_2O}$$

Calculate the number of moles of water in that mass.

$$64.8\ g \times \frac{1\ mole}{18\ g} = \mathbf{3.6\ moles\ H_2O}$$

Find the ratio of moles of $CoCl_2$ to moles of H_2O.

$$\frac{3.6\ moles\ H_2O}{0.60\ mole\ CoCl_2} = \frac{6\ moles\ H_2O}{1\ mole\ CoCl_2}$$

$$\therefore \mathbf{CoCl_2 \bullet 6H_2O}$$

Mass of filter paper	**1.501 g**	**You need to subtract this number from all the masses measured.**
Mass of LiCl tablet	1.204 g	Mass of tablet + filter paper − mass of filter paper 2.705 g − 1.501 g
Mass of precipitate after first drying	4.813	Mass of precipitate + filter paper − mass of filter paper 6.314 g − 1.501 g
Mass of precipitate after second drying	4.810	Mass of precipitate + filter paper − mass of filter paper 6.311 g − 1.501 g
Mass of precipitate after third drying	4.809	Mass of precipitate + filter paper − mass of filter paper 6.310 g − 1.501 g

485. Pb^{2+} *(aq)* + 2 Cl^- *(aq)* → $PbCl_2$ *(s)*
Net-ionic equations are a good way to express precipitation, single and double replacement, and neutralization reactions. The other ions present (Li^+ and NO_3^{2-}) during the reaction do not participate in the reaction (they are spectator ions). Because they do not participate in the reaction they are unnecessary to include in the reaction.

486. The wet precipitate will lose mass (due to the evaporation of water) until all the water is lost. It is standard practice to dry a precipitate until a constant mass (or as close to constant as possible) has been reached. **This ensures all the water has been evaporated from the sample.**

487. The experiment summary explicitly states that the 0.2500 M $Pb(NO_3)_2$ *(aq)* solution was added in excess, so we can safely assume the $[Li^+] < [NO_3^{2-}]$.

The determination of Li^+ is done indirectly, by precipitation of the Cl^- and inferring the amount of Li^+ from the mass of the precipitate. If all the Cl^- was not precipitated out, the analysis would be inaccurate (and incomplete).

488. (a) Identify the precipitate.

The precipitate is $PbCl_2$ (molar mass = 278.1).

Find the mass of the precipitate.

Choose the **mass of the precipitate** after the third drying, **4.809 g**. (Don't forget to subtract the mass of the filter paper; see table above.)

Calculate the number of moles of the precipitate.

$$4.809 \text{ g} \times \frac{1 \text{ mole PbCl}_2}{278.1 \text{ g}} = 0.01729 = \mathbf{1.729 \times 10^{-2} \text{ mole PbCl}_2}$$

(b) Because two moles of Cl^- were precipitate for every one mole of Li^+ in the tablet (LiCl versus $PbCl_2$), the number of moles of Li^+ ions is twice the number of moles of Cl^- in the precipitate.

$$0.01729 \text{ mole PbCl}_2 \times \frac{2 \text{ moles Cl}^-}{1 \text{ mole PbCl}_2} = 0.03458 \text{ mole Cl}^-$$

$$\therefore 0.03458 = \mathbf{3.458 \times 10^{-2} \text{ mole Li}^+}$$

(c) Convert moles of Li^+ to grams of Li^+.

$$0.03458 \text{ mole Li}^+ \times \frac{6.941 \text{ g Li}^+}{1 \text{ mole Li}^+} = 0.2401 = \mathbf{2.401 \times 10^{-1} \text{ gram Li}^+}$$

Divide the mass of lithium by the mass of the tablet. (Don't forget to subtract the mass of the filter paper that the tablet was weighed on; see table above.)

$$\frac{0.2401 \text{ g Li}^+}{1.204 \text{ g tablet}} = \mathbf{19.9\% \text{ Li}^+}$$

489. (a) The result would not change. The amount of water used to solubilize the tablet is irrelevant as long as there was enough to solubilize the tablet and not too much to solubilize a significant amount of $PbCl_2$. Most of what we call *insoluble* is actually a tiny, tiny bit soluble (depending on the K_{SP}), so too much water could keep enough of what you're trying to precipitate in solution to make your results significantly inaccurate.

(b) The result would not change. You might need to add more of the solution, but the amount of precipitate formed would be the same. The drop from 0.25 to 0.20 is not very large. See answer 489(a).

(c) Find the number of moles of lead nitrate needed.

In answer 488 you determined that there are 0.03458 mole Cl^-, which means we need half that many moles of Pb^{2+} ions to precipitate it out. (Each Pb^{2+} will remove 2 Cl^- from solution.)

$$0.03458 \text{ mole } Cl^- \times \frac{1 \text{ mole } Pb^{2+}}{2 \text{ moles } Cl^-} = 0.01729 = \mathbf{1.729 \times 10^{-2} \text{ mole } Pb^{2+}}$$

$$\therefore \mathbf{1.729 \times 10^{-2} \text{ moles } Pb(NO_3)_2 \text{ needed}}$$

Divide the number of moles of $Pb(NO_3)_2$ by the number of liters (10.00 mL = 0.0100 L).

$$\frac{0.01729 \text{ mole } Pb(NO_3)_2}{0.0100 \text{ L}} = \mathbf{1.729 \text{ M}}$$

(d) As calculated in part (c), 0.01729 mole Pb^{2+} ∴ 1.729×10^{-2} moles $Pb(NO_3)_2$ are needed.

The concentration, 0.25 M, can be expressed as 0.25 moles per liter of solution *or* as 1 liter of solution contains 0.25 moles. We'll use the latter to solve this problem:

$$\frac{1 \text{ L}}{0.25 \text{ mole}} \times 0.01729 \text{ mole} = 0.6916 \text{ L, or } \mathbf{69.16 \text{ mL}}$$

490. No, the assumption was not reasonable.

(a) Calculate **final volume** of solution.

Final volume = distilled water + $Pb(NO_3)_2$
= 50.00 mL + 62.16 mL
= **112.2 mL, or 0.1122 L**

Assume the final solution is saturated. You can then use the K_{SP} value to calculate the concentration.

$$PbCl_2\ (s) \leftrightharpoons Pb^{2+}\ (aq) + 2\ Cl^-\ (aq)$$
$$\text{Let } x = [Pb^{2+}] \text{ and } 2x = [Cl^-]$$
$$K_{SP} = 1.2 \times 10^{-5}$$
$$K_{SP} = [Pb^{2+}]\ [Cl^-]^2$$
$$1.2 \times 10^{-5} = (x)(2x)^2$$
$$= 4x^3$$
$$\therefore x = \mathbf{[Pb^{2+}]} = 0.01442 = \mathbf{1.442 \times 10^{-2}\ M}$$
$$\therefore 2x = \mathbf{[Cl^-]} = 0.02884 = \mathbf{2.884 \times 10^{-2}\ M}$$

Use the concentration and volume to **calculate the number of moles and masses of the ions.**

$$Pb^{2+} = \frac{0.01442 \text{ mole}}{\text{L}} \times 0.1122 \text{ L} = 0.001618 = \mathbf{1.618 \times 10^{-3} \text{ mole } Pb^{2+}}$$

$$0.001618 \text{ mole} \times \frac{207.2 \text{ g}}{\text{mole}} = 0.3352 = \mathbf{3.352 \times 10^{-1} \text{ grams } Pb^{2+}}$$

$$Cl^- = \frac{0.02884 \text{ mole}}{L} \times 0.1122 \text{ L} = 0.003236 = \mathbf{3.236 \times 10^{-3} \text{ mole } Cl^-}$$

$$0.003236 \text{ mole} = \frac{35.45 \text{ g}}{\text{mole}} = 0.1147 = \mathbf{1.147 \times 10^{-1} \text{ grams } Cl^-}$$

(c) You need to use the volume calculated in answer 489 because that is the minimum volume needed to precipitate out "all" of the Cl^-. Because the K_{SP} of $PbCl_2$ is approximately 10^{-5}, there will be nonnegligible amounts of Cl^- left in solution (3.236×10^{-3} mole Cl^-) because there was not enough Pb^{2+} to prevent Cl^- from re-solubilizing. The number of chloride ions is how you count lithium ions, so leaving chloride in solution produces an underestimation of the number of moles of lithium ions.

Total mass of $PbCl_2$ = 5.259 g

4.809 g $PbCl_2$ was recovered by precipitation

∴ 0.4499 g $PbCl_2$ left in solution (0.3352 g Pb^{2+} + 0.1147 g Cl^-).

8.55% left in solution (by mass)

$$\frac{0.4499 \text{ g left in solution}}{\text{Total mass of } 5.259} \times 100 = \mathbf{8.55\% \text{ solubilized}}$$

91.4% recovered by precipitation

$$\frac{4.809 \text{ g recovered}}{\text{Total mass of } 5.259} \times 100 = 91.4\% \text{ recovered}$$

The assumption of negligible solubility was not reasonable. **Almost 9% of the chloride ions were left in solution.**

(d) The original procedure used **excess $Pb(NO_3)_2$** to remove all the chloride ions from solution and prevent them from re-solubilizing by keeping the concentration of Pb^{2+} ions high (which precipitates all the remaining Cl^- as the equilibrium shifts to consume the excess Pb^{2+}). The "minimum volume of 0.25 M $Pb(NO_3)_2$ solution that would precipitate out all the Cl^- ions" (from question 489d) was calculated by adding 1 Pb^{2+} ion for every 2 Cl^- ions and assumed the solubility was negligible.

491. Yes, the K_{SP} of AgCl is sufficiently low (much lower than $PbCl_2$) to precipitate the chloride ions.

492. Yes. The mass of the tablet is 1.204 grams and the balance can measure to +/−0.001 grams; then the answer can be expressed in four significant figures, but the last number is uncertain. For questions like this, you must consider the quantity you are massing relative to the error of the balance. If you had more than one tablet, you could mass several of them at once and then divide by the number (for example, if 25 tiny tablets mass out to 1.5 grams, each weighs 60 mg, 1,500 mg total for 25 tiny tablets).

493. Calculate the total **mass of pigment.**

$$30.0 \text{ ml} \times \frac{1.00 \text{ g}}{\text{mL}} = \mathbf{30.0 \text{ grams}}$$

Convert mass of pigment **to moles** of pigment

$$30.0 \text{ g} \times \frac{1 \text{ mole}}{642 \text{ g}} = \mathbf{0.0467 \text{ mole}}$$

Use number of moles and volume of solution to **calculate molarity** of solution.

$$\frac{0.0467 \text{ mole}}{0.03 \text{ L}} = \mathbf{1.56 \text{ M}}$$

494. For this question it is helpful to create a table.

	0.00 M	0.10 M	0.20 M	0.30 M	0.40 M	0.50 M	0.60 M
Volume H_2O	5	4.68	4.36	4.04	3.72	3.40	3.10
Volume of original solution	0	0.32	0.64	0.96	1.28	1.60	1.90
Total volume	5	5	5	5	5	5	5

Sample calculation:
Solution 1 = original solution of pigment
Solution 2 = diluted solution for standard curve

$$\mathbf{M_1V^*_1 = M_2V^*_2}$$
$$(1.56 \text{ M})V_1 = (0.1 \text{ M})(5 \text{ mL})$$
$$\mathbf{V_1 = 0.32 \text{ mL}}$$

Total volume = 5 mL
Volume of original solution = 0.32 mL
Volume of water = total volume − volume of original solution
= 5 − 0.32 = **4.68 mL water added**

* In this formula, you can substitute mL for L if it makes the calculation easier, but remember the unit you plug in for volume is the unit of volume your answer will be in.

495. (a) The standard absorbance curve is shown in the graph below. A best-fit line is used for interpolation (predicting concentrations from absorbances and absorbances from concentrations between measured data points).

(b) $\mathbf{y = 1.6 + 0}$

The equation of the line gives the slope, which can be calculated from any two points on the curve. **The *y*-intercept should theoretically be zero**, since there should be no absorbance of a zero molar solution. The buffer is chosen so that it is "invisible" to the spectrophotometer at the specific wavelength you are measuring, so it should not affect absorbance. The 0 M sample gave a 0% absorbance in the data given, but if there were any absorbance for the 0 M sample, the *y*-intercept would be the percent absorbance at 0 M.

The most reliable way to interpolate an unknown value that is between two measured data points is to generate a linear equation between those two data points. If you need to generate a curve that will allow you to determine many data points, you'll need to generate an equation for the line that best fits all the data points you have. When choosing points for the slope, *choose two points on the best-fit line*. This will make your interpolations consistent. The *x* and *y* values of those points will be estimations, so there will be variation between students' answers. In addition, because the actual values do not create a perfect line, each student may draw a slightly different best-fit line. The value of the y-intercept is so close to zero it can be considered negligible, so I ignored it.

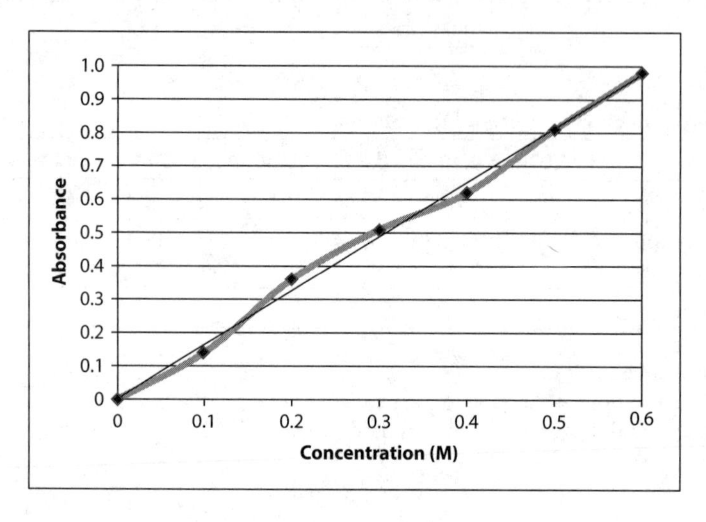

(c) The slope is the ratio of how much the absorbance changes with a change in concentration.

(d) The y-axis values are from 0 to 1, or 0 to 100%, so a **slope of 1.6** means that the absorbance increases 160% for a 1 M increase in concentration. The molarity increases in increments of 0.1 M, so a 0.16, or 16%, increase is expected for a 0.1 M increase in concentration.

Check the logic:
0 M = 0 absorbance

0.1 M ≈ 0.14 absorbance (0.16 predicted—this value checks out!)
0.2 M ≈ 0.36 absorbance (0.16 × 2 = 0.32—close enough!)

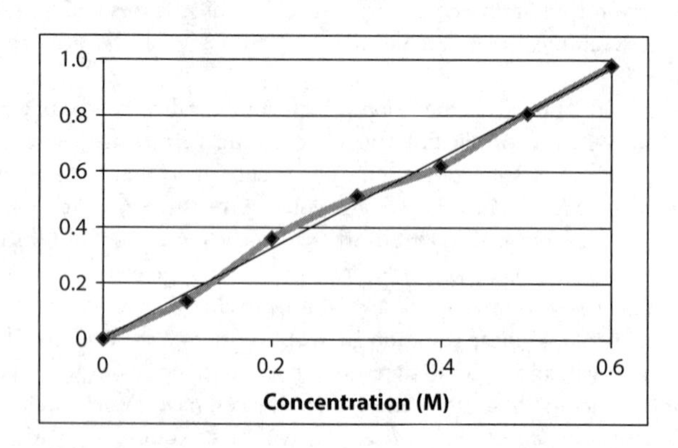

(e) Estimate by interpolation and calculate using the equation of the best-fit line. Remember that an absorbance of 1 is 100%.

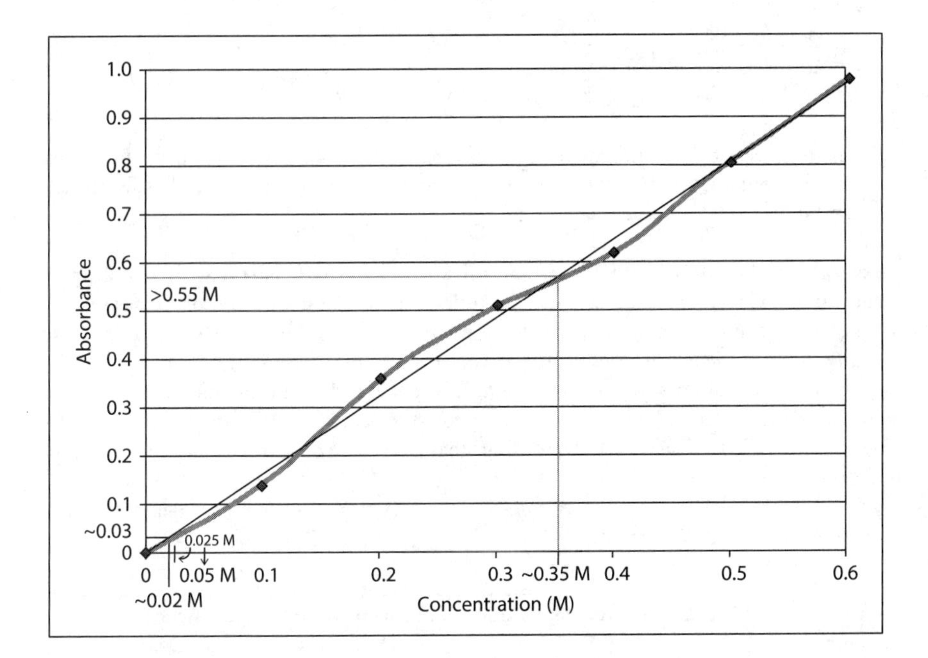

To calculate the percent absorbance of a 0.020 M concentration:

$y = mx$
$y = (1.62)(0.020)$
$\mathbf{y}$ **= absorbance = 0.032**
$0.032 \times 100 =$ **3.2%**

The answer matches the interpolation.
To calculate the percent absorbance of a 0.36 M concentration:
$y = mx$
$y = (1.62)(0.36)$
$\mathbf{y}$ **= absorbance = 0.58**
$0.58 \times 100 =$ **58%**

The answer matches the interpolation. Notice that the graph is too small to distinguish 0.36 M exactly, so interpolation really only gives you an estimate.

An alternative approach: You can use the actual data points to find the slope of the line between the two points that flank the value of interest.

For example, concentration = 0.36 M
Falls between 0.3 M, 0.51%, and 0.4 M, 0.62%

$$\text{Slope} = \frac{\Delta y}{\Delta x} = \frac{(0.62 - 0.51)}{(0.4 - 0.3)} = 1.1$$

$y = mx$
$y = (1.1)\,(0.36)$
y = absorbance = 0.396 ≈ **0.40**
$0.40 \times 100 = 40\%$

The purpose of the standard curve is to average out all the variations in the measured points. The construction usually involves very accurate dilutions and several trials of each concentration.

496. (a) An absorbance of 1 means **the dilution is too concentrated**. It may be just concentrated enough so that the best-fit curve predicts the concentration accurately, or it can be overconcentrated. The problem is, there's no way to know unless you **dilute the sample further**. Make sure you record the volume used for the second dilution!

(b) Generally, it's best to have your sample read somewhere in the middle of the range you're spectrophotometer is set to read in. If your absorbance is very close to zero or 100%, you're in the "fuzzy zone" where the readings may lose some of their accuracy.

497. The table below shows the concentration changes of A over time. Your values may vary due to differences in estimation.

Time (minutes)	Absorbance (%)	Concentration (M)	$\frac{\Delta \text{ concentration}}{\text{time}}$	$\frac{\Delta \text{ concentration}}{\text{time}}$
0	~23.0	0.34	$\frac{0.26\text{M}}{2 \text{ min}} \therefore \frac{0.13\text{M}}{\text{min}}$	
2	~6.0	0.08		$\frac{0.06\text{M}}{2 \text{ min}} \therefore \frac{0.03\text{M}}{\text{min}}$
4	~1.5	0.02		

The reaction rate will be expressed as a function of its stoichiometry. Because two moles of A are consumed for each mole of X_2 consumed, the reaction can be expressed as moles A consumed per minute, moles X_2 consumed per minute, or moles AX formed per minute (which will have the same value of moles of A consumed per minute).

From 0 to 2 minutes:

$\frac{0.13\text{M A}}{\text{min}}$ **or** $\frac{0.065\text{M X}}{2 \text{ min}}$ **consumed** **or** $\frac{0.13\text{MAX}}{2 \text{ min}}$ **formed**

From 2 to 4 minutes:

$\frac{0.03\text{M A}}{\text{min}}$ **or** $\frac{0.0015\text{M X}}{2 \text{ min}}$ **consumed** **or** $\frac{0.03\text{MAX}}{2 \text{ min}}$ **formed**

498. ~17.5%

You can **estimate a reaction order for A** from the original data (your assumption is that it's true). The reaction rate in the first minute of the reaction is 0.160 M per minute. In that time, the concentration of A is decreased by one-half. From 1 to 2 minutes, the rate is 0.08 M per minute, so the reaction rate was halved when the concentration of A was halved. Therefore, A is a first-order reactant. The **half-life is one minute.** If the initial absorbance was 35%, the concentration of A is 0.50 M, so in 1 minute, 0.25 M concentration of A would be left and the absorbance would be ~17.5%.

499. 22.5

Make an "ICE box," but assume you have all the X_2 you need to completely react with A. In other words, the complete consumption of 0.32 moles of A requires 0.16 moles of X_2.

	2 A	**X_2**	**2 AX**
Initial	0.32	0.16	0.00
Change	−0.30	−0.15	+0.30
Equilibrium	0.20	0.10	0.30

500. 38 grams H_2O, 56 grams CO, and 90.2 grams CO_2

The graph shows that 300 grams of the hydrate was present. (This represents approximately two moles of the hydrate, molar mass 146 g mol^{-1}.) The first drop in mass corresponds to the evaporation of water from the hydrate. The loss of mass corresponds to the mass of two moles of water (**~38 grams**).

The second drop corresponds to the conversion of calcium oxalate to calcium carbonate and carbon monoxide gas. The loss of CO gas results in a loss of approximately **56 grams**, the mass of two moles of carbon monoxide.

The third drop corresponds to the "loss" of CO_2. The mass lost is approximately **90.2 grams**, corresponding to two moles of CO_2.

Periodic Table of the Elements

May be used with all questions.

The periodic table

Key:

Atomic number
Symbol
Atomic weight (mean elative mass)

1	2		3	4	5	6	7	8	9	10	11	12	13	14	15	16	17	18
1 H 1.008																		2 He 4.002602(2)
3 Li 6.94	4 Be 9.012182(3)												5 B 10.81	6 C 12.011	7 N 14.007	8 O 15.999	9 F 18.9984032(5)	10 Ne 20.1797(6)
11 Na 22.98976928(2)	12 Mg 24.3050(6)												13 Al 26.9815386(2)	14 Si 28.085	15 P 30.973762(2)	16 S 32.06	17 Cl 35.45	18 Ar 39.948(1)
19 K 39.0983(1)	20 Ca 40.078(4)		21 Sc 44.955912(6)	22 Ti 47.867(1)	23 V 50.9415(1)	24 Cr 51.9961(6)	25 Mn 54.938045(5)	26 Fe 55.845(2)	27 Co 58.933195(5)	28 Ni 58.6934(4)	29 Cu 63.546(3)	30 Zn 65.38(2)	31 Ga 69.723(1)	32 Ge 72.63(1)	33 As 74.92160(2)	34 Se 78.96(3)	35 Br 79.904(1)	36 Kr 83.798(2)
37 Rb 85.4678(3)	38 Sr 87.62(1)		39 Y 88.90585(2)	40 Zr 91.224(2)	41 Nb 92.90638(2)	42 Mo 95.96(2)	43 Tc [97.91]	44 Ru 101.07(2)	45 Rh 102.90550(2)	46 Pd 106.42(1)	47 Ag 107.8682(2)	48 Cd 112.411(8)	49 In 114.818(3)	50 Sn 118.710(7)	51 Sb 121.760(1)	52 Te 127.60(3)	53 I 126.90447(3)	54 Xe 131.293(6)
55 Cs 132.9054519(2)	56 Ba 137.327(7)	57–70 *	71 Lu 174.9668(1)	72 Hf 178.49(2)	73 Ta 180.94788(2)	74 W 183.84(1)	75 Re 186.207(1)	76 Os 190.23(3)	77 Ir 192.217(3)	78 Pt 195.084(9)	79 Au 196.966569(4)	80 Hg 200.59(2)	81 Tl 204.38	82 Pb 207.2(1)	83 Bi 208.98040(1)	84 Po [209]	85 At [210]	86 Rn [222]
87 Fr [223.02]	88 Ra [226.03]	89–102 **	103 Lr [262.11]	104 Rf [265.12]	105 Db [268.13]	106 Sg [271.13]	107 Bh [270]	108 Hs [277.15]	109 Mt [276.15]	110 Ds [281.16]	111 Rg [280.16]	112 Cn [285.17]	113 Nh [284.18]	114 Fl [289.19]	115 Mc [288.19]	116 Lv [293]	117 Ts [294]	118 Og [294]

*lanthanoids	57 La 138.90547(7)	58 Ce 140.116(1)	59 Pr 140.90765(2)	60 Nd 144.242(3)	61 Pm [144.91]	62 Sm 150.36(2)	63 Eu 151.964(1)	64 Gd 157.25(3)	65 Tb 158.92535(2)	66 Dy 162.500(1)	67 Ho 164.93032(2)	68 Er 167.259(3)	69 Tm 168.93421(2)	70 Yb 173.054(5)
**actinoids	89 Ac [227.03]	90 Th 232.03806(2)	91 Pa 231.03588(2)	92 U 238.02891(3)	93 Np [237.05]	94 Pu [244.06]	95 Am [243.06]	96 Cm [247.07]	97 Bk [247.07]	98 Cf [251.08]	99 Es [252.08]	100 Fm [257.10]	101 Md [258.10]	102 No [259.10]

NOTES

NOTES

NOTES

NOTES